Adaptation and Resilience

Adaptation and Resilience

The Economics of Climate, Water, and Energy Challenges in the American Southwest

EDITED BY

Bonnie G. Colby and George B. Frisvold

Routledge
Taylor & Francis Group
LONDON AND NEW YORK

First published 2011 by RFF Press

2 Park Square, Milton Park, Abingdon, Oxfordshire OX14 4RN
711 Third Avenue, New York, NY 10017

First issued in paperback 2018

Routledge is an imprint of the Taylor & Francis Group, an informa business

ISBN: 978-1-933115-93-1 (hbk)
ISBN: 978-1-138-38019-6 (pbk)

Copyedited by Joyce Bond
Typeset by OKS Press Services
Cover design by Maggie Powell
Cover images: "Fresh snow Snowmass" © Michael Braun/istockphoto.com; "Electrical power lines" © Pgiam/istockphoto.com

Library of Congress Cataloging-in-Publication Data

Adaptation and resilience : the economics of climate, water, and energy challenges in the American Southwest/edited by Bonnie G. Colby and George B. Frisvold.
p. cm.
Includes bibliographical references and index.
ISBN 978-1-933115-93-1 (hardback : alk. paper) 1. Water resources development–Environmental aspects–Southwestern States. 2. Water-supply–Southwestern States–Management. 3. Climatic changes–Southwestern States. 4. Water-supply–Economic aspects–Southwestern States. I. Colby, Bonnie G. II. Frisvold, George B.
HD1695.S664A33 2010
333.9100979–dc22 2010035305

A catalogue record for this book is available from the British Library

About Resources for the Future *and* RFF Press

Resources for the Future (RFF) improves environmental and natural resource policymaking worldwide through independent social science research of the highest caliber. Founded in 1952, RFF pioneered the application of economics as a tool for developing more effective policy about the use and conservation of natural resources. Its scholars continue to employ social science methods to analyze critical issues concerning pollution control, energy policy, land and water use, hazardous waste, climate change, biodiversity, and the environmental challenges of developing countries.

RFF Press supports the mission of RFF by publishing book-length works that present a broad range of approaches to the study of natural resources and the environment. Its authors and editors include RFF staff, researchers from the larger academic and policy communities, and journalists. Audiences for publications by RFF Press include all of the participants in the policymaking process—scholars, the media, advocacy groups, NGOs, professionals in business and government, and the public. RFF Press is an imprint of **Earthscan**, a global publisher of books and journals about the environment and sustainable development.

The RFF Press Water Policy Series

Books in the RFF Press Water Policy Series are intended to be accessible to a broad range of scholars, practitioners, policymakers, and general readers. Each book focuses on critical issues in water policy with the mission to draw upon and integrate the best scholarly and professional expertise concerning the physical, ecological, economic, institutional, political, legal, and social dimensions of water use. The interdisciplinary approach of the series, along with an emphasis on real world situations and on problems and challenges that recur globally, are intended to enhance our ability to apply the full body of knowledge that we have about water resources—at local, country, regional, and international levels.

We welcome new contributions to the series. For editorial queries about the RFF Press Water Policy Series, please write to waterpolicy@rff.org

Contents

List of Figures

Contributors

Bonnie G. Colby is a professor of agricultural and resource economics at the University of Arizona. She is a coeditor or coauthor of *Arizona Water Policy: Management Innovations in an Urbanizing, Arid Region, Negotiations over Tribal Water Rights,* and *Braving the Currents: Resolving Conflicts over the Rivers of the American West.*

George B. Frisvold is a professor and extension specialist in the Department of Agricultural and Resource Economics at the University of Arizona. He is coeditor of *Global Environmental Change and Agriculture.*

Katherine Adams is a doctoral candidate in the Department of Agricultural and Resource Economics at the University of California–Davis.

Rosalind Bark is a resource economist with the Commonwealth Scientific and Industrial Research Organisation (CSIRO) in Adelaide, Australia.

Elisabeth Basta completed her master's degree in the Department of Agricultural and Resource Economics at the University of Arizona in December 2010.

Shailaja Deva is a credit policy and risk analyst for HSBC Bank in Elmhurst, Illinois.

Kyle J. Emerick is a doctoral candidate in the Department of Agricultural and Resource Economics at the University of California–Berkeley.

Lana Jones is an economist, educator, and writer based in southern Arizona.

Kazim Konyar is a professor in the Department of Economics at the California State University–San Bernardino.

Xudong Ma is a doctoral candidate in the Department of Applied Economics at the University of Minnesota.

Anand Murugesan is a doctoral candidate in the Department of Agricultural and Resource Economics at the University of Maryland.

Srinivasa Ponnaluru is a postdoctoral research associate at the IMPACT Center of the School of Economic Sciences at Washington State University.

Paulo Tanimoto is an economist and energy consultant with KEMA, an international energy consulting firm, in Madison, Wisconsin.

Acknowledgments

The authors acknowledge former and current research colleagues with the Climate Assessment for the Southwest (CLIMAS) program at the University of Arizona. This work was supported by that project through the National Oceanic and Atmospheric Administration's (NOAA) Climate Program Office, grant NA16GP2578.

Colby also appreciates her research colleagues on the University of Arizona Enhancing Water Supply Reliability research project for their contributions over the years of the project and the Bureau of Reclamation for its generous support and interest.

The work reported in this book was also supported by a NOAA Sector Applications Research Program (SARP) grant. The statements, findings, conclusions, and recommendations are those of the research team and do not necessarily reflect the views of NOAA, U.S. Department of Commerce, or the U.S. government Climate Program.

In addition, we acknowledge support provided through the U.S. Department of Agriculture's Multistate Research Project, W2190, Water Policy and Management Challenges in the West, and support from the Marvin and June Morrison Chair of Agribusiness and Resource Management at Arizona State University.

We thank our anonymous reviewers for their helpful input in improving the manuscript, and Chuck Howe for his generous and thoughtful Foreword.

We appreciate the collegiality and intellectual stimulation provided by our colleagues in Agricultural and Resource Economics at the University of Arizona. We are grateful for the expert and timely staff assistance in graphics and editing, accounting, and graduate student support provided by Nancy Bannister, Nancy Smith, and Elizabeth Tilley.

Any errors or omissions are the responsibility of the authors.

We dedicate this book to our children: George, Peter, and Korey (Bonnie); and Dana and Alec (George). Their generation undoubtedly will continue to wrestle with the challenges we discuss here. It's a good thing they are smarter than we are. They will need to be!

Foreword

Questions of adaptation to changing natural resource conditions extend back at least to the beginnings of the industrial revolution. The definition of resources and the confining conditions surrounding their use have changed over time. For example, by the time of Elizabeth I's reign in England, circa 1558, the competing demands on the nation's forests for shipbuilding, charcoal for smelting, and cleared land for agriculture created sufficient concern that restrictions were placed on the use of trees for fuel in areas close to the coast. Timber shortages pushed the smelters of iron, vital to the emerging industries, to attempt to substitute coal for charcoal, at which they were unsuccessful until Abraham Darby developed a new process using coke in 1709. A century and a half later, economist and philosopher Stanley Jevons (1865) worried about the exhaustion of England's coal. Earlier, Thomas Malthus (1798) had raised concerns about the tendency of the human population to increase faster than the food supply provided by agriculture.

Michael Graham's fabled "whale oil story" (1973), covering the era from medieval whaling off the coast of France in the ninth century to the U.S. discovery of petroleum as a resource in the 1860s, clearly illustrates the broad systems context within which changing demands, invention, and substitution can lead to resilience in the face of changing scarcities. Harold Barnett and Chandler Morse's classic *Scarcity and Growth* (1963) again illustrated this process, arguing that discovery and inventiveness had effectively prevented natural resource scarcity, with the possible exceptions of the agricultural and forestry sectors. However, Barnett and Morse also suggested that the environmental impacts of natural resource extraction and use threatened to increase costs and limit resource use.

Attention to water as an important resource in the United States was initially in the context of fisheries and transportation. Albert Gallatin's brilliant report on roads and canals (1808) not only emphasized the importance of internal navigation to the westward expansion of the country, but also outlined the basic principles of cost-benefit analysis for comparing alternative public investments. Theodore Roosevelt's establishment in 1902 of the Reclamation Service (later the Bureau of Reclamation) established irrigated agriculture as the base of westward expansion. In 1907, Roosevelt appointed the Inland Waterways Commission to develop policies and procedures for multiple-purpose river development (Hays 1959). Water quality as a dimension of water's value did not become a significant policy issue until the 1953 *Report of the Mid-Century Conference on Resources for the Future*, and water quality legislation was not passed until the early 1970s.

Although climate itself cannot be counted as a resource, it certainly defines an important part of the framework within which all natural resources are extracted and used. The variability of climate initially received attention in the context of drought. Advances in paleohydrology have clarified the existence of long swings in the water environment, such as in precipitation and streamflow. While these studies have greatly expanded the databases for understanding climate variation, they have also called into question the methods of hydrologic analysis that have been based on the stationarity of hydrologic regimes. We now recognize that climate is changing over the long term. The work of the Intergovernmental Panel on Climate Change has made it clear that for all practical planning purposes, we are faced with climate change, especially in arid regions like the U.S. Southwest (Watson et al. 1996; McCarthy et al. 2001).

This volume represents a collection of studies that adds to this long tradition of natural resource and environmental concern by analyzing some of the implications of climate variability and change for the southwestern United States. The vulnerabilities of particular agricultural areas and crops are identified, along with potential impacts on water and electric power demands. Implications for outdoor recreation—in particular, winter sports and national park visitation—are also investigated.

Major mechanisms of adjustment to the changing environment are evaluated through theoretical formulations and empirical case studies. The greater use of weather and climate information in agriculture, which accounts for more than 80% of total consumptive water use, promises to improve the choice of technologies and crops, especially those related to water conservation. Curiously, the authors point out that under some conditions, increased irrigation efficiency may not actually free up water for use in other sectors.

Increasing short- and long-term flexibility in the allocation of existing water supplies through water markets or marketlike procedures is a major adaptive mechanism. Case studies in Arizona, New Mexico, and California draw lessons from the effective functioning of these water market arrangements. Transaction costs and third-party impacts of water transfers impede and complicate the transfer process.

The findings and recommendations in this book will be useful to federal, state, and regional policymakers who set the legal and regulatory framework for the more effective use of climate information and efficient functioning of water markets.

Chuck Howe
Professor Emeritus of Economics,
University of Colorado at Boulder

REFERENCES

Barnett, Harold, and Chandler Morse. 1963. *Scarcity and Growth: The Economics of Natural Resource Availability.* Baltimore: Johns Hopkins University Press.

Gallatin, Albert. 1808. *Report of the Secretary of the Treasury on the Subject of Public Roads and Canals.* Washington, DC: R.C. Weightman.

Graham, Michael. 1973. *A Natural Ecology.* Manchester, UK: Manchester University Press.

Hays, Samuel P. 1959. *Conservation and the Gospel of Efficiency: The Progressive Conservation Movement,* 1890–1920. Cambridge, MA: Harvard University Press.

Jevons, W. Stanley. 1865. *The Coal Question: An Inquiry concerning the Progress of the Nation, and the Probable Exhaustion of Our Coal-Mines.* London: Macmillan and Company.

Malthus, Thomas Robert. 1798. *An Essay on the Principle of Population.* London: J. Johnson, in St. Paul's Church-yard.

McCarthy, J.J., O. Canziani, N.A. Leary, D.J. Dokken, and K.S. White, eds. 2001. *Climate Change 2001: Impacts, Adaptation and Vulnerability.* Contribution of Working Group II to the Third Assessment Report of the Intergovernmental Panel on Climate Change. Cambridge, UK: Cambridge University Press.

Resources for the Future. 1954. *The Nation Looks at Its Resources: Report of the Mid-Century Conference on Resources for the Future, Washington, D.C., December 2, 3, 4, 1953.* Washington, DC: Resources for the Future.

Watson, R.T., M.C. Zinyowera, and R.H. Moss, eds. 1996. *Climate Change 1995: IPCC Second Assessment Report: Scientific-Technical Analyses of Impacts, Adaptations, and Mitigation of Climate Change.* Cambridge, UK: Cambridge University Press.

Acronyms and Abbreviations

ac-ft	acre-feet
AEPCO	Arizona Electric Power Cooperative
AR	autoregressive
AR(1)	first-order autocorrelation
BB&B	Bed, Board and Booze
CAP	Central Arizona Project
CCP	countercyclical payment
CES	constant elasticity of substitution
CIMIS	California Irrigation Management Information System
cm	centimeters
CPC	Climate Prediction Center
CV	contingent valuation
DF	degrees of freedom
EIS	environmental impact statement
EMP	elasticity of the marginal product
ENSO	El Niño Southern Oscillation
EPA	U.S. Environmental Protection Agency
EQIP	Environmental Quality Incentive Program
ERS	Economic Research Service (of USDA)
ET	evapotranspiration
FRIS	Farm and Ranch Irrigation Survey
FTE	full-time equivalent
GCM	General Circulation Model
GW	gigawatts

IID	Imperial Irrigation District
IPCC AR4	Intergovernmental Panel on Climate Change Fourth Assessment Report
kPa	kilopascals
kWh	kilowatt-hours
LDP	loan deficiency payment
m	meters
m^2	square meters
m^3	cubic meters
MA	moving-average
MAPE	mean absolute percentage error
max.	maximum
MEI	Multivariate ENSO Index
MGM2	Money Generation Model 2
min.	minimum
MJ	megajoules
mm	millimeters
m/s	meters per second
MSE	mean squared error
MVP	marginal value product
MW	megawatts
MWD	Metropolitan Water District (of Southern California)
NEPA	National Environmental Policy Act
NHP	national historical park
NHS	national historic site
NM	national monument
NMem	national memorial
NOAA	National Oceanic and Atmospheric Administration
NP	national park
NPS	National Park Service
NRA	national recreation area
NRCS	Natural Resources Conservation Service (of USDA)
NROVC	net returns over variable costs

NRTW	net returns to water
NWS	National Weather Service
PCSE	panel-corrected standard error
PDO	Pacific Decadal Oscillation
PMP	Positive Mathematical Programming
ppm	parts per million
PRE	proportionate reduction in error
PVID	Palo Verde Irrigation District
QSA	Quantification Settlement Agreement
REAP	Regional Environment and Agriculture Programming
RMSE	root mean squared error
SAS	statistical analysis software
SBC	Schwartz Bayesian criterion
SDCWA	San Diego County Water Authority
SM	Southern Mountain (region)
SNWA	Southern Nevada Water Authority
SPI	Standard Precipitation Index
SS	skill score
SST	sea-surface temperature
std. dev.	standard deviation
TSCS	time-series–cross-section
USARM	U.S. Agricultural Resources Model
USBR	U.S. Bureau of Reclamation
USDA	U.S. Department of Agriculture
USFS	USDA Forest Service
USGS	U.S. Geological Survey
USLE	universal soil loss equation
WTP	willingness to pay

CHAPTER 1

The Climate-Water-Energy Nexus in the Arid Southwest

Bonnie G. Colby and George B. Frisvold

This book explores the challenges that climate change and climate variability pose for water and energy managers and users, communities, and policymakers in the arid U.S. Southwest and demonstrates the application of economic methods to analyze and respond to these challenges. It provides valuable tools for both those interested in resource management and climate change and those seeking to understand how economic methods can be used to understand contemporary social problems and design appropriate responses. Throughout the book, we apply a resilience framework (B. Nelson et al. 2007). The concept of adaptation focuses on how individual economic agents or institutions respond to variability or longer-term change in climate. A resilience framework follows a broader systems approach, going beyond considering adaptive behaviors of individual actors or institutions. Resilience is a measure of a system's ability to undergo changes while still maintaining its fundamental structure and function. Individual adaptation strategies can be assessed in terms of individual incentives and effects, but they also need to be considered in terms of broader systematic effects.

The book uses a broad definition of the Southwest. Most chapters focus on Arizona, New Mexico, or Southern California, or combinations of these areas. Problems of water and energy management are not confined within fixed boundaries, however, and thus in some cases, the scope of our analysis extends to others of the seven Colorado River Basin states (Wyoming, Utah, Colorado, Nevada) or to Mexico.

Figure 1.1. *The Colorado River Basin and Surrounding Areas*

THE INTERRELATIONSHIP OF WATER AND ENERGY RESOURCES

Water and energy use are inextricably linked, a phenomenon now widely referred to as the "water-energy nexus" (Larson et al. 2007; Goldstein et al. 2008). Large amounts of energy are used in the extraction, conveyance, treatment, and distribution of water. At the same time, significant portions of southwestern water supplies are used in energy production. Energy and water issues customarily have been the concern of separate federal and state agencies, disparate public and private utilities, and differing user groups and other stakeholders. Because of this, electricity production planning has neglected to consider impacts on regional water needs, and planning for water supply and delivery and wastewater treatment has not taken into account strains on regional electricity supplies and costs. Additionally, environmental effects are neglected in both water and energy planning.

A U.S. Department of Energy report (DOE 2006) to Congress on the interdependency of energy and water in the United States highlights how regional water use is inextricably linked to energy use and vice versa. The population of the Southwest has been projected to increase substantially, and hence a large portion of growth in U.S. electricity demand over the next 25 years is likely to occur in the Southwest, where water supplies are limited. Climate change models also indicate that water supplies will both decrease on average and exhibit more year-to-year variability. Increases in the scarcity and costs of water and energy could become constraining factors on economic and population growth in the arid southwestern states. Various energy technologies can reduce energy-related water demand, but resource managers and policymakers must jointly consider energy and water development and the relative costs of each resource. Integrated coordination across the energy and water sectors will be essential.

Figure 1.2 shows water consumption for various stages of energy production, and Figure 1.3 shows energy consumption for various water conveyance and treatment needs, both based on studies from the western United States. Actual values will be location-specific, but the figures highlight the significance of the water-energy interrelationship.

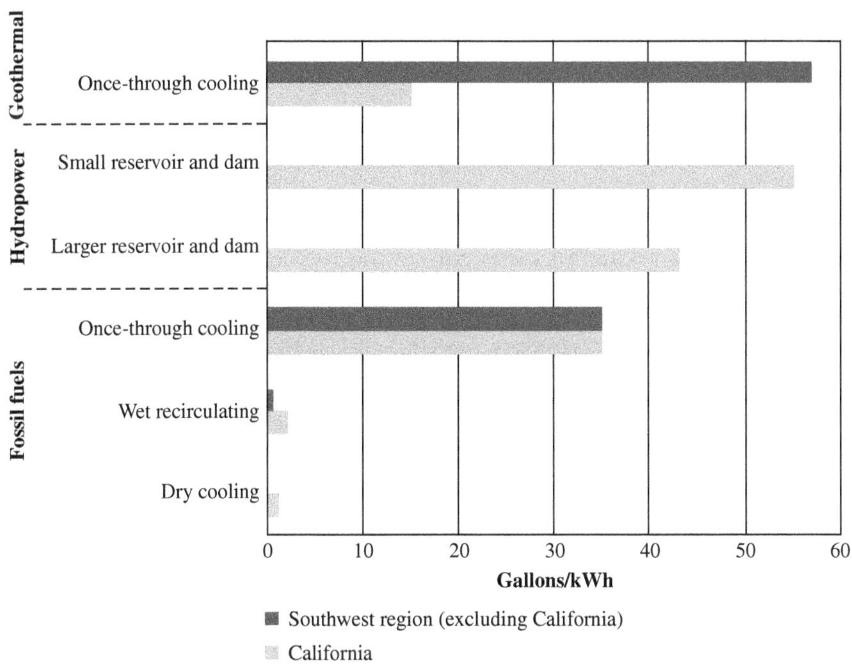

Figure 1.2. *Water Demand for Energy Production in California and the Southwest*

Source: Adapted from Larson et al. (2007)

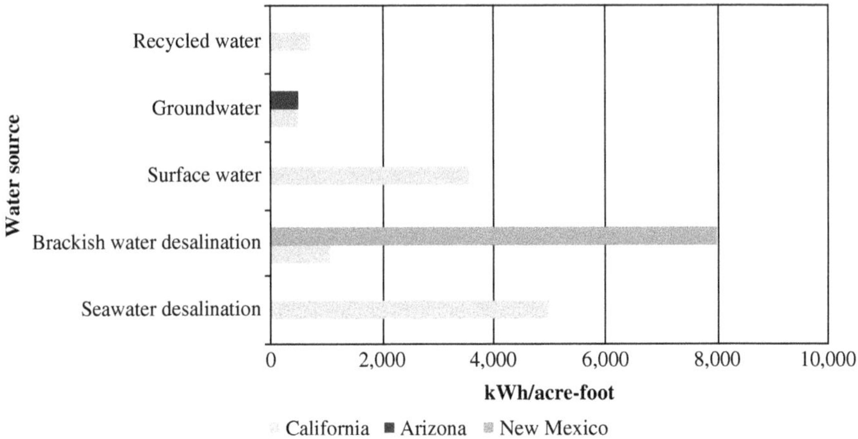

Figure 1.3. *Energy Demand for Water Treatment and Transport in Select States*

Source: Adapted from Wilkinson (2007)

Water requirements for energy production vary greatly with the primary energy source and with conversion and cooling technologies. Bioenergy from irrigated crops, hydropower, and thermoelectric plants with once-through cooling have the largest water requirements per kilowatt-hour (kWh). Dry-cooled thermoelectric facilities, solar photovoltaic, bioenergy from waste, and wind turbines have the lowest water requirements.

Energy requirements related to water use are highly location-specific but are significant. The California Energy Commission (CEC 2005) estimates that nearly 20% of California's electricity demand and more than 30% of the state's natural gas demand are linked to water use. Energy is used in various stages of the water use cycle: in diverting surface water and pumping groundwater, in conveying water from its source to points of use, and in treating water prior to use, particularly for potable uses. End residential users consume energy with in-home softeners or filters, and agricultural users through irrigation systems. Collecting and treating wastewater also consume significant amounts of energy. Wastewater treatment plants in California use 500 to 1,500 kWh per acre-foot (ac-ft) of water treated (Larson et al. 2007).

CLIMATE CHANGE PROJECTIONS AND IMPLICATIONS FOR RESOURCE MANAGEMENT

Climate change is predicted to produce significant and challenging changes in the hydrologic cycle of the U.S. Southwest. Models predict that snow levels will retreat upward by 500 feet in elevation for each increase of 1 degree Celsius (1.8 degrees Fahrenheit), and floods and large storms are expected to increase in size and frequency, overwhelming flood control systems and making floodplains and low-elevation coastline areas vulnerable to significant damage (D.R. Nelson et al.

2007). Sea levels are anticipated to rise 7 to 23 inches by 2100, with negative effects on freshwater supplies, estuaries and wetlands, and coastal development and infrastructure (IPCC 2007). Another study projects a sea level rise of 20 to 55 inches by 2100, threatening coastline areas as well as groundwater and surface water located near the coast and vulnerable to saltwater intrusion (Rahmstorf 2007). Higher temperatures will affect fish habitat, with studies estimating that up to 38% of stream reaches suitable for cold-water fish may be too warm to provide habitat by 2090 (D.R. Nelson et al. 2007).

All of these changes will challenge both the water supply infrastructure and the elaborate body of laws and policies that govern water management and use. These were designed for seasonal timings and magnitudes of runoff from historical temperature, precipitation, and snowmelt patterns. Climate change and variability affect all elements of water management, from watershed catchment areas to storage, conveyance systems, and wastewater treatment. Regional water systems in the Southwest are already stressed by overdraft and contamination of groundwater, saltwater intrusion in coastal aquifers, overappropriation of surface water, and population growth with increased water demand.

Under current stressors of finite water supplies and increasing demand, water providers seek additional reliable sources from surface water, groundwater, recycled water, and desalinated water. Given the implications of climate change for water and energy resources, it will become increasingly important to consider how various water supply alternatives compare when energy costs are considered.

Energy use associated with surface water varies considerably with location and topography of source and destination. Surface water is pumped long distances and over mountain ranges in many areas. For instance, conveying water from Northern to Southern California consumes 3,000 kWh/ac-ft because of the need to pump the water over the high lift of the Tehachapi Mountains. Even gravity conveyance systems typically require pumping into and out of reservoirs (Cohen 2007).

Energy requirements for extraction and delivery of groundwater depend on depth to the groundwater and pump efficiencies. In California, energy requirements for groundwater pumping may range from 292 kWh/ac-ft along the central coast to 740 kWh/ac-ft in the Central Valley (Cohen 2007). In southern Arizona, energy demands for pumping may range from 146 to 730 kWh/ac-ft, given lifts of 100 to 500 feet and assuming a 70% energy-efficiency pump (Seasholes 2008).

Energy required for treatment of wastewater and transport to the point of use ranges from 325 to 1,000 kWh/ac-ft in California (CEC 2005). Wastewater reuse has become more common and may be an energy-efficient supply alternative in some areas (Cohen 2007).

Seawater desalination as a water supply has been limited because of the high cost associated with its high energy consumption, about 40% of overall desalination costs. However, costs vary widely with plant design and the quality of the source water. Energy demand estimates for proposed California seawater desalination plants range from about 4,400 to 5,500 kWh/ac-ft. A brackish inland water source may require less energy; two such facilities in Southern California require only 405 and 1,700 kWh/ac-ft (Cohen 2007). When considered in New Mexico in 2004, a range of 9,775 to 16,293 kWh/ac-ft was estimated (NM ENV 2004).

All interests involved in water supply planning and financing must factor in the predicted impacts of climate change. Energy-conserving options are likely to prove the best investments under the more variable conditions anticipated with climate change. Wastewater reuse and conservation can help regions meet demand for reliable water supplies, while simultaneously saving on energy costs and reducing emissions that contribute to climate change (D.R. Nelson et al. 2007). Water and energy managers will need to consider the significant and unevenly distributed effects of climate change across locations, direct water users, and other stakeholders. Climate change is likely to decrease the yields from surface storage and diversion projects. Projected changes in the hydrological cycle will make reservoir management more challenging as a result of more intense trade-offs between storing water against drought and releasing stored water prior to the flood season to prevent downstream flood damages. Under projected increased severity of both drought and flood periods, it will be more difficult to store adequate water to meet delivery obligations without also increasing flood risk (D.R. Nelson et al. 2007).

Many argue that increased investment in water and energy efficiency represents a "no regrets" approach to addressing climate change impacts (D.R. Nelson et al. 2007). Water conservation results in both water and energy savings, as does energy conservation. Traditionally separate arenas, water agencies and energy utilities can develop joint programs to encourage customers to conserve both water and energy. Although such investments need to be scrutinized for cost-effectiveness and unintended socioeconomic impacts, they can reduce carbon emissions and free up water supplies to benefit water-dependent aquatic ecosystems. Local, state, and federal planning and coordination among agencies need to incorporate climate vulnerability analyses and impacts in management tools.

A regional water- and energy-efficient portfolio that is well suited to address climate variability demands the right balance of primary energy sources and generating and cooling technologies, as well as an optimal mix of surface water, groundwater, and recycled water, along with conservation programs for both water and energy. Energy utilities should focus on water-efficient electricity generation, and water utilities need to focus on energy-conserving water conveyance and treatment options. A careful life-cycle assessment of electricity generation alternatives with a focus on water use will clarify the full water requirements of electricity generation. Similarly, a life-cycle assessment of water supply options can highlight the energy implications (Larson et al. 2007).

Specific use sectors, such as municipalities, manufacturing plants, and orchards and vineyards, place a high value on reliability in both their energy and water supplies. Integration of energy and water planning must consider the varying reliability needs and willingness to pay among different use sectors. In addition, environmental impacts need careful consideration. Extensive solar panel areas, for instance, may reduce pollution and generate energy using little water, but significantly affect habitat and regional biodiversity. While some common principles may apply across locations, impacts will be region-specific. As the current emphasis on devoting cropland to producing biofuels illustrates, the production of dedicated energy crops may be more economically and environmentally sound in some regions than others.

CLIMATE, CLIMATE VARIABILITY, AND CLIMATE CHANGE

People and institutions adapt to climate, climate variability, and climate change. Climate refers to long-run average patterns of temperature, precipitation, and other weather variables. The climate of the Southwest is hot and dry, with bimodal rainfall patterns of winter storms and summer monsoons. Cropping and recreation patterns have evolved over time on account of the pervasive climate of the region. Irrigators have adjusted crop mixes on account of long growing seasons and high water requirements. The area along the mainstem of the Lower Colorado River has grown into a national center of year-round fruit and vegetable production. At the same time, few recreationists venture into low desert regions at the height of summer, when temperatures can exceed 110 degrees F.

Interannual climate variability refers to regular deviations in any given year from "normal" climate. Water management institutions have evolved in the Southwest to smooth water deliveries in response to variability. River systems are heavily managed with reservoir systems allowing lakes to be drawn down in drought years and replenished in wet years. Groundwater use is also heavily regulated in some areas. In agriculture, growers can adopt irrigation technologies and water management practices that afford them flexibility in response to year-to-year variations in temperature or precipitation. Water and energy managers can also use weather forecast information to better balance energy and water demands with supplies.

Climate change refers to long-term changes in both averages (such as average temperature or rainfall) and climate variability (year-to-year or seasonal fluctuations and probabilities of extreme events). Many strategies to adapt to climate variability such as flooding or droughts may aid in adaptation to climate change. For example, irrigation practices that allow producers to adapt to consecutive years of drought may help them adapt to a future where what are now considered drought conditions become the new normal conditions. Market mechanisms that facilitate temporary water leasing across regions or uses may pave the way for more long-term dry-year transfer contracts in the future. In other cases, adaptation thresholds may be exceeded, so that previously successful minor adjustments in behavior prove insufficient to cope with long-term changes in climate.

In 1995, the Powell Consortium completed the research project Severe, Sustained Drought in the Southwestern United States (SSD study).[1] Although resilience terminology was not used at the time, this project can be viewed as an early application of a resilience framework. It considered the question of how the Colorado River Basin would adapt to a multiyear megadrought. Original analysis was based on tree-ring data from the 1579–1600 megadrought, considering what would happen if such a drought recurred. However, the regional water management system proved remarkably resilient to this scenario.

Researchers then reordered the years of drought occurrence to increase the number of consecutive years of below-average Colorado River flows and thus the severity of the drought impacts. Even with this change, the Colorado River management system proved quite resilient in the face of prolonged drought because of its extensive storage capacity—approximately four years of average flows, according to Rajagopalan et al. (2009). Consumptive uses, particularly those in the Lower Basin, were well protected; the greatest losses were felt by nonconsumptive uses (hydropower and water-based recreation), followed by Upper Basin consumptive uses.

The SSD study made several important contributions. First, it highlighted the importance of impact assessments of climate-sensitive sectors of the regional economy, including agriculture, water-based recreation, and hydropower generation. Second, it showed the importance of the climate-water-energy nexus. Specifically, there were important trade-offs in maintaining reservoir levels for hydropower and maintaining consumptive (primarily agricultural and urban) uses of water. Third, it illustrated the role of reservoir management in buffering water users from effects of prolonged drought and the ability of that system to maintain system resilience. And fourth, it demonstrated how reallocating water uses across activities and locations could greatly mitigate the costs of prolonged drought, indicating the value of water transfer mechanisms as a drought adaptation strategy.

Kenney et al. (2010) have pointed out, however, that the SSD study significantly underestimated future water use in central Arizona. The SSD study was completed just as the Central Arizona Project (CAP) was also being completed. Today the CAP delivers 1.5 million ac-ft of water to central Arizona, about three times more water than the SSD predicted (Kenney et al. 2010). This additional demand places a greater burden on the Colorado River management system. Because CAP water deliveries have the most junior water rights among Lower Colorado River users, central Arizona users are the most vulnerable in the event that the U.S. Bureau of Reclamation declares a shortage on the river and begins to curtail deliveries. New shortage sharing rules negotiated among the seven basin states are intended to provide flexibility to address some of these vulnerabilities (Rajagopalan et al. 2009).

This book revisits some of the themes of the SSD study. Projections of future climate in the region suggest warmer temperatures and less runoff, similar to conditions considered in the SSD scenarios. The book examines the role of water reallocation mechanisms as an adaptive response to water shortages, focusing on market-based mechanisms, specifically short-term water lease markets, large-scale agricultural-urban water transfer agreements, and dry-year option contracts. It also provides new estimates of sector impacts of climate change, considering agriculture and outdoor recreation. The climate-water-energy nexus is an important consideration throughout the book. Maintaining regional hydropower production is an important consideration in regional water management, affecting institutional strategies to adapt to climate change. Unlike the SSD study, however, this book devotes more attention to the role of modern irrigation technologies and information systems as mechanisms to adapt to climate change.

ORGANIZATION OF THE BOOK

This book is divided into three main parts. Chapters 2 through 4 examine the role of a new generation of innovative market mechanisms that facilitate transfers of water between uses and regions. The use of such innovative mechanisms, beyond the "buy and dry" approach common for much of the past three decades, has increased considerably in the western United States, and such voluntary incentive-based arrangements will likely play a crucial role in regional climate adaptation. Chapters 5 through 7 examine sector impacts of climate change, focusing on irrigated agriculture and outdoor recreation, including national park visitation and skiing. Chapters 8 through 11 examine the role of information and technology in climate change adaptation, paying particular attention to the role of climate and weather information in agricultural and energy management decisions.

Increasingly, economists work as part of multidisciplinary research teams. This is especially true of agricultural, environmental, energy, water, and climate research. Colleagues in other disciplines, as well as public- and private-sector decision-makers, require measurements of costs and benefits of economic, environmental, and policy changes. Close and fruitful interactions depend on good communication of economic methods and ways of thinking about problems. Economists in turn rely on colleagues in other disciplines for knowledge of physical, environmental, and biological relationships. Insights from anthropology, psychology, and geography also enhance economists' understanding of human behavior.

For noneconomists, we provide here an overview of the economic methods applied in this book to examine implications of climate variability and change for water, energy, agriculture, and recreation. Readers may see which methods might be useful in their own work and be more informed when reviewing studies that apply these economic methods. For each chapter in the book, Table 1.1 summarizes the research questions addressed, the methods and data used to answer those questions, and the main findings. Within the individual chapters, we also point out where and how economists benefit from knowledge from other disciplines. Many chapters include appendixes that provide additional methodological detail for readers particularly interested in the methodologies applied.

Part I: Voluntary Water Transfers as Adaptation Mechanisms

Chapters in Part I of this book consider the scope for voluntary water-trading mechanisms to reallocate water as a climate change adaptation and the effects of climate on the functioning of these markets. Chapter 2 examines water lease markets in California and New Mexico over three decades. Water leases are evaluated in a supply-demand framework in which price and quantity of water leased are determined simultaneously. Climate factors play an important role in determining supply-and-demand relationships. A major finding is that spatial scale is important, with local hydrologic conditions having more of an effect than regional climate on water transfer prices. Lease price also depends on what the lessee intends to do with the water (e.g., whether it is for agriculture, urban use, or environmental restoration).

Table 1.1. *Research Problems, Methods of Analysis, and Results by Chapter*

Chapter	*Research problem*	*Methods of analysis*	*Results*
2	Effects of climate variables on water lease prices; lease price and quantity may be simultaneously determined	Multivariate regression analysis of data on individual transactions; test and correction for simultaneity bias	Leases are effective at reallocating water from low- to high-value uses and smoothing water supply reliability during droughts; price and quantity are determined simultaneously; water lease price is driven more by local hydrologic conditions than by regional climate
3	Effects of technology, economic, and policy variables on large-scale agricultural-urban water transfers; outcomes include water transferred, transfer price benefits to sellers and buyer, and third-party impacts	General cooperative game theory model to derive general results; empirically based numerical simulation model of IID-SDCWA/MWD transfer agreement	Irrigation district collective bargaining increases amount of water transferred; improved irrigation efficiency does not necessarily increase transfers or conserve water; environmental mitigation requirements are highly sensitive to model assumptions
4	Develop and illustrate methodology to identify price ranges at which agricultural water users would be willing to lease water for other uses during dry periods	Residual (farm budget) approach; estimation of net returns to water above variable costs; calculation of value of water in producing different crops	Leasing water is most viable for cotton and wheat, more costly for alfalfa, and prohibitively costly for lettuce; exercise price increases if lease option is exercised after planting; minimum supply prices are estimated
5	Region-level impacts of large reductions in regional water supplies or increases in water prices on agricultural land use, production, prices, farm employment, and farm income	Water price and availability changes simulated using a Positive Mathematical Programming (PMP) model	Region's agriculture (especially specialty crops) is resilient to supply shocks; production of central Arizona crops is vulnerable; water price increases are more costly than water quantity reductions
6	Effects of climate, water availability, environmental variables, economic variables, and population on visits to southwestern National Park System; effects on local economies of climate change–induced changes in visits	Multivariate time-series–cross-section data analysis of visits to southwestern parks; input-output analysis mapping changes in visits to changes in local sales, personal income, and employment	Summer warming decreases visits, but winter warming increases them (except in low desert parks); reduced reservoir levels reduce water-based recreation; effects of population growth and gas prices are measured; effects on local economies are quantified

7	Effects of climate change on southwestern snow-based recreation; economic viability of investments in snowmaking	Multivariate regression model of skier visitation; financial model of returns to investment in snowmaking	El Niño is an important positive predictor of ski visits; ENSO forecasts could increase returns to snowmaking
8	Effects of farm size on use of water management information, irrigation scheduling, investments to conserve water or energy, and conservation program participation	Univariate, categorical data analysis of data from Arizona and New Mexico; nonparametric measures of association; Cochrane-Armitage trend test	Small farms less likely to investigate irrigation improvements and use management-intensive irrigation scheduling; larger farms use more data sources, especially private ones
9	Impacts of farm size, climate, energy costs, water sources, and soil characteristics on choice of irrigation technology	Multivariate regression analysis of state-level adoption data stratified by farm size; data for 17 westernmost contiguous states	Sprinkler irrigation adoption increases with farm size, soil erodibility, energy costs, rainfall, and temperatures; high temperatures and long growing seasons decrease adoption rates
10	Use of weather and climate information to improve electricity load forecasts; use of improved specification of seasonality	Time-series econometric forecasting model; quantitative evaluation of forecast accuracy (MAPE, RMSE, and forecast skill)	New models that include weather and climate data outperform traditional forecasting models that do not; temperature is the most important weather variable
11	Influences of Arizona farmer and rancher personal characteristics and operation on use of weather data for management decisions	Multivariate discrete choice and count data analysis of farm-level data	Production diversification within agriculture encourages data use; off-farm income discourages use; risk attitudes affect data use, but Internet access does not

Notes: **IID-SDCWA/MWD = Imperial Irrigation District and San Diego County Water Authority/Metropolitan Water District; ENSO = El Niño Southern Oscillation; MAPE = mean absolute percentage error; RMSE = root mean squared error**

The findings in Chapter 2 add to our knowledge base on temporary and voluntary water leases as a climate adaptation strategy. Persons and organizations charged with securing reliable water supplies for cities and for environmental restoration in the face of increased supply uncertainty can anticipate increased per-unit costs when prices are negotiated in the midst of a dry period rather than in advance; increases over time in water leases as a mechanism to secure water supplies; and different lease costs for different types of water users.

The transactions analyzed in Chapter 2 are not the anonymous perfectly competitive market of basic economics textbooks. Water markets are often what economists call "thin." There are few buyers and sellers, and the special needs of each and their relative bargaining power matter. Indeed, some scholars have been critical of the perfectly competitive market paradigm being applied to water transactions. Evaluations of water markets have been criticized for paying insufficient attention to the bilateral nature of water market transactions, asymmetries in bargaining power, negative third-party impacts,[2] transaction costs, and harm to the environment (Colby 1990; Bauer 2004; Merrett 2003).

Chapter 3 applies game theory to address these neglected issues by considering negotiations over a large-scale agricultural-urban water transfer between an irrigation district and a large municipal water provider. Such negotiations are similar to collective-bargaining negotiations between a large labor union and a large employer. The difference is that the bargaining parameters are the price and quantity of water transferred, rather than jobs and wages. In perfectly competitive market models, individuals exert little influence on market outcomes. In contrast, game theory focuses on cases where institutions may act strategically. Economic actors can influence outcomes and know they can do so. Analysis of perfectly competitive markets focuses on efficiency of market-based outcomes and often downplays the distribution of gains from trade. Game theory, in contrast, explicitly considers the exercise of monopoly power, distribution, and often highly asymmetric benefits of exchange. By dealing with this "dark side" of exchange, game theory can overcome some of the distrust other social scientists, environmental advocates, and environmental managers often have of economics and reliance on market-oriented outcomes.

Chapter 3 illustrates that collective bargaining by irrigation district members will lead to more, not less, water transferred to an urban buyer (although district members receive a better price for their water). Thus large-scale water transfer agreements may be insufficient to prevent negative third-party impacts unless those affected parties are brought into negotiations. In times of drought, urban buyers may demand more water, but irrigators may require water for crops. Thus transfers may not be a panacea for cities wishing to supplement water supplies in times of drought. Spatial scale of drought is important. One needs to consider hydrologic conditions in both the sending and receiving locations of water transfers.

The chapter develops a numerical simulation model based on the Quantitative Settlement Agreement (QSA) involving the Imperial Irrigation District (IID) and San Diego County Water Authority (SDCWA), the United States' largest agricultural-urban water transfer agreement. Results suggest that without specific mitigation requirements to protect return flows to the Salton Sea, improved irrigation efficiency increases consumptive use of water. Although the current mitigation requirements under the QSA appear sufficient to protect return flows and reduce consumptive use, this result is highly sensitive to small changes in model parameters. This suggests that ongoing hydrologic monitoring and adaptive environmental management will be needed to obtain better estimates of the parameters that influence results the most and to mitigate as-yet-unanticipated damages. Even with these features in place, the transfer agreement will reduce available water for ecosystem services and agriculture across the border in Mexico. These individual adaptations to water scarcity may not further the resilience of the system as a whole.

Chapter 4 considers the economic informational needs for negotiating prices with farmers to acquire senior agricultural water when needed to increase the reliability of water supplies for urban or environmental uses. Dry-year transfers provide cities and habitat restoration programs the flexibility to lease water only when supply augmentation is necessary. Instead of incurring the economic and political costs of purchasing permanent water rights, temporary transfers allow for year-to-year decisions about leasing additional water. Temporary water leases and irrigation forbearance could spare regional economies from significant economic losses in sectors facing climate-related supply variability.

Because irrigated agriculture accounts for the vast majority of consumptive water use in the western United States, agreements to make agricultural water available to others during extended drought will be essential. The cost of acquiring water from agriculture depends fundamentally on the value of water in agricultural production. Information about the returns to water in agriculture is valuable for establishing a baseline for water transfer negotiations.

Chapter 4 also considers the value of water in agricultural production in western Arizona, where producers have senior water rights to Colorado River water and therefore could be reliable suppliers of water under dry-year option contracts. A common method for valuing irrigation water, and the one applied in the chapter, is the residual or farm budget method. This method constructs models of representative farms with production characteristics typical of a given area. Data generally obtained from farmer and extension agent interviews are used to calculate average net returns to water (NRTW) per acre for each crop. NRTW provides a baseline to inform water negotiations, as it represents the maximum value an irrigator can pay for water while still covering the variable costs of production.

Results from the NRTW exercise suggest that water used for cotton and wheat production have the lowest returns and would be the first acres fallowed to supply water under option contracts. Taking alfalfa out of production is more costly, while taking high-valued lettuce out of production would be prohibitively costly. The cost of forbearing crop production varies significantly with the timing of water deliveries under option contracts. Requiring water deliveries after field crops have been planted or at times or in quantities that interfere with production of high-value crops will drive up the price of transferred water.

Part II: Sector Impacts of Climate Change

Chapters in the second part of this book examine the effects of climate change on vulnerable sectors and areas of the Southwest by looking at impacts on three climate-sensitive activities: agriculture, national park visitation, and skiing. The chapters consider different approaches to estimating climate change impacts and various mechanisms that agricultural and outdoor recreation sectors may use to adapt.

Water management in the Southwest is a story of systemwide resilience but local vulnerability to climate change. The region has extensive reservoir, canal, and river management systems that buffer municipal water users against drought. Reservoirs are managed to protect hydropower supplies. Allowing reservoir levels to be drawn down in drought years stabilizes water supplies for urban and hydropower use but increases the vulnerability of communities dependent on water-based recreation. While irrigators along the mainstem of the Colorado River have senior water rights, protecting them from water shortages, central Arizona irrigators have the most junior rights to Lower Colorado River water. Not only will central Arizona irrigators be among the first to feel impacts of areawide water shortages, but they also face some of the region's greatest urban development pressure.

Chapter 5 considers effects of pervasive water shortages in southern mountain states, using a rationing model approach to evaluating effects of shortages. The rationing model is an extension of the residual valuation method employed in Chapter 4. In both cases, there is little scope for agricultural producers to adapt to water shortages, other than fallowing irrigated land. Chapter 5 compares rationing model results with those from an agricultural production model. The agricultural model allows several types of adaptation. Growers can change their crop mixes from irrigated to dryland production or switch from high- to low-water-using crops. They may also reduce water application intensity (deficit irrigation). Changing cropping patterns and irrigation intensity can greatly reduce the costs of water shortages to agriculture. While high-value specialty crops of the type grown along the Colorado River mainstem are little affected, model results suggest significant reductions in acreage and water use for crops grown in central Arizona. Further, commodity programs based on supporting producer prices do a poor job at protecting producers from drought risk.

Chapter 6 examines the effects of climate change on national park system visitation. Analysis proceeds in two stages. First, the authors use multivariate regression analysis to quantify annual visits to parks, which are affected by climate variables, gasoline prices, regional economic and population growth, and park attributes. Regression results are then combined with input-output analysis to trace changes in visits to changes in recreation-based sales, local incomes, and local employment in park gateway communities. Management of Lakes Mead and Powell has important implications for water-based recreation and the communities that rely on it. Water levels in these reservoirs are as much the product of human management decisions as of long-term hydrologic conditions. Reservoir managers must consider trade-offs among providing water for agriculture, cities, hydropower, and water-based recreation. To evaluate trade-offs and water shortage scenarios, economists rely on climatologists, hydrologists, and water agency managers to understand the size and nature of potential supply shortages, as well as the objectives and constraints of water management institutions. Economists in turn can provide water managers with quantitative estimates of the economic consequences of different decisions that affect reservoir levels and supplies available for various water uses.

Chapter 7 considers the implications of climate change for the Southwest's ski industry. It links climate prediction science with recreation demand analysis, developing an econometric model of skier visitation and a financial model that examines returns to investment in snowmaking. The econometric model predicts visitation at an Arizona ski destination using climate cycle and snowfall data. The El Niño phase in the Southwest's climate—a cooler, higher winter precipitation cycle—is a significant and positive predictor of skier visits. Total seasonal snowfall and ski resort days open explain 77% of the variation in skier visitation.

Snowmaking is currently the most common ski industry climate change adaptation strategy. To assess payoffs to investing in snowmaking, Chapter 7 considers trade-offs between snowmaking effectiveness in improving ski season reliability and the significant expense of making snow. Costs, excluding the capital costs of investing in the specialized equipment, range from 15% to 25% of total resort annual operating costs. The future of snowmaking as an adaptation strategy needs careful consideration, given global warming projections, as snowmaking is not technically viable at temperatures higher than −2 degrees C (Scott et al. 2003) and may not be financially viable at cooler temperatures, depending on input costs. Snowmaking requires large volumes of water, which can be costly and legally complex to secure in the Southwest.

Returns to snowmaking investment depend on winter season climate types—El Niño (wet), La Niña (dry), and neutral seasons—and snowmaking capacity, the share of skiable area that can be covered. Snowmaking costs and water requirements vary by season type, as do predicted average skiable days. Model results suggest that investment in snowmaking yields positive returns except in the case of a dry (La Niña) season with only half capacity. These models can enable winter recreation businesses to make better-informed decisions using seasonal climate forecasts. Such decisions include whether to increase advertising or seasonal labor hiring to capitalize on strong El Niño conditions or whether to increase snowmaking in La Niña years.

Part III: Information, Technology, and Adaptation

The third part of this book examines relationships among climate, technology, and information in water and energy management. In particular, it looks at the role of information and information technology in climate change adaptation.

Chapter 8 considers the role of farm scale in irrigator use of information for water management. It also examines barriers to adopting water- and energy-conserving improvements. The chapter begins by developing a simple theoretical model of an irrigator's demand for water management information. This model is used to develop a set of hypotheses about the relationship between scale of operation and the demand for information. Although multivariate regression is the "bread and butter" of empirical economic analysis, it is not the only means to evaluate data or hypotheses. This chapter addresses the problem of what to do with data that are not amenable to multivariate regression. The USDA's Farm and Ranch Irrigation Survey (FRIS), conducted roughly every five years, is the most extensive and nationally comprehensive data collected on irrigator behavior in the United States. Yet results of this survey are published in aggregate form, and access to individual data is limited, to protect respondent confidentiality. Aside from USDA economists, who have conducted interesting and important analyses, economists have not applied econometric analysis to this valuable data source. Aggregate FRIS data is published as cross-tabulations showing 2 × 2 relationships between variables. The USDA's Economic Research Service has made available more detailed FRIS cross-tabs that report a host of irrigation variables by different farm sales classes (USDA ERS 2004).

Chapter 8 uses cross-tabulated data from this special tabulation to test several hypotheses about the relationship between farm size and irrigation water management. It employs some different statistical methods specifically designed for analysis of cross-tab data, using two measures of association: the Goodman-Kruskal (1980) gamma (γ) coefficient and the Cochrane-Armitage trend test. Researchers often have access only to cross-tab data. This chapter demonstrates that it is possible to do considerably more with such data than use them for mere description. The statistical techniques employed are a powerful tool for exploratory data analysis. Although the tests are univariate (i.e., they consider effects of only one explanatory variable in isolation), they are nonetheless quite informative. For example, results suggest that smaller-scale irrigators were less likely to investigate irrigation improvements to conserve water or energy, use management-intensive methods for irrigation scheduling, participate in government conservation programs, or rely on private specialists or consultants for irrigation information.

Results also suggest that larger farms use more data sources, especially private sources, to make irrigation management decisions. There were also significant differences between irrigators in Arizona and New Mexico. In New Mexico, in contrast to Arizona, agriculture is characterized by a large number of small-scale operators. This dates back to differences in land tenure systems established at the time of Spanish colonization. This illustrates the importance of historical factors in economic analysis and suggests there is scope for combining anthropological and economic approaches for explaining differences in local water management systems.

Chapter 9 is concerned with how climate and other variables affect farmer choice of irrigation technology. Although a substantial literature exists on the economics of irrigation technology choice), few studies have examined the influence of climate. This is because most analyses are based on data for a single year and a single irrigation district or relatively small production region. This means there is insufficient variation in climate across space or in weather across time to measure their effects. This chapter adds to the few exceptions with sufficient geographic scope to consider the role of climate (Negri and Brooks 1990; Caswell et al. 2001; Mendelsohn and Dinar 2003).

Chapter 9 makes use of data from special cross-tabulations of the USDA Farm and Ranch Irrigation Survey to conduct a western regionwide analysis of irrigator behavior (USDA ERS 2004). The chapter examines irrigator choice of sprinkler versus gravity flow irrigation, with observations stratified by state and farm sales class for the 17 westernmost contiguous states. Whereas field application efficiencies usually range from 40% to 65% for gravity systems, they are more typically 75% to 85% for sprinkler systems (Aillery and Gollehon 2003). Multivariate regression analysis is used to examine the roles of economic and environmental variables in irrigation technology choice. The authors find that reliance on sprinkler irrigation increases with several factors: farm size, potential for sheet and rill erosion, number of months with freezing temperatures, energy costs, and reliance on groundwater. Results also suggest that sprinkler adoption increases and then decreases with temperature, with very high temperatures discouraging sprinkler adoption. Scope for irrigators to adapt to a warmer climate by switching from gravity to sprinkler technology may be limited in warm, arid regions of the West. However, because sprinkler systems are a land-augmenting technology, enhancing the water-holding capacity of soils, they may allow conversion from dryland to irrigated production in some areas.

Chapter 10 develops econometric models of electricity loads using data from a southwestern electric utility. The authors evaluate how including weather and climate information improves utility load forecasts. Electricity cannot be efficiently stored, so the amount to be generated and delivered must be estimated in advance. Forecasts are essential in the electric sector to predict timing and location of demand sufficiently in advance so that commensurate supplies can be delivered when and where demanded. Improved forecasts of electricity demand allow power generation to occur more cost-effectively and can be an important component of climate change adaptation. Improved load forecasts allow more precise use of generating facilities and can reduce pollution and result in more reliable and cost-effective provision of electricity.

The chapter employs time-series models. A time series is a sequence of data points based on observations at successive time intervals. Time-series forecast models seek to predict future observations based on the known series of past observations. The authors develop a model to forecast hourly electricity loads and a simple medium-term model to assess the feasibility of incorporating climate information in planning. They then compare the performance of models that use only past observations with those that also include weather and climatic information. The accuracy of forecasts is evaluated using methods from econometrics and applied meteorology.

Time-series models of electricity demand are challenging because of three important temporal patterns: an annual peak in the summer and another in the winter; a daily electricity use pattern that changes from unimodal in the summer to bimodal in the winter; and a weekly "seasonality" in which weekday demands differ from weekend demands. The authors develop an innovative modeling technique to address these issues. Temperature proved to be the most influential climate-weather factor in explaining electricity demand. Including variables for humidity, precipitation, and wind speed further reduced the percentage of prediction error. Although the improvements in forecasting were small in percentage terms, such improvements could save millions of dollars annually.

Chapter 11 continues the theme of information use, focusing on the use of weather information for agricultural decisionmaking. Smit and Skinner (2002) identified better use of weather data as an important climate change adaptation strategy. Although weather data can have value in decisionmaking, it can be costly to apply. Weather information may thus be considered in a supply-and-demand framework. While substantial public investments have been made in expanding the supply of climate and weather information services, the demand for information is less well understood. This chapter develops a simple economic model of the demand for weather information. This model is then used to inform empirical analysis of primary survey data from Arizona farmers and ranchers. The authors examine what types of weather data agricultural producers use and what decisions they use weather data for.

Much of the literature on the value of weather data focuses on estimating the value of information to users (Stewart 1997; Wilks 1997). However, Freebairn and Zillman (2002) pointed out that a large share of potential users does not use meteorological services in decisionmaking. The data use model produces negative economic information of sorts. By distinguishing between users and nonusers of data, it distinguishes between agricultural producers that expect a net benefit from using weather information and those that do not. Although we cannot obtain estimates of the value of weather information, we can identify the prevalence and characteristics of producers who do not derive expected benefits from weather data.

Regression analysis suggests that production diversification *within* agriculture encourages weather data use, as does the importance a producer places on government risk management programs. Off-farm income discourages weather data use, but Internet access has no significant effects. Nearly 40% of Arizona survey respondents did not have Internet access, which is similar to the national average for farmers. These findings have important implications, given that government agencies, extension agencies, and universities are devoting significant resources to web-based weather information services.

This work also informs studies by climate and atmospheric scientists on the economic value of climate and weather forecasts, which have combined decision theory with climate science (Katz and Murphy 2005). Although provision of information enhances the ability of others to adapt to climate change, it is not in itself an adaptation (Smit and Skinner 2002). Supplying information is not enough; it is also necessary to understand whether and how that information is used.

CONCLUSIONS

Looking at the questions addressed and economic methods used in this book, it is striking that the economics of information figures so prominently. Although use of improved information holds great promise for risk management and adaptation to climate change, significant barriers to information use remain. While the chapters in this volume illustrate that an economic framework can be usefully applied to study of the "supply and demand" of climate information services, they also highlight the importance of hydrology and climate sciences in designing effective economic responses to climate variability and change.

Other important themes addressed are the roles of technological and institutional innovation as mechanisms to adapt to climate change. Technological innovations include improved irrigation technologies, snowmaking, and irrigation scheduling systems. The primary institutional innovation is use of more flexible market instruments to voluntarily reallocate water as an adaptation to the increased supply variability associated with climate change. This book illustrates several methods to evaluate these market instruments.

Finally, the methods in the book allow for a more complex assessment of vulnerabilities and reliance in the face of climate change. In many cases, the Southwest as a whole is quite resilient to potential impacts of climate change. Amid this reliance, however, methods used in this book are able to identify which sectors, locations, resources, and populations remain vulnerable to climate change.

NOTES

1. Results of the study were published together in volume 31, issue 5 (1995) of *Water Resources Bulletin* (now *Journal of the American Water Resources Association*). Young (1995) provides an overview of the study.

2. An example of a third-party impact is as follows: A farmer diverts 1,000 units of water for irrigation, with 400 units running back to a river for downstream use by other farms. Suppose the farmer now sells all 1,000 units to a city. The farmer and city may benefit from the exchange, but downstream users are deprived of 400 units of return flows. The downstream users (the third parties) are worse off from the exchange.

REFERENCES

Aillery, M., and N. Gollehon. 2003. Irrigation Water Management. In *Agricultural Resources and Environmental Indicators, 2003*, edited by Ralph Heimlich. Agriculture Handbook No. AH722. Washington, DC: U.S. Department of Agriculture, Economic Research Service, Ch. 2.2.

Bauer, C.J. 2004. Results of Chilean Water Markets: Empirical Research since 1990. *Water Resources Research* 40: W09S06, doi:10.1029/2003WR002838.

Caswell, Margriet, Keith Fuglie, Cassandra Ingram, Sharon Jans, and Catherine Kascak. 2001. *Adoption of Agricultural Production Practices: Lessons Learned from the U.S. Department of Agriculture Area Studies Project*. Agricultural Economic Report No. 792. Washington, DC: U.S. Department of Agriculture, Economic Research Service, Resource Economics Division.

CEC (California Energy Commission). 2005. California's Water–Energy Relationship, www.energy.ca.gov/2005publications/CEC-700-2005-011/CEC-700-2005-011-SF.PDF (accessed October 10, 2010).

Cohen, R. 2007. The Water Energy Nexus. *Southwest Hydrology* 6 (5):16–19.

Colby, B.G. 1990. Transactions Costs and Efficiency in Western Water Allocation. *American Journal of Agricultural Economics* 72 (5):1184–1192.

DOE (U.S. Department of Energy). 2006. Energy Demands on Water Resources, Report to Congress on the Interdependency of Energy and Water. www.sandia.gov/energy-water/docs/121-RptToCongress-EWwEIAcomments-FINAL.pdf (accessed October 10, 2010).

Freebairn, J.W., and J.W. Zillman. 2002. Economic Benefits of Meteorological Services. *Meteorological Applications* 9:33–44.

Goldstein, C., R. Newmark, C. Whitehead, E. Burton, J. McMahon, G. Ghatikar, and D. May. 2008. The Energy-Water Nexus and Information Exchange: Challenges and Opportunities. *International Journal of Water* 4 (1–2):5–24.

Goodman, L.A., and W.H. Kruskal. 1980. *Measures of Association for Cross Classification*. New York: Springer-Verlag.

IPCC (Intergovernmental Panel on Climate Change). 2007. *Climate Change 2007: Synthesis Report*. Valencia, Spain: IPCC.

Katz, R.W., and A.H. Murphy, eds. 2005. *Economic Value of Weather and Climate Forecasts*. Cambridge, UK: Cambridge University Press.

Kenney, D., A. Ray, B. Harding, R. Pulwarty, and B. Udall. 2010. Rethinking Vulnerability on the Colorado River. *Journal of Contemporary Water Research & Education* 144:5–10.

Larson, D., C. Lee, S. Tellinghuisen, and A. Keller. 2007. California's Energy Water Nexus: Water Use in Electricity Generation. *Southwest Hydrology* 6 (5):20–21, 30. www.sahra.arizona.edu/swhydro/archive/V6_N5/feature3.pdf (accessed October 11, 2010).

Mendelsohn, R., and A. Dinar. 2003. Climate, Water, and Agriculture. *Land Economics* 79:328–341.

Merrett, S. 2003. The Urban Market for Farmers' Water Rights. *Irrigation and Drainage* 52:319–326.

Negri, D.H., and D.H. Brooks. 1990. Determinants of Irrigation Technology Choice. *Western Journal of Agricultural Economics* 15 (December):213–223.

Nelson, B., M. Schmitt, R. Wilkinson, R. Cohen, and N. Ketabi. 2007. In Hot Water: Water Management Strategies to Weather the Effects of Global Warming. Natural Resources Defense Council. www.nrdc.org/globalWarming/hotwater/execsum.pdf (accessed August 28, 2008).

Nelson, D.R., W.N. Adger, and K. Brown. 2007. Adaptation to Environmental Change: Contributions of a Resilience Framework. *Annual Review of Environment and Resources* 32:395–419.

NM ENV (New Mexico Environment Department). 2004. *New Mexico Brackish Groundwater Assessment Program Workshop: Report of Findings and Recommendations*. Albuquerque, NM: NM ENV.

Rahmstorf, S. 2007. A Semi-Empirical Approach to Projecting Future Sea-Level Rise. *Science* 315 (January):368–370, doi:10.1126/science.1135456.

Rajagopalan, B., K. Nowak, J. Prairie, M. Hoerling, B. Harding, J. Barsugli, A. Ray, and B. Udall. 2009. Water Supply Risk on the Colorado River: Can Management Mitigate? *Water Resources Research* 45:W08201, doi:10.1029/2008WR007652.

Scott, D., G. McBoyle, and B. Mills. 2003. Climate Change and the Skiing Industry in Southern Ontario (Canada): Exploring the Importance of Snowmaking as a Technical Adaptation. *Climate Research* 23:171–181.

Seasholes, K. 2008. *Historical/Hydrologic Overview of Tucson Active Management Area (AMA)*. Paper presented to the Tucson/Pima County Water and Wastewater Study Oversight Committee. July 2008, Tucson.

Smit, B., and M.W. Skinner. 2002. Adaptation Options in Agriculture to Climate Change: A Typology. Climatic Change. *Mitigation and Adaptation Strategies for Global Change* 7:85–114.

Stewart, T.R. 1997. Forecast Value: Descriptive Decision Studies. In *Economic Value of Weather and Climate Forecasts*, edited by R.W. Katz and A.H. Murphy. Cambridge, UK: Cambridge University Press, 147–181.

USDA ERS (U.S. Department of Agriculture, Economic Research Service). 2004. Data: Western Irrigated Agriculture. http://ers.usda.gov/Data/WesternIrrigation/ShowTables.asp (accessed August 28, 2008).

Wilkinson, R. 2007. The Energy Intensity of Water Supplies. *Southwest Hydrology* 6 (5):30.

Wilks, D.S. 1997. Forecast Value: Prescriptive Decision Studies. In *Economic Value of Weather and Climate Forecasts*, edited by R.W. Katz and A.H. Murphy. Cambridge, UK: Cambridge University Press, 109–145.

Young, R.A. 1995. Coping with a Severe, Sustained Drought on the Colorado River: Introduction and Overview. *Water Resources Bulletin* 31:779–788.

PART I

VOLUNTARY WATER TRANSFERS AS ADAPTATION MECHANISMS

CHAPTER 2

Temporary Water Transactions and Climate Change Adaptation

Bonnie G. Colby, Elizabeth Basta, and Katie Adams

Access to water supplies in many areas of the western states is governed by prior appropriation, a "first-in-time, first-in-right" water rights system. Irrigators across the West hold senior water rights, while municipalities and newer uses such as instream flows and recreation typically have junior priority. In the event of shortage, irrigators with senior water rights will receive their full allocations, while junior rights holders may be shorted. Lower-priority water users face high economic and political costs during water shortages. However, these losses can be mitigated through well-structured negotiated temporary transfers of water. Temporary transfers are not suitable to provide long-term supplies for population growth (as a temporary borrowing of water from an established user), but they can serve as a "buffer" over subsequent periods of drought. Further, temporary transfers are less costly in both financial and political terms than permanent acquisitions of agricultural land and water, and they generally engender less heated opposition over potential third-party impacts.

In every political jurisdiction, transfers are subject to public agency oversight with the intent of protecting other rights holders and public interests. The approval process, however, can be lengthy and costly and may discourage desirable transfers. Some states, such as Oregon and Colorado, have recognized that bureaucratic delays create costs and uncertainties, and thus have tried to mitigate the problem by assuring more rapid decisions on temporary water transfer applications (Landry and Anderson 1999).

This chapter provides background on the role for temporary transactions as an adaptation response to climate-related changes in water supply and demand. Temporary transfers of water among water-using sectors and locations not only are a valuable adaptation response by individual actors (farms, cities, environmental instream habitat restoration programs), but also provide a resilience strategy for regional economies and water management systems. We look at actual water volumes leased and pricing trends and develop econometric models of relationships between climate and negotiated water prices in the southwestern states. The analysis is based on hundreds of temporary water transactions occurring over two decades in California and New Mexico. These states were selected because of an active history of temporary water transfers and their susceptibility to water supply disruptions linked to climate variability.

VOLUME TRENDS

One useful way to consider the role of temporary water transfers is to examine the volumes of water involved in such transfers over a period of time. Figures 2.1 and 2.2 provide this information for the states of California and New Mexico over three decades, setting the context for the econometric analysis that follows. Our source for data on volumes transferred through leases is *Water Strategist*, a monthly publication focusing on water transactions, policy, and law updates.

Examining the graphs of quantities of water leased for California and New Mexico highlights the vast difference between the two, as the scale of water leased in California is about 10 times that of New Mexico. California's peak volume of annual quantity of water leased is approximately 1 million acre-feet (ac-ft), while New Mexico has a maximum of approximately 120,000 ac-ft. As another indicator of differences in scale, the total reported yearly water use for each state

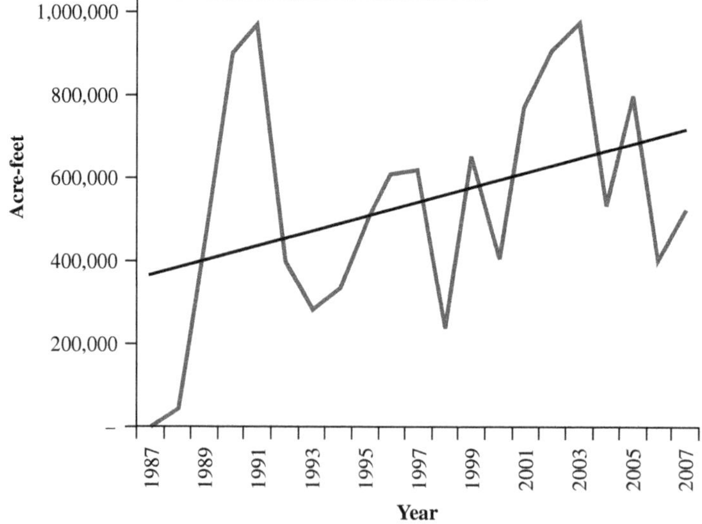

Figure 2.1 *Quantities of Water Leased in California*

Source: Water Strategist (various dates)

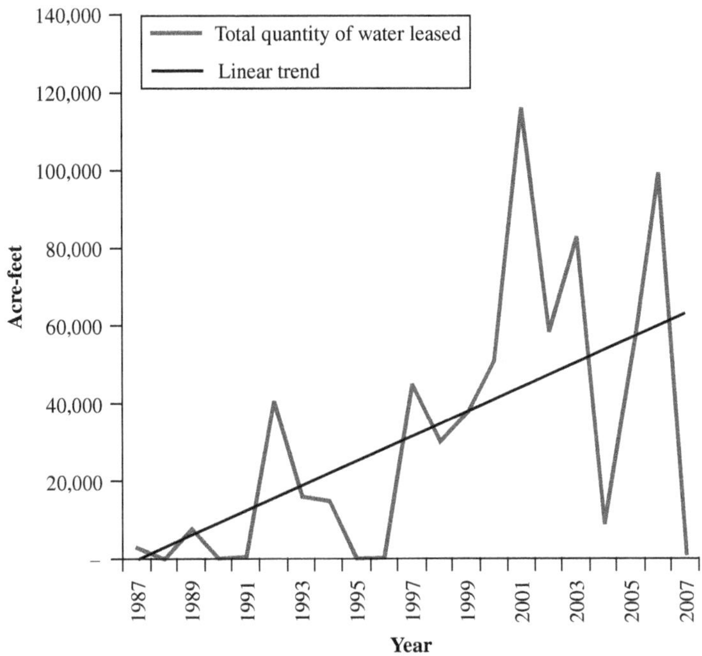

Figure 2.2 *Quantities of Water Leased in New Mexico*

Source: Water Strategist (various dates)

(not including environmental uses) is also about 10 times higher for California. Reported yearly total water use for each state, for agriculture and urban uses, is about 40 million ac-ft in California and 4 million ac-ft in New Mexico. California also reports annual environmental use totals, which add about another 40 million ac-ft in an average year to the total volume used in the state (CA DWR, 2005; NMOSE 2005).

Examples of dry-year lease transfers in the past for each state can be seen through a corresponding volume increase on each state's graph. The severe California drought in the early 1990s spurred government involvement and led to the creation of the California Drought Bank (*Water Strategist*). The bank served to encourage and facilitate voluntary water transfers and ensure that water rights holders would retain their permanent rights. The yearly volumes of water leased in 1990 and 1991 both exceeded 900,000 ac-ft and correspond to the first notable volume increase on the California graph. The water leased for these two years alone is 16% of the total amount of water leased over the two decades.

More recently, in New Mexico, the Bureau of Reclamation entered into multiple one-year leases with irrigators in an effort to enhance stream flows for the Middle Rio Grande silvery minnow, a fish listed and protected under the Endangered Species Act. Corresponding water volumes leased in 2005 and 2006 were about 45,500 and 99,500 ac-ft, respectively. This volume increase is also clearly seen on the New Mexico graph and is about 22% of the total amount of water leased over the two decades.

Over the past two decades, leases in each state transferred substantial volumes of water and served as a vehicle for shifting water from lower-value to higher-value uses. The total volume of water traded over the 20-year period is more than 11 million ac-ft in California and more than 660,000 ac-ft in New Mexico. In each state, average annual volume traded represents less than 1% of annual water use.

DATA

The dataset consists of 259 temporary water transfers occurring in California and New Mexico between 1987 and 2005. The primary sources of transaction data used in the analysis are three private publications: the *Water Strategist, Water Market Update,* and *Water Intelligence Monthly.* These subscription publications provide monthly coverage of marketing, legislation, litigation, and financial information on water resources. These sources do not, however, report transactions between users in the same water district. Such transfers, which are a regular component of farmers' water management strategies, can involve substantial volumes of water (Howitt and Hanak 2005).

Occasionally, the information reported on a specific transaction was limited, ambiguous, or raised questions about the legitimacy of the transaction as a true *market* transaction. We employed the following criteria in the sample selection process: When purchase price per acre-foot was reported as a dollar range, we assigned the midpoint to the transaction; we did not include transactions of less than $5 per acre-foot in New Mexico or $10 in California, as we felt they were too low to be based on supply-and-demand factors (we set the cutoff higher in California because of generally higher reported market values for water there); and we also omitted transactions for which the reported price per acre-foot was not set, varied, or was unclear. Transactions that included land or other infrastructure in addition to a water right were considered on an individual basis. If it was not obvious that the transaction occurred solely for the acquisition or use of the water right, the transaction was deleted. Transactions were also eliminated if the length of the lease was over 40 years, because we would expect transfers longer than this to take on the characteristics of a permanent transfer.

Explanatory Variables Related to Water Source and Transaction Characteristics

Following are brief descriptions of the explanatory variables related to water source and transaction characteristics, compiled based on information provided in the three sources named above. If applicable, we also state the hypothesized sign of parameter.

Price. The price, *lnAdj_Price*, is the natural log of the real price per acre-foot paid for leased water. All prices are adjusted to 2008 US$ using the Consumer Price Index. If the life of the lease is more than one year, the annual price per acre-foot is used to ensure comparative prices across leases of various lengths.

Date. For each temporary transfer, the date is a record of the month and year the transaction was reported. It should be noted, however, that because of a lag in reporting, this date does not correspond to the date the transaction was actually formalized. As a result, select variables are lagged to account for the discrepancy between the actual transaction date and the date it was reported.

The year a transaction occurred is used to construct a trend variable to assess whether a statistically significant trend up or down is occurring over time in the price of leased water. Over time, water is becoming a scarcer resource (Griffin 2006). Factors such as population growth are shifting the demand curve for water outward, which increases the price of water. However, a negative time trend may also be reasonable. This is because as a lease market matures, transaction costs may be reduced, and potential sellers may become more comfortable with the process and confident their long-term water rights will be protected. These factors in turn may increase the number of sellers in the market, causing the supply curve to shift out and price to decrease.

Quantity. The number of acre-feet of water transferred is captured in the variable *lnaf*, which is the natural log of total acre-feet transferred per lease transaction. Presumably, some of the transaction costs inherent in leasing water, such as locating a partner and gaining approval for the transfer, would not significantly increase with the quantity of water leased, and thus economies of scale come into play. Economies of scale, or a price-per-unit decrease as the number of units increases, would suggest a negative relationship between the quantity of water leased and its price per acre-foot.

Original Use. This variable indicates how the water was originally used. Major categories include reclaimed, municipal, industrial, and storage/surplus.

New Use. The new use variable indicates how the leased water is put to use. Major new use categories include agriculture, environmental, municipal (landscape and golf course irrigation, as well as new development), recreation, industrial, and storage.

Because the marginal value of water varies across uses, in relatively immature lease markets, the lease price of water leased for lower-value uses, such as low-value crop production, would be expected to be less than the lease price for relatively higher-value uses, such as municipal or environmental. In more mature lease markets, however, the market would operate so as to reduce variation in lease prices. (In a "textbook" perfectly functioning market, the equi-marginal principle would lead to a convergence of price across types of uses. However, water leases in the western United States do not yet approach the model of a frictionless market.)

Type of Water. The type of water leased is categorized as surface water, groundwater, or effluent. Consumers' tastes, preferences, and aversion to risk may come into play in determining whether the type of water has a positive or negative impact on price. Because groundwater is less susceptible to the immediate effects of drought than surface water and thus lessens the risk of shortage, people may value groundwater more highly than surface water or effluent and therefore have a higher willingness to pay for groundwater supplies. In terms of quality, effluent is generally considered inferior in quality to surface water and groundwater, so consumers' willingness to pay for reclaimed water may be relatively lower.

Length of Lease. This variable represents the number of years of the life of a lease. The sign of its coefficient is difficult to hypothesize. On one hand, lessees may have a higher willingness to pay for a longer lease, because of the longer assured supply. On the other hand, a shorter-term lease may represent an effort to meet an immediate need, and hence the willingness to pay might be higher.

Explanatory Variables Related to Climate and Regional Economy

In addition to the variables compiled based on information related to water and transaction characteristics, a number of other factors are hypothesized to influence the price of leased water. These variables are briefly described here.

Economic Indicators. A state or region's economic vitality and population are also expected to have some impact on the price of water, and we collected a number of economic indicators from the U.S. Census Bureau and Bureau of Economic Analysis. These include population and personal income, defined as income earned from all sources, including participation in production as well as from government and business transfer payments (BEA 2006). In general, increases in these economic indicators would shift the demand curve out, and thus the expected sign on these parameter estimates would be positive. However, given the temporary nature of water leases and the reasons people lease water, the positive relationship between economic vitality/population growth and price may not be so straightforward.

Drought Index. Relative dry and wet conditions are captured using the Standard Precipitation Index (SPI), which was developed to monitor drought and is a measure of the probability of precipitation for a given time period (NOAA 2003; WRCC 2001). The SPI can capture short-term droughts; droughts of three to six months, which affect soil moisture and small rivers and ponds (and thus agricultural production); and longer-term droughts of six months to several years, which have implications for major river, aquifers, and other large bodies of water. SPI data were obtained from the Western Region Climate Center database and assigned to each lease based on the year, month, and climate region in which the transaction took place. One SPI value was assigned per transaction if the water remained in the same climate region. For water transfers across climate regions, two SPI values were assigned to each transaction, based on both the climate region where the water transfer originated and the climate region where the water was ultimately transferred.

The index ranges from −4 to 4, with −4 representing extremely dry conditions and 4 extremely wet. In general, as the SPI goes down and hydrologic conditions become dryer, the supply function for water in a specific region would shift inward. This would imply a negative parameter estimate on the SPI variable.

During drought, this may be further compounded by an increase in demand for water, as junior rights holders find themselves in a shortage situation and thus enter the lease market, shifting out the demand curve.

Climate Index. A newly developed climate index, the Multivariate ENSO Index (MEI), is also explored in this analysis. The El Niño Southern Oscillation (ENSO) index is linked to yearly climate variability (Wolter 2004). The MEI is a weighted average of sea level pressure, the east-west and north-south components of the surface wind, sea surface temperature, surface air temperature, and the total fraction of cloudiness in the sky. The MEI is calculated on a sliding bimonthly basis, and all seasonal values are standardized with respect to each season and to the 1950–1993 reference period to keep the MEI comparable (Wolter 2006). Because the climate and hydrologic implications of the MEI vary from region to region, the expected sign varies as well. In general, in the southwestern United States, a positive MEI value (El Niño) suggests a wet winter, and a negative MEI value (La Niña) suggests a dry winter. Thus in the Southwest, as the MEI increases corresponding to wetter conditions and an outward shift of the supply curve, the price of leased water is expected to decrease.

Lagged Values of SPI and MEI. Lagged values of both SPI and MEI were also included. Negotiations and the water transfer approval process can take many months to finalize, and the month a transaction was finalized does not generally correspond to the date the transfer was reported. Thus a lag effect of 6 and 12 months for the SPI and MEI are assessed.

METHODOLOGY

We estimate a derived demand model of water lease markets for New Mexico and California, with water lease prices as the dependent variable (Pittenger 2006). Price and quantity are often simultaneously determined, meaning price can be a determinant of quantity, while quantity can also be a determinant of price. When variables are determined simultaneously, they are both dependent variables, also termed endogenous variables, which leads to what is known as endogeneity bias. To test for endogeneity, we perform the Hausmann-Wu test on each model to determine whether price and quantity are endogenous. If quantity is found to be endogenous in the price equation, this would, if unaddressed, lead to inconsistency and bias in standard ordinary least squares (OLS) regression analysis (Wooldridge 2003). The remedy applied here is called a two-stage least squares (2SLS) regression analysis. In the first stage, instrumental variables (which are assumed to be exogenous) are used to predict the value of quantity. Supply-side variables generally are used as instruments for quantity, and demand-side variables are used in the price models. In the second stage, the predicted values of quantity are then substituted back into the price equation as an explanatory variable. Because the Hausmann-Wu test indicates that quantity is endogenous in both models, we use the 2SLS in the analysis of lease prices for both California and New Mexico.

After estimating each model using 2SLS, we then conduct the Breusch-Pagan test and White's test for heteroskedasticity to determine whether the variance is constant. If the tests indicate a nonconstant variance, the variance-covariance matrix is used to construct Huber-White robust standard errors. Correcting for nonconstant variance is critical so that valid hypothesis tests can be carried out and the statistical significance of each variable can be accurately assessed.

ECONOMETRIC ANALYSIS OF WESTERN WATER LEASES: MODELS AND RESULTS

As each state has a distinct water supply system, its own specific laws governing water rights and transactions, and different patterns of demand, a cross-state comparison of water markets is of interest. The next sections provide a brief overview of each state's pertinent water policies and transfer history. Following each description, the model, summary statistics, and regression results for the state models of the price of leased water are presented.

California

The distribution of California's water supply system in both time and space gives rise to the need for reallocation. Over 75% of the demand for water is generated in the southern portion of the state, while more than 71% of average annual runoff originates in the northern half (Yolles 2000). Further, there is great seasonal variability in precipitation, with 75% of average annual precipitation occurring between November and March. The summer months, when agricultural and urban water demands are at their peak, are quite dry, and the state is prone to multiyear droughts.

To address seasonal and annual variability in water supply and increasing demand due to rapid population growth, California has historically relied on water supply storage and conveyance systems. Today, however, the state's extensive water infrastructure is inadequate, and water markets have received much attention as a mechanism to stimulate the efficient reallocation of water to meet demand.

Water rights in California are use rights, meaning the actual ownership of the water remains with the state (BLM 2001). California's water rights are regulated by both the riparian doctrine and the prior appropriation system. The riparian doctrine establishes a use right as a result of landownership adjacent to a surface water source, and no permit is necessary. Appropriative rights, however, require an application and approval from the State Water Resources Control Board. Riparian rights are generally senior to appropriative rights.

California has experienced growth and development, albeit limited, of its water market since the mid-1980s. Although the state is making strides toward creating an institutional environment more conducive to efficient water marketing, it still lacks a vigorous water market to effectively address the situation of fluctuating supplies and changing patterns of demand.

Between 1987 and 2005, California water users leased approximately 6,485,000 ac-ft of water at an average price of \$140/ac-ft (2008 US\$). The average volume per transaction over the period was 30,591 ac-ft.

Tests indicate endogeneity in the price model, so we apply 2SLS techniques. These involve the use of instrumental variables and estimation of quantity using these instrumental variables. Variable descriptions, summary statistics, and regression results for all instrumental variable equations can be found in Appendix 2.1. Using the predicted values of quantity (denoted *lnafhat*) from the first-stage regression (using instrumental variables), the second-stage equation is modeled and described as follows:

$$\begin{aligned} lnAdj\_Price = {} & \beta_0 + \beta_1\, lnafhat + \beta_2\, Trend + \beta_3 MEI6L6 + \beta_4 SPI24orig \\ & + \beta_5\, ag\_NewUse + \beta_6\, enviro\_NewUse + \beta_7\, other\_NewUse \\ & + \beta_8\, BasinDummy + \beta_9\, effluent + \beta_{10}\, grnd\_water + \beta_{11}\, trans\_per\_year \end{aligned}$$

The positive, significant coefficient on the quantity variable (quantity predicted from instrumental variable estimation), *lnafhat*, indicates that an increase in the quantity corresponds to a higher price per acre-foot. This result likely reflects more controversy and higher transaction costs associated with leasing larger quantities of water. For example, third-party impacts may be greater for larger transfers, and gaining state approval for a transfer may be more cumbersome if the proposed transfer involves a large quantity of water. The variable trend, however, is not significant at the 10% level, indicating that prices do not have a clear increasing or decreasing trend over time.

Both the drought and climate indices are significant with the expected signs. The parameter estimate on the MEI is positive, meaning that as the MEI increases, the price of leased water correspondingly increases. The implication of a negative or positive MEI is variable through California (Wolter 2004). A positive MEI value

Table 2.1 *Description of Variables Used in the California Water Price Model*

Name	*Description*
lnAdj_Price	The natural log of the real price per acre-foot, 2008 US\$
lnafhat	Natural log of the predicted value of acre-feet after instrumental variable estimation
Trend	A time trend equal to 1 if year = 1987, 2 if year = 1988, etc.
MEI6L6	6-month MEI lagged six months
SPI24orig	24-month SPI for the climate region of origin
ag_NewUse	Binary variable equal to 1 if new use is agricultural
enviro_NewUse	Binary variable equal to 1 if new use is environmental
other_NewUse	Binary variable equal to 1 if the new use falls into some other category
BasinDummy	Binary variable equal to 1 if the transfer was intrabasin
effluent	Binary variable equal to 1 if the transferred water was effluent
grnd_water	Binary variable equal to 1 if the transferred water was groundwater
trans_per_year	The number of transactions that occurred within a year

Table 2.2 *Summary Statistics for the California Water Price Model*

Variable name	*Mean*	*Std. dev.*	*Variance*	*Min.*	*Max.*
lnAdj_Price	4.55	0.78	0.61	2.62	6.35
lnafhat	9.22	0.89	0.79	5.96	11.86
Trend	12.34	5.19	26.91	2.00	19.00
MEI6L6	0.47	0.81	0.65	−1.13	2.51
SPI24orig	−0.16	0.90	0.81	−2.08	2.23
ag_NewUse	0.22	0.41	0.17	0.00	1.00
enviro_NewUse	0.22	0.41	0.17	0.00	1.00
other_NewUse	0.11	0.32	0.10	0.00	1.00
BasinDummy	0.75	0.43	0.19	0.00	1.00
effluent	0.12	0.33	0.11	0.00	1.00
grnd_water	0.18	0.38	0.15	0.00	1.00
trans_per_year	13.74	5.71	32.61	1.00	27.00

indicates an El Niño event in Southern California, suggesting a wet winter. In Northern California, however, the index has the opposite interpretation: a positive MEI implies a dry winter. Intuitively, interpreting the MEI coefficient as it is linked to climate events in Northern California makes more sense. That is, as the MEI goes up and winter conditions are drier, the price of leased water correspondingly increases. The fact that the majority of temporary transfers

Table 2.3 *Regression Results for the California Water Price Model*

Variable name	*Parameter estimate*	*Robust standard error*	*Chi-square*	*Robust P-value*
Intercept	2.955	0.929	3.18	0.0017
lnafhat	0.189	0.093	2.04	0.043
Trend[a]	0.002	0.012	0.14	0.8878
MEI6L6	0.123	0.062	1.99	0.0479
SPI24orig	−0.110	0.058	−1.9	0.0592
ag_NewUse[b]	−0.553	0.127	−4.36	< 0.0001
enviro_NewUse[b]	−0.357	0.126	−2.84	0.005
other_NewUse[b]	0.298	0.172	1.73	0.0851
BasinDummy[a,c]	−0.137	0.146	−0.94	0.3475
effluent[d]	0.498	0.212	2.35	0.0196
grnd_water[d]	0.687	0.142	4.84	< 0.0001
trans_per_year[a]	−0.011	0.009	−1.25	0.2137
Observations	212			
R^2	0.33			

Note: Dependent variable is log of adjusted price per acre-foot

[a] Insignificant at 10% level

[b] Base group is *muni_NewUse*

[c] Base group is interbasin transfer

[d] Base group is surface water

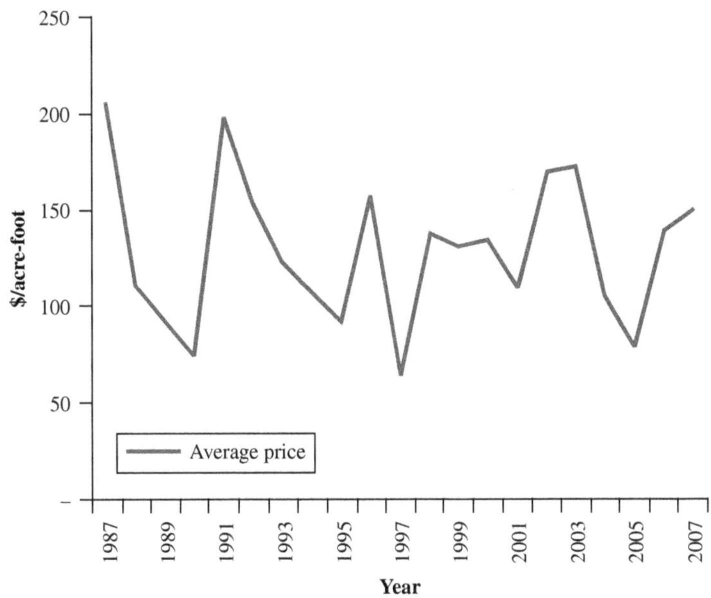

Figure 2.3 *Average Water Lease Prices in California (2008 US$)*

Source: *Water Strategist* (various dates)

originate in Northern California, as does much of Southern California's water supply, adds legitimacy to this interpretation. Similarly, the negative coefficient on the SPI, which is specific to the climate region in which the transfer originated, implies a sensitivity of the price of water to drought conditions.

The negative and significant sign on *ag_NewUse* suggest that, all else constant, the price per acre-foot of water is lower for agricultural leases relative to the reference group, municipal use. This result is indicative of agriculture's relatively lower-valued use of water and represents an opportunity for efficient reallocation. The coefficient on *enviro_NewUse* is also negative and significant. This may be attributed to the large number of environmental purchases made by the federal government to satisfy environmental protection mandates in California.

The negative coefficient on *BasinDummy* indicates that the price of intrabasin transactions is less than that of interbasin transfers, all else equal. This result may be capturing the higher transactions costs of moving water between basins. The positive coefficient on *grnd_water* is a signal that groundwater is more highly valued than surface water. This result may suggest that buyers and sellers value the availability or predictability of a water source. Because of general overallocation problems in California, as well as drought, which affects surface water levels before groundwater, the supply of surface water may not be as reliable as groundwater. The significant, positive coefficient on effluent, however, is more difficult to explain. Effluent is generally perceived as inferior in quality to surface water, and therefore we would expect its market price to be lower.

But the results here suggest a higher price for effluent than for surface water, all else constant.

New Mexico

Water rights in New Mexico are governed by the prior appropriation doctrine, and water transactions between users occur with relative ease, given the supportive nature of laws and regulations surrounding water transfers. Surface water rights are prioritized in reference to 1907, the year surface water was declared public (Brookshire et al. 2004). Surface water rights established post-1907 are considered permitted and therefore cannot be transferred. However, surface rights that were developed before 1907 all share a common priority date and are treated as owned property that can be leased out or sold. This standardized priority date creates a uniformity of water rights, which lends itself to the development of a water market, because transaction costs are lower and a single market clearing price can be established (Brookshire et al. 2004; Howe and Goemans 2003).

Both groundwater and surface water transactions are subject to the approval of the state engineer. The majority of temporary water transfers in New Mexico occur to meet municipal, recreation, or instream flow needs to support endangered species. The primary supplier of water for temporary transfers is agriculture. Between 1987 and 2005, there were 47 reported leases of water in New Mexico, at an average price of approximately $58/ac-ft (2008 US$).

The Hausmann-Wu test for endogeneity revealed that price and quantity are indeed endogenous in the New Mexico lease market, so 2SLS is employed to correct for bias and inconsistency. The R^2 for the first stage instrumental variable equation is 0.51. Using these estimates for *lnaf*, the second-stage least squares equation is modeled as follows:

$$\ln Adj\_Price = \beta_0 + \beta_1 \ln afhat + \beta_2 Trend + \beta_3 MEI24l6 + \beta_4 SPI12L6 + \beta_5 ag\_NewUse + \beta_6 enviro\_NewUse + \beta_7 other\_NewUse + \beta_8 Albuq$$

The explanatory variables included in the New Mexico second-stage price model are described and summarized below.

Table 2.4 *Descriptions of Variables in the New Mexico Water Price Model*

Name	*Description*
Trend	A time trend equal to 1 if year = 1987, 2 if year = 1988, etc.
lnafhat	Natural log of predicted value of acre-feet after IV estimation
MEI24l6	12-month MEI lagged 6 months
SPI12L6	12-month SPI lagged 6 months for climate region
ag_NewUse	Binary variable equal to 1 if the water was originally used for agriculture
enviro_NewUse	Binary variable equal to 1 if the water was originally used for environmental purposes
other_NewUse	Binary variable equal to 1 if the water was originally used for some other purpose
Albuq	Binary variable equal to 1 if the transaction involved the Albuquerque metro area

Table 2.5 *Summary Statistics for the New Mexico Water Price Model*

Variable name	*Mean*	*Std. dev.*	*Variance*	*Min.*	*Max.*
Trend	10.66	5.42	29.40	2.00	19.00
lnafhat	7.50	1.45	2.11	4.42	10.89
MEI24l6	0.26	0.68	0.46	− 1.25	2.51
SPI12L6	0.32	0.85	0.72	− 1.73	1.64
ag_NewUse	0.02	0.15	0.02	0.00	1.00
enviro_NewUse	0.53	0.50	0.25	0.00	1.00
other_NewUse	0.28	0.45	0.20	0.00	1.00
Albuq	0.47	0.50	0.25	0.00	1.00

Tests for heteroskedasticity indicate that the New Mexico price model may be biased because the variance is not constant. Thus asymptotic standard errors are calculated and used to recalculate heteroskedasticity robust test statistics.

The 2SLS regression results indicate that a statistically significant time trend is not associated with the price of leased water in New Mexico. Visual inspection of a simple plot of average price versus year confirms that the price of leased water has fluctuated fairly randomly between 1987 and 2007, without a clear trend up or down.

The number of acre-feet transferred is highly significant in predicting the price of leased water in New Mexico. A potential explanation for this positive price-quantity relationship is that transaction costs are higher as the quantity increases. Generally, it is more difficult and time-consuming to receive authorization for a transfer involving more water because of concern over third-party impacts and environmental effects, with higher transaction costs for larger transfers.

Table 2.6 *Regression Results for the New Mexico Water Price Model*

Variable name	*Parameter estimate*	*Robust standard error*	*Chi-square*	Robust P-value
Intercept	2.242	0.497	20.32	< 0.0001
Trend[a]	0.016	0.025	0.38	0.5355
lnafhat	0.232	0.061	14.33	0.0002
MEI24l6	0.341	0.099	11.85	0.0006
SPI12L6	−0.456	0.093	24.04	< 0.0001
ag_NewUse[b]	−0.426	0.248	2.95	0.086
enviro_NewUse[b]	− 1.202	0.274	19.26	< 0.0001
other_NewUse[a,b]	−0.365	0.244	2.24	0.1345
Albuq	0.844	0.125	45.84	< 0.0001
Observations	47			
R^2	0.71			

Note: Dependent variable is log of adjusted price per acre-foot

[a] Insignificant at 10% level

[b] Base group is *muni_NewUse*

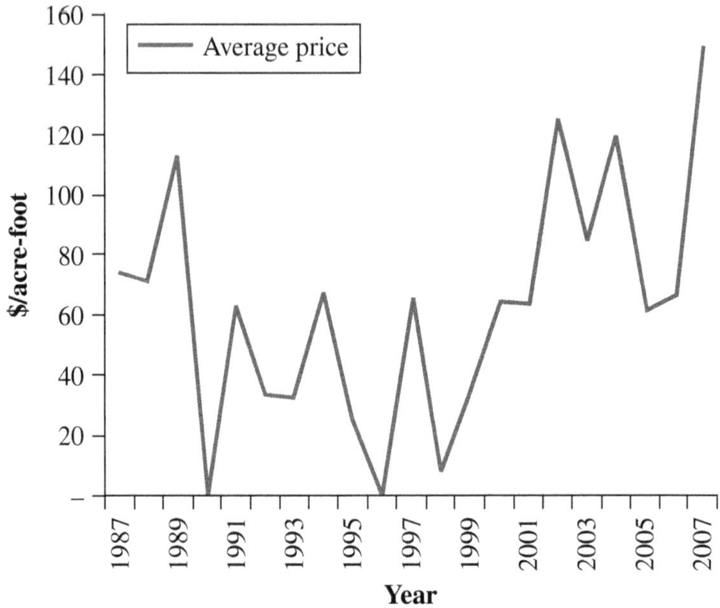

Figure. 2.4 *Average Water Lease Prices in New Mexico (2008 US$)*

Source: *Water Strategist* (various dates)

The climate and drought indices are both significant, though the sign on the climate index, MEI, again is not expected. The SPI, however, does take on the expected sign in New Mexico, indicating a premium for leased water when hydrologic conditions are dry.

The agricultural and environmental new use variables are significant, with the expected sign. In comparison with water leased for municipal purposes, water leased for agriculture or the environment is predicted to be less expensive. This likely reflects the relatively higher use value of municipal water than, for instance, the value of water used to irrigate low-value crops. Water leases in the Albuquerque metropolitan area are also predicted to carry a higher premium than transactions in the rest of the state, all else constant.

CONCLUSIONS

The potential for temporary water transfers to be an effective climate change adaptation mechanism in the face of increased hydrologic variability is evident in our analysis of both California and New Mexico water leases. Over the past three decades, water users have been able to take advantage of the ability to move water from low-value to high-value uses during times of drought. Looking at the total volumes of leased water traded over time in each state, we can link the response of yearly quantity increases to periods of corresponding drought. With respect to prices, economic theory suggests that when water is in short supply, the price will go up to reflect its scarcity value. Results of this analysis conclude, in both models evaluated, that the price of leased water has the anticipated inverse relationship with water supply conditions. That is, when water supply is less abundant as a result of drought, its price per acre-foot rises. This finding suggests that specialized types of conditional water transfers, negotiated in advance of actual shortages, would be a prudent strategy for cities and environmental programs wanting to secure reliable supplies.

It is important to note that the econometric models are based on assumptions that may create bias in the results. The transaction data available may not be representative of all temporary transactions in these two states, yet no more complete data are available. There is a possibility of omitted variable bias in the models as they are now specified. A number of other attributes also could have an effect on the price of water, such as the priority date of water leased. Data are not available on this characteristic for the transactions analyzed in this chapter.

Although water leases are particularly conducive to minimizing drought-induced water supply variability, perhaps of more interest is what the results suggest about which drought indicators are systematically related to price trends. This analysis indicates that the SPI may better reflect the effect of climate and drought on the price of water than the MEI. Whereas the MEI is indicative of the large-scale climate cycle, the SPI is specific to intrastate climate divisions, thus allowing for location-specific analysis of the impact of drought on the price of leased water. The MEI is significant in both models, but the sign is inconsistent. The significance and expected sign of the coefficient on the SPI variable in each model suggest that the price of leased water is driven more by local hydrologic conditions than by the overall regional climate.

This research contributes to the understanding of lease water markets as a climate adaptation and resilience mechanism in two ways. First, yearly total quantities of water leased in California and New Mexico provide evidence of high volumes traded during times of drought and serve as examples of the flexibility benefits of temporary water transfers. And second, statistical analysis of lease water transactions in the two states provides insight into the determinants of the price of leased water. This information is useful in providing water users and managers with a clear concept of the determinants of the price and quantities of water leased under different climate conditions. The findings suggest that negotiating water transfer contracts to improve supply reliability prior to a shortage rather than waiting to make arrangements during an actual drought crisis would result in lower per-unit prices. Continued research into the relationships determining prices and quantities of water leases will remain vital as we continue to learn more about, and adapt to, climate change.

APPENDIX 2.1: INSTRUMENTAL VARIABLE ESTIMATION: STATISTICS AND RESULTS

The primary findings of this econometric modeling of water leases in two states are presented in the main chapter text. However, some readers may benefit from reviewing the details of the instrumental variable estimation presented below.

California

Table 2.7 *Description of Variables Used in the California Water Quantity Model*

Name	*Description*
lnaf	Natural log of acre-feet
Trend	The real 3-month average price of alfalfa at the state level
adj_alfalfa	Natural log of real, annual personal income at the state level
indust_OrigUse	Binary variable equal to 1 if the water was originally used for industrial purposes
munic_OrigUse	Binary variable equal to 1 if the water was originally used for municipal purposes
storage_OrigUse	Binary variable equal to 1 if the water was originally in storage
reclaim_OrigUse	Binary variable equal to 1 if the water was originally reclaimed water
lnincome	Natural log of annual personal income at the state level
MEI12L6	12-month MEI lagged 6 months
SPI24orig	24-month SPI in the climate region of origin
BasinDummy	Binary variable equal to 1 if the transfer was intrabasin
years_lease	Number of years of the life of the lease
effluent	Binary variable indicating the type of water transferred was effluent
grnd_water	Binary variable indicating the type of water transferred was groundwater
north	Binary variable indicating the transfer originated in Northern California
droughtbank	Binary variable indicating the transfer occurred in a droughtbank year
trans_per_year	Number of lease transactions that occurred in each year
lnpopulation	Natural log of annual state population

Table 2.8 *Summary Statistics for the California Water Quantity Model*

Variable name	*Mean*	*Std. dev.*	*Variance*	*Min.*	*Max.*
lnaf	9.22	1.73	2.99	3.71	13.12
Trend	12.34	5.19	26.91	2.00	19.00
adj_alfalfa	128.21	20.37	414.90	92.97	172.17
indust_OrigUse	0.04	0.19	0.04	0.00	1.00
munic_OrigUse	0.22	0.41	0.17	0.00	1.00
storage_OrigUse	0.31	0.46	0.22	0.00	1.00
reclaim_OrigUse	0.02	0.15	0.02	0.00	1.00
lnincome	20.85	0.13	0.02	20.60	21.01
MEI12L6	0.40	0.72	0.52	−1.10	2.49
SPI24orig	−0.16	0.90	0.81	−2.08	2.23
BasinDummy	0.75	0.43	0.19	0.00	1.00
years_lease	2.79	6.21	38.54	0.00	40.00
effluent	0.12	0.33	0.11	0.00	1.00
grnd_water	0.18	0.38	0.15	0.00	1.00
north	0.11	0.32	0.10	0.00	1.00
droughtbank	0.20	0.40	0.16	0.00	1.00
trans_per_year	13.74	5.71	32.61	1.00	27.00
lnpopulation	17.32	0.07	0.00	17.14	17.40

Table 2.9 *Regression Results for the California Water Quantity Model*

Variable name	Parameter estimate	Standard error	T-stat	Pr > |t|
Intercept	−996.618	402.840	−2.47	0.0142
Trend	−0.673	0.320	−2.1	0.0369
adj_alfalfa	0.003	0.008	0.47	0.642
indust_OrigUse	0.224	0.619	0.36	0.718
munic_OrigUse	−0.209	0.311	−0.67	0.5028
storage_OrigUse	0.410	0.265	1.55	0.123
reclaim_OrigUse	1.193	0.821	1.45	0.1481
lnincome	−8.831	4.368	−2.02	0.0446
MEI12L6	−0.040	0.180	−0.22	0.8244
SPI24orig	−0.096	0.161	−0.59	0.5536
BasinDummy	−0.942	0.267	−3.52	0.0005
years_lease	0.018	0.018	0.99	0.3239
effluent	−1.752	0.377	−4.65	< 0.0001
grnd_water	−0.206	0.342	−0.6	0.548
north	−0.216	0.371	−0.58	0.5616
droughtbank	−1.219	0.532	−2.29	0.0231
trans_per_year	−0.005	0.026	−0.19	0.8457
lnpopulation	69.224	22.983	3.01	0.0029
Observations	212			
R^2	0.26			

Note: Dependent variable is natural log of acre-feet

New Mexico

Table 2.10 *Description of Variables Used in the New Mexico Water Quantity Model*

Name	Description
lnaf	Natural log of acre-feet
adj_alfalfa	The real 3-month average price of alfalfa at the state level
lnincome	Natural log of real, annual personal income at the state level
trans_per_year	Number of lease transactions that occurred in each year
MEI12L6	12-month MEI lagged 6 months
SPI12L6	12-month SPI lagged 6 months for climate region
munic_OrigUse	Binary variable equal to 1 if the water was originally used for municipal purposes
storage_OrigUse	Binary variable equal to 1 if the water was originally in storage
reclaim_OrigUse	Binary variable equal to 1 if the water was originally reclaimed water
years_lease	Number of years of the life of the lease
jan–nov	Binary variables indicating in which month the transaction was reported

Table 2.11 *Summary Statistics for the New Mexico Water Quantity Model*

Variable name	*Mean*	*Std. dev.*	*Variance*	*Min.*	*Max.*
lnaf	7.50	2.04	4.17	3.89	11.51
adj_alfalfa	149.70	16.78	281.46	122.00	201.57
lnincome	17.45	0.55	0.30	14.99	17.79
trans_per_year	4.49	2.46	6.04	1.00	9.00
MEI12L6	0.23	0.74	0.55	−1.10	1.63
SPI12L6	0.18	0.73	0.53	−1.08	1.98
munic_OrigUse	0.04	0.20	0.04	0.00	1.00
storage_OrigUse	0.51	0.51	0.26	0.00	1.00
reclaim_OrigUse	0.15	0.36	0.13	0.00	1.00
years_lease	4.04	5.61	31.43	1.00	20.00
jan	0.23	0.43	0.18	0.00	1.00
feb	0.09	0.28	0.08	0.00	1.00
mar	0.06	0.25	0.06	0.00	1.00
jun	0.11	0.31	0.10	0.00	1.00
sep	0.06	0.25	0.06	0.00	1.00
oct	0.13	0.34	0.11	0.00	1.00
nov	0.21	0.41	0.17	0.00	1.00

Table 2.12 *Regression Results for the New Mexico Water Quantity Model*

Variable name	*Parameter estimate*	*Standard error*	*T-stat*	*Pr > |t|*
Intercept	8.964	11.932	0.75	0.4584
adj_alfalfa	−0.095	0.026	−3.64	0.001
lnincome	0.751	0.671	1.12	0.2724
trans_per_year	0.049	0.192	0.25	0.8014
MEI12L6	−0.798	0.556	−1.44	0.1613
SPI12L6	−0.623	0.592	−1.05	0.3007
munic_OrigUse	0.403	1.509	0.27	0.7914
storage_OrigUse	1.416	1.026	1.38	0.1778
reclaim_OrigUse	−2.664	1.761	−1.51	0.1409
years_lease	−0.089	0.068	−1.31	0.1998
jan	−0.308	1.163	−0.26	0.793
feb	0.254	1.383	0.18	0.8555
mar	−0.081	1.566	−0.05	0.9589
jun	0.171	1.321	0.13	0.898
sep	0.830	1.436	0.58	0.5676
oct	−1.069	1.274	−0.84	0.4081
nov	−0.947	1.474	−0.64	0.5256
Observations	4			
R^2	0.51			

Note: Dependent variable is natural log of acre-feet

REFERENCES

BEA (Bureau of Economic Analysis). 2006. Glossary. www.bea.gov/bea/glossary/glossary_p.htm (accessed March 3, 2006; site discontinued).

BLM (Bureau of Land Management). 2001. California Water Rights Fact Sheet. Western States Water Laws. www.blm.gov/nstc/WaterLaws/california.html (accessed April 15, 2006).

Brookshire, David S., Bonnie Colby, Mary Ewers, and Philip T. Ganderton. 2004. Market Prices for Water in the Semiarid West of the United States. *Water Resources Research* 40:W09S01, doi:10.1029/2004WR003394.

CA DWR (California Department of Water Resources). 2005. *California Water Plan Update.* Sacramento: CA DWR.

Griffin, Ronald C. 2006. *Water Resource Economics: The Analysis of Scarcity, Policies, and Projects.* Cambridge, MA: MIT Press.

Howe, Charles W., and Christopher Goemans. 2003. Economic Efficiency and Equity Considerations in Regional Water Transfers: A Comparative Analysis of Two Basins in Colorado. www.economics.hawaii.edu/research/seminars/02-03/01-17.pdf (accessed April 15, 2006).

Howitt, Richard E., and Ellen Hanak. 2005. Incremental Water Market Development: The California Water Sector, 1985–2004. *Canadian Water Resource Journal* 30 (1):73–82.

Landry, Clay, and Terry Anderson. 1999. *Market Transfers of Water.* Oakland, CA: Pacific Institute.

NMOSE (New Mexico Office of the State Engineer). 2005. New Mexico Water Use Data 2005 by County. www.ose.state.nm.us/publications_technical_reports_wateruse_county05.html (accessed April 16, 2009).

NOAA (National Oceanic and Atmospheric Administration). 2003. Standard Precipitation Index. www.ncdc.noaa.gov/paleo/drought/drght_spi.html (accessed March 3, 2006).

Pittenger, Katherine A. 2006. Economic Arrangements to Improve Dry-Year Supply Reliability in the Lower Colorado River Basin. Master's thesis, University of Arizona.

Water Strategist. Various dates, 1990–2010. Water Strategist: Analysis of Water Marketing, Finance, Legislation, and Litigation, www.waterstrategist.com/ (accessed September 22, 2010).

Wolter, Klaus. 2004. Multivariate ENSO Index. www.cdc.noaa.gov/ENSO/enso.mei_index.html (accessed April 30, 2006).

———. 2006. Multivariate ENSO Index (MEI). 2006. www.cdc.noaa.gov/people/klaus.wolter/MEI/mei.html (accessed April 30, 2006).

Wooldridge, Jeffrey M. 2003. *Introductory Econometrics: A Modern Approach.* 2nd ed. Mason, OH: South-Western.

WRCC (Western Regional Climate Center). 2001. Explanation of Terms. www.wrcc.dri.edu/spi/explanation.html (accessed April 15, 2006).

Yolles, Peter L. 2000. *Update 2000: Progress and Limitations in Developing a Water Market in California.* Point Richmond, CA: Western Water Company.

CHAPTER 3

Applying Bargaining Theory to Western Water Transfers

George B. Frisvold and Kyle J. Emerick

In the West, transfers from agriculture have become an important means to meet growth in urban water demand. In an exhaustive survey of water market trends, Brewer et al. (2008) found that agricultural-urban transfers are increasing over time, as are prices paid for transferred water. Moreover, reliance on permanent transfers and long-term leases has been increasing, relative to short-term leases. Water transfers (both leases and sales) out of agriculture have exceeded 3 million acre-feet (ac-ft) in the 12 westernmost contiguous U.S. states (Libecap et al. 2005).

This chapter focuses on large-scale agricultural water transfers between irrigation districts and municipal water providers, for several reasons. First, agriculture is by far the dominant supplier of water for transfers. Second, large-scale transfers can account for a disproportionate share of total water transferred in the West. The quantity of water transferred annually in agricultural-urban trades averaged about 3,000 ac-ft per transfer per year (based on figures reported by Brewer et al. (2008) from 1987 to 2005). Recent and contemplated large-scale transfers dwarf this average. The Metropolitan Water District (MWD) of Southern California has entered into a number of bilateral agreements with different irrigation districts, with transfers of 50,000 to 100,000 ac-ft of water per agreement per year. The Quantification Settlement Agreement reached between the Imperial Irrigation District (IID) and the San Diego County Water Authority/Metropolitan Water District (SDCWA/MWD) will transfer up to 200,000 ac-ft of water per year after 10 years, with the option of increasing this amount to 303,000 ac-ft per year in later years (IID 2004). The Central Arizona Groundwater Replenishment District is looking to meet over 200,000 ac-ft in future demand from leases from Indian tribes and from irrigation districts along the Colorado River (Jacobs and Megdal 2004). Tijuana has considered construction of an aqueduct to transfer agricultural water from the Mexicali Valley to the coast.

There has even been discussion of a binational aqueduct that would transport Colorado River water to San Diego, Tijuana, and Ensenada (Michel 2002; Mumme and Lybecker 2006). Nevada and Utah are negotiating over a pipeline project that would move 132,000 ac-ft out of the rural Snake Valley along their borders to Las Vegas and other urban areas.

Third, the sheer size and scope for strategic behavior in large-scale transfers require a fundamental rethinking of what is meant by "water markets." Merrett's (2003) discussion of rural-urban water transfers is a useful starting point. The author pointed out that water markets do not operate as the perfectly competitive markets of economic textbooks, represented in simple supply-and-demand graphs. Rather, "the absolutely predominant form of transaction is the bilateral deal." He argued that because markets are thin, there is significant scope for exercise of monopsony power by buyers or monopoly power by sellers. Asymmetries in technical capacity and bargaining power are important. Negative effects of transfers on third parties (externalities) also are usually important. Merrett concluded, "The market equilibrium approach is rarely applicable," and "theory of the neoclassical type does not well represent social processes" of water transfers. He conceded that a "rural monopoly–urban monopsony approach" might be useful, and that "concepts of a farmer's minimum release price and an urban actor's maximum bid price are still appropriate, as is the account of how these prices are shaped" (2003, 319, 326).

A goal of this chapter is to illustrate that "theory of the neoclassical type" extends well beyond the introductory economics perfect-competition framework. The real-life complexities Merrett introduced—bilateral monopoly, asymmetric capacity and bargaining power, and negative externalities of transfers—can be readily addressed in a neoclassical framework using relatively simple game-theory concepts that have become standard in modern microeconomics.[1] In addition, the minimum release price and maximum bid price, important to Merrett's discussion, are fundamental parts of a game-theoretic bargaining solution.

Developments in the strategic bargaining approach have made game theory more applicable to policy evaluation and design. The earlier, axiomatic approach to bargaining relies on developing a set of convincing properties that a bargaining solution would (or should) have and then to determine whether such a solution is achievable. The focus is the mathematical properties of the solution, abstracting from the bargaining process itself. Player attributes, institutional rules, and the precise bargaining environment are implicit, while the economic efficiency of outcomes is emphasized. Often (as in the classic case of bilateral monopoly) the distribution of payoffs to bargaining parties is discussed somewhat vaguely in terms of differences in "bargaining power," without exploring the source from which such power derives.

In contrast, under the strategic bargaining approach, the bargaining process itself along with player attributes are explicitly considered (Binmore et al. 1986). For example, a player's bargaining power is strengthened by a low discount rate (ability to defer a payoff), the ability to make and respond to offers quickly (relative to other players), and the ability to accept risk (being less risk-averse than other players).

The strategic approach has features that make it naturally more appealing. While also being highly mathematical, it is couched in the language of playing games (in terms of players and moves), which is more accessible than discussion of axioms. The explicit consideration of the environment and rules of negotiations highlight how much institutions matter. Consideration of repeated games further illustrates the importance of reputation, reciprocity, institutional memory, and history of negotiations. Analysis of perfectly competitive markets focuses on efficiency of market-based outcomes and often downplays the distribution of gains from trade. Game theory, in contrast, is well equipped to explore issues such as the exercise of monopoly or monopsony power, distribution, and often highly asymmetric benefits of exchange. By dealing with this "dark side" of exchange, game theory can overcome some of the distrust other social scientists, environmental advocates, and environmental managers often have of economics and reliance on market-oriented outcomes. There is some evidence that a game-theoretic approach to analysis of water markets has been received more readily than "perfect market" approaches (Cortez Lara et al. 2009; Mumme and Lybecker 2006).

In the next section of the chapter, we introduce a bilateral monopoly model of water transfers. The seller is an irrigation district, and the buyer is a large municipal water provider. Negotiations are treated as an alternating-offer game that, under certain assumptions, reduces to the well-known Nash cooperative solution. We pay special attention to the role of negative third-party impacts and mitigation requirements on the transfer agreement. Bilateral water transfer agreements have considerable scope to generate externalities. By reducing local water used for irrigation, they can reduce subsequent groundwater recharge and surface water return flows. This in turn can take water away from other local water users or reduce water that maintains riparian habitats. Negotiated terms of large-scale transfers thus will be sensitive to laws governing mitigation of these external costs. Transboundary externalities are an additional complication, affecting international relations between the United States and Mexico. We also examine the effects of changes in irrigation technology and efficiency on the water transfer. It is often assumed that improved irrigation efficiency will increase water availability and facilitate transfers (e.g., USDA NRCS 2003). However, as we illustrate, improved efficiency does not necessarily reduce on-farm water demand or basinwide consumptive use. In fact, it can increase both under quite plausible conditions.

Next, we present and discuss numerical simulations based on the Quantitative Settlement Agreement involving IID and SDCWA, the largest agricultural-urban water transfer agreement made to date in the United States. The simulations demonstrate the utility of the theoretical model and aid in quantifying the nature of third-party effects of the transfer. While the transfer agreement provides significant benefits to contracting parties, it also has scope for negative externalities affecting the ecosystem of California's Salton Sea as well as wetlands and agriculture across the border in Mexico. Although mitigation requirements have been included in the agreement to address the Salton Sea, simulation results suggest that the adequacy of these requirements is tenuous and continued environmental monitoring and adaptive management are warranted. We conclude by noting that mitigation efforts to address externalities imposed on Mexico have yet to be undertaken and suggest that the concept of interlinked games might be a useful way to address outstanding transboundary conflicts between the United States and Mexico.

WATER TRANSFERS AS BILATERAL MONOPOLY

As Merrett (2003) noted, water transfers often have characteristics of imperfect competition. A single buyer can exercise considerable monopsony power. Whereas negotiations between a large buyer and a single small irrigator may be one-sided, irrigation districts can wield enormous bargaining power. For example, the Imperial Irrigation District alone accounts for more than 62% of Southern California's consumptive use of Colorado River water. Combined, the Imperial, Palo Verde, and Coachella Valley Irrigation Districts use 78% of Southern California's Colorado River water.

While large municipal water providers control large segments of urban markets, irrigation districts, when their members bargain collectively, can exert great countervailing power. Thus large transfer agreements, rather than looking like a competitive supply-and-demand graph, resemble more closely collective-bargaining negotiations between the United Auto Workers and Ford Motor Company. Frisvold and Caswell (1995) adapted McDonald and Solow's (1981) model of union-employer wage and employment negotiations to irrigation district–municipal water provider negotiations. As a labor union collectively supplies labor at a contracted wage, the irrigation district collectively supplies water at an agreed transfer price. The problem is similar to a classic bilateral monopoly problem, with some important institutional differences that create deviations from a textbook bilateral monopoly solution. First, the municipal water provider is not a profit-maximizing buyer, but a regulated public utility. The provider maximizes consumer benefits minus charges for cost recovery of acquiring new water supplies. The provider may also engage in two-part pricing, so that consumers pay a fixed rate plus a per-unit fee for water. This per-unit fee need not equal the marginal cost of acquiring water. Such two-part pricing for urban water is common in the United States and much of the developing world (Dinar and Subramanian 1998). Public utility–irrigation district negotiations may also apply to energy markets. For example, concentrated thermal solar facilities are seeking out agriculture to supply water for electricity generation (Woody 2009).

Irrigation District's Water Supply Problem

The source of water for rural-urban transfers is an irrigation district that receives a fixed allotment of water, X. The district can enter into a contract to transfer x acre-feet of its allotment to a municipal water provider for a fixed price of w per acre-foot of water.

If an agreement is reached, the district may be required to allocate a portion of its water allotment, m, to mitigate damages to other water users or to the environment that result from the transfer. Irrigation water not taken up by the crop or lost to evaporation may become available to other water users as return flows or groundwater recharge. Data from Shiklomanov (2000) suggest that globally, 70% of irrigation withdrawals are consumed and 30% are recycled. Return flows may also replenish riparian habitat and be critical for environmental protection. By sending water out of the basin, a transfer agreement may reduce return flows or recharge. This can lead to third-party damages to other water users, damages to the environment, or both.

The district's problem is to allocate water between rural-urban transfers and irrigation to maximize overall district profits, π, subject to any mitigation requirements. To simplify analysis, one can think of the district as made up of many homogeneous irrigators. We abstract from the differences among irrigators, intradistrict bargaining, and the distribution of gains among irrigators (Rosen and Sexton (1993) provide an interesting analysis of these issues).

Following Caswell and Zilberman (1986), agricultural production q is a function of effective water e used by the crop, $q = f(e)$, where $f'(e) > 0$ and $f''(e) < 0$. Effective water $e = h\,[X - (x + m)]$ depends on irrigation efficiency, h, which measures the percentage of water used by the crop. If there is no sale to the municipal provider, the district's entire allocation of water, X, is used for irrigation, and no water is required for mitigation. The cost of production is a constant c per acre-foot of water used for irrigation. We also assume quantity rationing of irrigation district water. In the United States and Mexico, districts often receive water at subsidized rates. Because of low water prices, irrigation demand for water is frequently quantity-rationed rather than price-rationed (Moore 1991; Moore and Dinar 1995).

Municipal Provider's Water Demand Problem

A municipal water provider supplies an urban market. The provider can supply water at a low cost, P^0, but faces a capacity constraint, C (Figure 3.1). If urban water demand curve is D^0, then demand can be met at price P^0. If demand increases to D^1, however, the municipal provider would have to either raise the price or impose quantity rationing. Politically, these may be unattractive options. For example, in June 1976, a "reform" coalition of four city council members in Tucson, Arizona, voted to increase water rates in response to demand growing beyond existing capacity. Public response to water rate increases was swift and negative. By November, three of the council members were removed in a special recall election, with the fourth having resigned prior to the election (Gelt et al. 1998).

To avoid rationing or price increases, the provider can seek out additional supplies by leasing water or purchasing water rights from agricultural sources. Urban total willingness to pay for agricultural water is represented by the area under D^1 to the right of the capacity constraint, C (Figure 3.2).

The provider is a regulated public utility whose objective is to maximize consumer benefits of new water supplies acquired from agriculture, $V(x)$, subject to a break-even constraint. The provider must cover the cost of acquiring water by charging fees to urban customers. Total fees collected must equal wx, the amount paid to the irrigation district. Fees can be collected from urban users as a combination of fixed payments and per-unit charges. Urban water consumers in the United States and many parts of the developing world are typically charged this type of two-part fee for water, with marginal payments lower than the marginal cost of supplying water (Dinar and Subramanian 1998). The municipal water

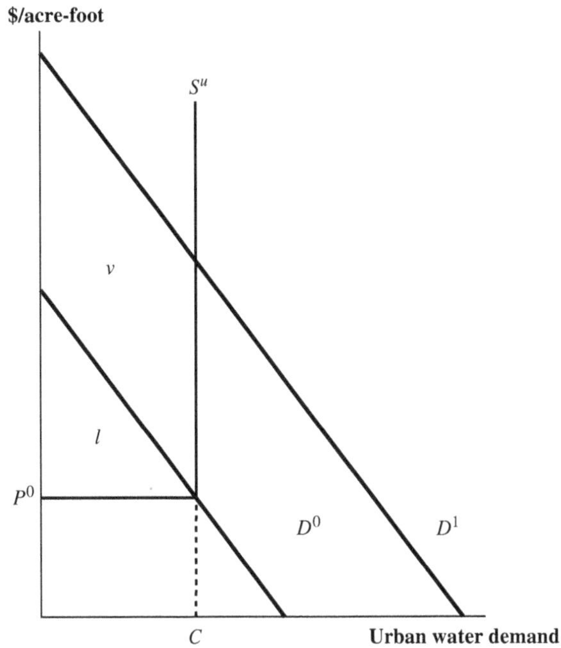

Figure 3.1 *Urban Water Demand Facing Municipal Water Provider*

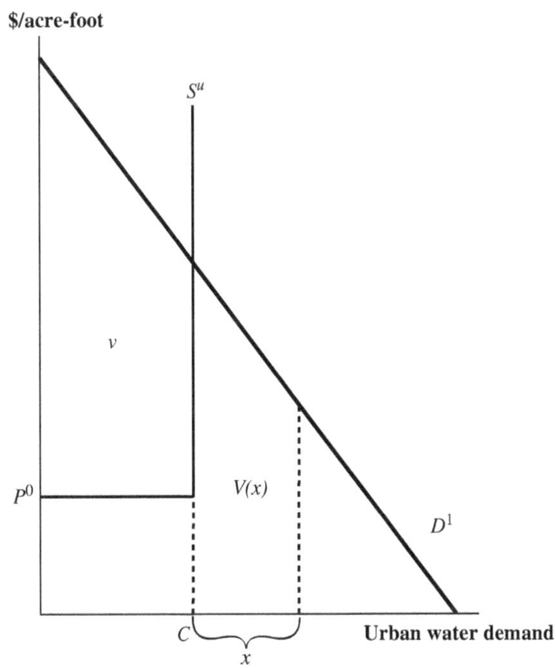

Figure 3.2 *Urban Willingness to Pay for Transferred Water*

provider's objective is to negotiate a quantity and price of transferred water to maximize $V(x) - wx$. If no transfer agreement is reached, $V = 0$.

Efficient Contracts

Rural-urban water transfer agreements can be thought of as contracts that jointly specify x (the quantity transferred) and w (the transfer price), subject to any mitigation requirement, m. (They can also specify water quality; see Frisvold and Caswell 1995). The contract curve is the set of efficient contracts between the irrigation district and municipal provider. Efficient contracts are $x - w$ pairs that, once agreed upon, cannot be altered to make one party better off without making the other worse off.[2] No deviations from efficient contracts can simultaneously make both parties better off.

Mathematically, the contract curve is the locus of points of tangency between the iso-payoff curves of the district and the municipal provider (see Appendix 3.1). These iso-payoff curves are like indifference curves and show which combinations of x and w yield a given level of return to the district and the provider. Figure 3.3 shows the iso-payoff curves along with disagreement points and the contract curve. The contract curve is the vertical line AB in $w - x$ space. There is a single efficient quantity of water to be transferred determined independently of w. The transfer price, w, determines the distribution of gains between the irrigation district and the

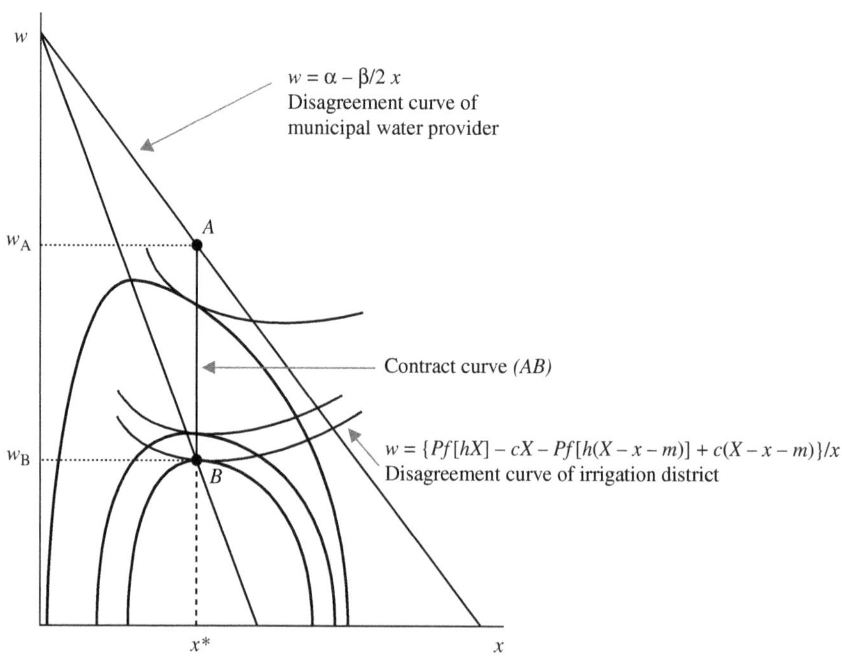

Figure 3.3 *Irrigation District and Municipal Water Provider Payoff Contours, Disagreement Curves, and the Contract Curve*

municipal water provider. At a transfer price of w_A, the irrigation district captures all the gains from the transfer; at w_B, the municipal provider captures all the gains from the trade. Transfer price w_A represents the provider's maximum willingness to pay for x^* acre-feet of water, and price w_B represents the minimum payment that the irrigation district would be willing to accept to sell x^* acre-feet of water. The negotiated value of w^* could be specified as the outcome of some yet-to-be-determined bargaining process. Although different game-theoretic solutions could be used to solve for w^*, many solutions, such as the Nash (1953) or Kalai-Smorodinsky (1975) solution, require efficiency as a criterion. For such solutions, a negotiated agreement will yield x^* as in Figure 3.3. Different bargaining solutions (that require efficiency) may differ in their solution to w^*, but not x^*.

A NASH BARGAINING MODEL FOR WATER TRANSFERS

This section considers negotiations between the municipal water provider and the irrigation district as a two-person Nash bargaining game. The Nash cooperative bargaining framework is relatively straightforward to implement. Yet the Nash solution can closely approximate noncooperative games, such as strategic models of sequential offers, if those games have high enough discount rates or the period between offers is sufficiently short (Binmore et al. 1986; Chae 1993; Krishna and Serrano 1996).

The bargaining process may represent a noncooperative sequential game, where two parties have the opportunity to exploit some mutually advantageous opportunity, but there is some risk that negotiations will break down and the opportunity will be lost. For example, during negotiations between IID and SDCWA, federal and state agencies offered to fund some mitigation programs as part of a negotiated deal. However, when negotiations temporarily broke down, the U.S. Department of Interior moved to reduce IID's water allotment by 275,900 ac-ft and California's overall use of Colorado River water by 800,000 ac-ft.

The Nash solution is found by choosing x and w to maximize the Nash product, N:

$$N = [\pi(X - (x + m), w) - \pi^0]^\gamma \; [V(x) - wx - V^0]^{1-\gamma} \tag{3.1}$$

The term π^0 represents the payoff to the irrigation district if bargaining breaks down. In this case, the district applies all of its water allotment, X, to irrigation. If negotiations break down, the municipal provider receives no agricultural water, so $V^0 = 0$. The parameter γ ranges from 0 to 1 and measures the bargaining strength of the irrigation district relative to the municipal provider. Including the bargaining power parameter γ allows for asymmetries in the bargaining game. The Nash product can be rewritten as

$$N = [\pi(X - (x + m), w) - \pi(X)]^\gamma \; [V(x) - wx]^{1-\gamma} \tag{3.2}$$

The negotiated transfer price $w^\star$ can be written as

$$w^\star = \gamma w_A + (1 - \gamma) w_B \tag{3.3}$$

where w_A is the provider's maximum willingness to pay for $x^\star$ and w_B is the minimum payment that the irrigation district would be willing to accept to sell $x^\star$ acre-feet of water. As the bargaining power of the irrigation district increases (γ tends toward 1), the transfer price approaches the municipal provider's maximum willingness to pay. Conversely, as the bargaining power of the municipal provider increases (γ tends toward 0), the transfer price approaches the price that is the irrigation district's minimum willingness to accept for $x^\star$.

FACTORS AFFECTING WATER TRANSFERS

In this particular model, the Nash bargaining solution is recursive. The optimal quantity of water transferred, $x^\star$, from the contract curve (Figure 3.3), can be determined independently of $w^\star$. It can be shown (see Appendix 3.1) that the quantity of water transferred does the following:

- increases as urban demand increases ($dx^\star/d\alpha > 0$ and $dx^\star/d\beta < 0$);
- increases as the costs of agricultural production increase ($dx^\star/dc > 0$);
- decreases as agricultural output price increases ($dx^\star/dP < 0$);
- increases as the irrigation district's water allotment increases ($dx^\star/dX > 0$);
- decreases as the mitigation requirement increases ($dx^\star/dm < 0$); and
- may increase or decrease as irrigation efficiency increases ($dx^\star/dh$? 0).

The first five results are intuitive. Increases in urban demand encourage rural-urban transfers, whereas factors increasing the profitability of irrigated agriculture discourage transfers. Given the diminishing marginal productivity of effective water, an increase in the water allotment of the district will increase the amount of water transferred.

Turning to environmental and agricultural policies, the mitigation requirement acts as a tax on water transfers, so increasing m reduces the quantity transferred. Higher agricultural prices increase returns to irrigation and reduce transfer. A policy implication is that a movement from coupled to decoupled farm support payments would encourage larger water transfers. Decoupling farm support programs would lead to a decline in P with lump-sum transfers being made to the irrigation district. In the United States, there has also been greater decoupling of farm program payments. Yet the prices of cotton, rice, and sugar beets and sugar cane are still supported through federal programs. Alfalfa and other types of hay also receive indirect demand stimulus through U.S. dairy price supports. Together, these crops account for a third of irrigation water applied in U.S. border states (USDA NASS 2003).

Increasing the cost of irrigated production, c, will increase the quantity of water transferred. Irrigation water is generally subsidized. The signs $dx^\star/dP < 0$ and

$dx^*/dc > 0$ imply that reallocation of water to urban uses can be achieved simply by removing current market distortions (output and input subsidies). As we discuss at greater length in Chapter 5, the Bureau of Reclamation provides nearly a third of surface water deliveries for irrigation in the 17 western states. Prices irrigators pay for water from these federal projects are often well below the full cost of providing the water. By subsidizing irrigation water, the federal government can discourage rural-urban water transfers.

The impacts of drought on the quantity of water transferred are multidimensional. Drought can affect both the demand for and supply of water. It can cause the regular water supplies of a city to contract, increasing demand for transferred water (α increases, β decreases, or both). On the supply side, drought may also reduce water available to the irrigation district (X decreases). A district with junior rights is more likely to have its allotments reduced during drought. Consequently, urban water providers may not be able to negotiate larger transfers in response to severe, prolonged drought if rural supplies also decline.

The comparative static results suggest that increasing irrigation efficiency (increasing h) need not increase water transfers. Economists have long recognized that improving irrigation efficiency can, under certain conditions, increase agricultural demand for water (Boggess et al. 1993; Caswell and Zilberman 1986). Increasing irrigation demand reduces the district's incentives to transfer water. Mirroring earlier results of Caswell and Zilberman, we find that whether improved irrigation technology increases or decreases agricultural water use depends crucially on the elasticity of the marginal product (EMP) of effective water with respect to the irrigation efficiency parameter. EMP measures how responsive the crop is to further irrigation. EMP is denoted as $\varepsilon(h)$, where

$$\varepsilon(h) = f''(e)\ (e/f'(e)) \tag{3.4}$$

The specific comparative static derivative for the effects of irrigation efficiency on quantity transferred is written as

$$\begin{aligned} dx^*/dh = &- [Pf'(e)\ (1+m/x)\ (1+\varepsilon)] \\ &/\ [\beta - h^2\ (1+m/x)^2\ Pf''(e)] \end{aligned} \tag{3.5}$$

The denominator is positive, so when $|\varepsilon| > 1$ (high EMP case), $dx^*/dh > 0$, and when $|\varepsilon| < 1$ (low EMP case), $dx^*/dh < 0$. Although beyond the scope of this chapter to explore in detail, this result when combined with the results of Caswell and Zilberman (1986) suggests that increased irrigation efficiency may actually discourage transfers in areas with deep wells, poor-quality land, and low output prices.

This result has important implications for agricultural conservation policy, specifically technology adoption subsidies to improve irrigation efficiency. The 2002 U.S. Farm Bill expanded the Environmental Quality Incentive Program (EQIP), which provides subsidies for adoption of agricultural conservation practices, to \$5.8 billion from 2002 to 2007. EQIP payments were budgeted to improve irrigation efficiency on nearly 25 million acres, primarily in the arid western United States, with projected reduction of agricultural water applications of 5.4 inches per acre. According to the U.S. Department of Agriculture, "Any water saved would be available for alternative uses such as by municipalities, utility generation, and wildlife habitat restoration" (USDA NRCS 2003). Assuming 20% loss in storage and transmission, the program would free 8.9 million ac-ft of water per year over 10 years. This is sizable considering that total water withdrawals by public suppliers, industry, and domestic self-suppliers in the 17 westernmost contiguous states were 25.1 million ac-ft/year in 2000 (Hutson et al. 2004).

This ambitious goal begs the question, will improving irrigation efficiency actually conserve agricultural water and encourage transfers? Previous studies have also considered the possibility that subsidizing irrigation technology adoption can backfire, working counter to water conservation goals (Huffaker and Whittlesey, 1995, 2000, 2003; Peterson and Ding 2005; Ward and Pulido-Velazquez 2008).

Without imposing more structure on the general model, it is not possible to derive unambiguous signs for comparative static results regarding w^*. An exception is the bargaining power parameter, γ. The price of transferred water w^* increases with the bargaining power of the irrigation district and decreases with the bargaining power of the municipal provider. Because x^* is determined independently of w^*, an increase in γ will unambiguously increase the payoff to the irrigation district.

WATER TRANSFERS AND EXTERNALITIES

We now consider how rural-urban water transfers affect third parties. Irrigation water not taken up by the crop or lost to evaporation may become available to other water users as return flows or groundwater recharge. Return flows may also replenish riparian habitat. Unused water is available for these other uses under the Nash bargaining solution $R^* = (1 - h)(X - (x + m)) + m$, where m is the quantity of water dedicated to mitigation. Water transfers reduce R^*, whereas mitigation increases it. Reductions in return flow thus capture third-party damages. Other damages would include lower demand for agricultural labor and other purchased inputs in the region.

There are (at least) two possible institutional responses to limit third-party damages. One would be to allow downstream water users or groups concerned about environmental impacts into the coalition with the irrigation district. Southern California's IID has representatives from the local community, not just irrigators, as voting members, but the neighboring Palo Verde Irrigation District (PVID) does not. In the recent water transfer to San Diego, IID negotiated for mitigation payments to compensate for pecuniary externalities, whereas water transfers from PVID have not included such provisions. However, including more heterogeneous actors may greatly increase transactions costs (Rosen and Sexton 1993).

A regulatory body may also set the mitigation requirement, m, such that return flows are not lower than those with no transfers, $m = m^0 = x(1 - h)/h$, where m is strictly decreasing in h. Holding transferred quantities x constant, improved irrigation efficiency (greater h) reduces the extent of mitigation required to prevent third-party damages. Conversely, if h is low, water transfers have the potential to create significant third-party damages, absent mitigation requirements.

Yet when h is low, the amount of water needed for mitigation is significant relative to the amount of water transfers. If $h < 0.5$, then $m^0 > x^*$; to prevent a reduction in return flows, more water would have to be dedicated to mitigation than would be transferred.

What might this mean for water policy? One might argue that to limit third-party damages, one should encourage investments to increase h, prior to encouraging water markets. However, a complication arises here, because increasing h has a direct negative effect on return flows and an indirect effect (via dx/dh) that may be positive or negative. These results, and the numerical simulations in the next section, suggest that limiting third-party damages of water transfers via technological improvements alone will be difficult, and that changes in mitigation requirements, m, will need to be continually monitored and adjusted over time.

NUMERICAL SIMULATION: THE QUANTIFICATION SETTLEMENT AGREEMENT

We now apply the Nash bargaining model to the recent Quantification Settlement Agreement (QSA), which transfers water from IID to SDCWA and is the largest rural-urban water transfer agreement in U.S. history. The QSA is a multiyear agreement, with payments and transfers increasing over time that could reach 0.3 million ac-ft of water per year. In 2005, 30,000 ac-ft were transferred at \$276/ac-ft. Our numeric simulation is calibrated to price and quantity accounting relationships observed in 2005. Choosing a single year abstracts from the multiyear nature of the agreement. However, the contract curve in each year depends only on exogenous parameter values for that year, and the static model is sufficient to illustrate key results.

Under the agreement, IID was also required to provide 15,000 ac-ft in 2005 for environmental mitigation of impacts to the Salton Sea, California's largest lake. The Salton Sea National Wildlife Refuge is a critical stop on the Pacific Flyway for migrating birds, including several state and federally listed endangered and threatened species. Agricultural drain water from the Imperial Valley accounts for about 75% of the freshwater inflow to the sea. Colorado River water transferred by IID directly to SDCWA reduces agricultural drain water feeding the sea. Total return flows, R, from IID depend on total diversions for agriculture, $X - x - m$, irrigation efficiency, h, and the quantity of water allocated to mitigation, m:

$$R = (1 - h)(X - x - m) + m \tag{3.6}$$

Consumptive use of water in the system, C, is simply total available water minus system return flows, $C = X - R$.

Table 3.1 reports parameter values and accounting relationships used in the numerical simulations. Appendix 3.2 discusses how parameter values were developed. Given model assumptions, the minimum price that the irrigation district would accept to transfer 30,000 ac-ft of water, W_B, was estimated to be \$41.41 per ac-ft. Based on Draper et al. (2003), the district's marginal willingness to pay for additional water was assumed to be \$23.40 at the disagreement point.[3] IID solicits bids from its members to participate in a land-fallowing program that frees up water for transfer and mitigation, offering a flat payment of \$60/ac-ft to participants. At this payment rate, IID can obtain more water than the minimum needed and has instituted a policy to ration participation. This suggests that the figure of \$41.41, lying between \$23.40 and \$60, is reasonable. SDCWA's estimated maximum willingness to pay for 30,000 ac-ft, W_A, was \$315.46. This rate is based on what SDCWA, a water wholesaler, charges smaller retail providers for treated water deliveries net of treatment and delivery costs for IID water. Given estimates of W_A, W_B, and an observed transfer price, w, of \$276 per ac-ft, the parameter measuring relative bargaining power, γ, was 0.85. This suggests asymmetric bargaining power favoring IID.

Table 3.2 reports estimates of the effects of changing irrigation efficiency on water transfers and transfer price. The system irrigation efficiency for IID was assumed to be 70% ($h = 0.7$), approximating previous estimates (Rosen and Sexton 1993). Around the baseline value, the simulations show that $dx/dh > 0$ and $dw/dh < 0$. In the IID case, improving irrigation efficiency appears to facilitate transfers. Table 3.2 shows, however, that if h is low, a small increase in h can reduce incentives to transfer water. Increasing h from 0.35 to 0.45 reduces incentives to transfer water. This has important implications for technology subsidy policies aimed at facilitating transfers by increasing h.

Table 3.3 shows that transfers tend to increase overall consumptive use relative to a no-transfer baseline. This occurs because transfers come at the expense of return flows. The effect of transfers on overall consumptive use is lessened as h increases, however. The QSA environmental mitigation provisions are intended to prevent reductions in return flows reaching the Salton Sea. This table shows how a combination of transfers with mitigation leads to a small decrease in consumptive

Table 3.1 *Values Used in Numerical Simulation of Effects of the Quantitative Settlement Agreement and Irrigation Technology Change on Water Use*

Variable name	*Variable description*	*Parameter or accounting value*
X	Irrigation district water allotment	3.2 million ac-ft
x	Quantity of water transferred	30,000 ac-ft
m	Quantity of water used for environmental mitigation	15,000 ac-ft
h	System irrigation efficiency	0.7
$Pf(X)$	Gross revenues of irrigation district without transfers	\$1.198 billion
c	Variable cost of production	\$115/ac-ft
w_B	Irrigation district's minimum acceptable price to transfer 30,000 ac-ft	\$41.41/ac-ft
w_A	Municipal provider's maximum acceptable transfer price to acquire 30,000 ac-ft[a]	\$315.46/ac-ft
w	Negotiated transfer price	\$276/ac-ft
α	Municipal provider's inverse demand intercept for additional sources of water[a]	\$600
β	Municipal provider's inverse demand slope parameter	0.01897
γ	Bargaining power coefficient	0.856

[a]Parameters account for treatment and delivery costs

Table 3.2 *Impact of Irrigation Efficiency on Water Transfers, Transfer Price, and Consumptive Use*

Irrigation efficiency h	Transfer price w	Water transferred (ac-ft) x	Change in consumptive water use resulting from transfer: With mitigation[a]	Change in consumptive water use resulting from transfer: Without mitigation
0.35	$ 304.60	27,738	12,780	18,049
0.40	$ 309.86	27,316	10,390	16,412
0.45	$ 311.97	27,143	8,179	14,955
0.50	$ 310.95	27,219	6,110	13,639
0.55	$ 306.79	27,544	4,145	12,427
0.60	$ 299.54	28,116	2,246	11,280
0.65	$ 289.25	28,935	377	10,162
0.70[b]	$ 276.00[b]	30,000[b]	− 1,500[b]	9,035
0.75	$ 259.86	31,310	− 3,422	7,861

[a]15,000 ac-ft of water devoted to mitigation
[b]Baseline simulation values

use (increase in return flows). Mitigation transfers of 15,000 ac-ft appear sufficient to maintain return flows for $h > 0.66$.

Table 3.3 shows the impacts on consumptive use of simultaneously introducing transfers and increasing irrigation efficiency. Here, a baseline with a lower h and no transfer can be compared with an alternative scenario with transfers combined with a higher h. In the early phases of the QSA, water conservation is to be achieved primarily by land fallowing. In later phases, however, it is expected that increasing irrigation efficiency is to play a more prominent role in reducing agricultural water use. In this table, consider a movement from no trades with $h = 0.70$ to one with trades and $h = 0.75$. Here, transfers increase consumptive use by more than 156,000 ac-ft. Now, consider a movement from no trades with $h = 0.65$ to one with trades and $h = 0.75$. Here, consumptive use increases by more than 300,000 ac-ft, even with 15,000 ac-ft set aside to supplement return flows.

The results from Tables 3.2 and 3.3 suggest that current mitigation provisions appear adequate to achieve stated short-run objectives, but that future mitigation

Table 3.3 *Impacts of Transfers and Irrigation Efficiency on Consumptive Use*

Irrigation efficiency h	Water transferred (ac-ft) x	Consumptive use[a] C
0.65	0	2,080,000[b]
0.65	28,935	2,080,377
0.70	0	2,240,000[b]
0.70	30,000	2,238,500
0.75	0	2,400,000[b]
0.75	31,310	2,396,578

[a]15,000 ac-ft of water devoted to mitigation
[b]Result without transfer agreement

requirements are highly sensitive to both base levels of h and changes in h. The numerical simulations imply that close environmental monitoring and water accounting will be necessary to recalibrate mitigation requirements to protect the Salton Sea. The QSA includes provisions for detailed and precise water accounting, as well as an adaptive management plan for environmental mitigation. Our simulation results suggest that these types of provisions will be necessary to meet the long-term environmental objectives of the QSA.

LINING OF THE ALL-AMERICAN CANAL

A key element to increase future transfers is the lining of the All-American Canal, which transports water from the Colorado River to the Imperial Valley. The unlined canal, built on sandy soils, loses water to seepage. In 2007, construction began on a concrete-lined canal parallel to the original to capture 67,700 ac-ft of water, most of which will be sent to SDCWA. The project is jointly financed by the U.S. Bureau of Reclamation, state of California, San Diego County Water Authority, Coachella Valley Water District, and Imperial Irrigation District (SDCWA 2010).

Lining the canal, however, will impose external costs on Mexico (Zamora Arroyo et al. 2006; Maganda 2005; Ries 2008). Groundwater seepage from the canal recharges the aquifer below the Mexicali Valley that supplies irrigated agriculture on the Mexican side of the border. Saille et al. (2006) estimate that canal lining would reduce the total amount of water available in the most affected area of Mexicali Valley by 14%. Also, because water from seepage is less saline than other water in the aquifer, canal lining would also increase the concentration of dissolved salts in the aquifer. The increase in soluble salts could lead to a 9% reduction in agricultural production and a 13% increase in energy costs (Saille et al. 2006). Canal seepage also supports the Andrade Mesa wetlands in Mexico (Dibble 2005), which consist of 500 acres of marsh and more extensive areas of mesquite bosque and provide habitat for endangered birds and migratory birds on the Pacific Flyway.

In the context of the model, one may distinguish between diverted water and delivered water. Let X in the model represent delivered water—water that actually reaches the irrigation district when water is diverted. Now, let $X = (1 - L)X'$, where X' represents surface water initially diverted and L represents the percentage of conveyance loss when transferring water. In this case, canal lining reduces L and increases X. The previous model results hold. Increasing X increases the quantity of water transferred and benefits both the irrigation district and the water provider. However, from Mexico's point of view, it receives agricultural benefits $A(L)$ and environmental benefits $E(L)$, where $A'(L) > 0$ and $E'(L) > 0$.

Historically, the United States' position has been that canal seepage represents U.S. surface waters allocated to it by the 1944 Water Treaty with Mexico, and that environmental mitigation under the Endangered Species Act (or other U.S. environmental laws) is not required for impacts beyond the U.S. border. Mexico has argued that the United States is taking action that adversely affects its groundwater, and that a protocol to the 1944 treaty requires that the countries formally consult with each other prior to undertaking such a project. However, a requirement for consultation is not necessarily a requirement for mitigation. The United States and Mexico have never signed a comprehensive agreement to manage transboundary groundwater resources.

In August 2006, the U.S. Ninth Circuit Court, in response to a lawsuit brought by Mexican business groups and environmental groups from both countries, issued a temporary injunction halting the lining of the All-American Canal (Dibble 2006). However, in December of that year, Congress passed the Tax Relief and Health Care Act of 2006, which included a section ordering that the canal lining project be completed "without delay" and stating that the 1944 treaty was the sole vehicle for addressing the relevant transboundary environmental issues (Pub. L. No. 109-432; Ries 2008). The Ninth Circuit Court subsequently dismissed all plaintiff claims (Cortez Lara et al. 2009; Ries 2008).

Bennett et al. (1997) noted that game-theoretic solutions to unidirectional externalities tend toward "victim pays" outcomes. In the All-American Canal lining case, this implies that Mexico's only option is to offer the United States some form of side payment to defer lining. Bennett et al. found "victim pays" regimes unsatisfactory because they run counter to the "polluter pays" principle accepted in the international community, and because countries may wish to avoid appearing to be weak negotiators.

Linking negotiation issues may be an alternative to side payments or accepting externalities. In interconnected games, negotiations over separate issues are joined in a repeated game. Each country's action in one game is conditional on the outcome of another. This allows for equilibrium solutions not attainable in isolated games and may yield higher joint payoffs. Interlinked solutions may also avoid side payments when isolated solutions do not (Folmer et al. 1994; Bennett et al. 1997). Ragland (1995) examined how Mexico was able to enhance its bargaining position vis-à-vis the United States in disputes over allocation of Colorado River water. Mexico successfully linked negotiations over the Colorado River (where it was the downstream, disadvantaged country) with negotiations over allocation of waters of the Rio Bravo/Rio Grande (where it was the upstream, advantaged country). By linking negotiations, Mexico was able to achieve a larger allocation of the Colorado's waters. Given that the United States and Mexico share 17 groundwater basins (Hall 2004), Mexico may consider such linkage over transboundary groundwater allocations.

Frisvold and Caswell (2000) and Frisvold (2009) suggested that an interconnected game framework might offer a solution to the All-American Canal dispute as well as other U.S.-Mexico water disputes. This suggestion appears to have gained some resonance among other social scientists and border scholars (e.g., Mumme and Lybacker 2006; Cortez Lara et al. 2009).

IMPLICATIONS FOR ADAPTATION AND RESILIENCE

Climate change and population growth in the Southwest will alter the relative values of water applied to competing uses. Reallocating water from lower- to higher-value uses can greatly reduce costs of drought or climate change. This chapter has examined large-scale, rural-urban water transfers as voluntary mechanisms to reallocate water. Given limits to increasing water supplies and growing urban water demands, municipal water providers have turned more and more to agriculture to supplement their water supplies.

How much water might be transferred in response to drought will depend on the geographic scope of the drought. One must consider climate in both water-supplying and water-demanding areas. Drought can simultaneously contract existing urban water supplies and increase urban water demands. Both factors would increase demand for new water transfers. However, on the supply side, drought in rural areas may simultaneously reduce agricultural water supplies and increase crop water requirements. These last two factors would discourage transfers. If the future climate throughout large areas of the Southwest becomes hotter and drier, as some studies suggest (e.g., Seager et al. 2007), the demand for transferred water will increase. But farmers would demand higher transfer prices to supply water to cities. If drought is more localized, or if water can be transferred over longer distances, there is a greater chance of transferring water from water-abundant rural areas to drought-stricken urban areas. As Chapter 2 demonstrated with water leases, local hydrologic conditions have important implications for transfer prices.

Water transfers can be a functional adaptation mechanism, helping both buyers and sellers directly involved in the exchange. But what do such transfers mean in considering a broader resilience framework (Nelson et al. 2007)? In other words, how do they help the region as whole maintain the basic structure and function of its economic activities? First, agriculture can serve as a form of insurance for urban water demanders. Agriculture possesses a residual reserve of water that cities may acquire, for a price. For agriculture, large water payments can be an important source of income. This chapter has focused on more permanent transfers. Chapter 4 examines the scope for more flexible dry-year option contracts, which make transfer contingent on seasonal weather and water availability.

From a resilience perspective, it is important to look beyond those agents immediately involved in water transfers. Although bilateral transfers can benefit those directly involved in the exchange, this chapter illustrates that they can have negative implications for downstream users or the environment. In particular, the QSA has potentially negative consequences for the Salton Sea, the ecosystem of the Colorado River Delta, and Mexican farmers. This raises questions about whether transfers improve the resilience of the system as a whole. If, in our example, "the system" includes Mexico, this is a debatable point.

A resilience framework also calls for a broader, systemwide evaluation of adoption of modern irrigation technologies. Improved irrigation technology can improve *on-farm* water application efficiency. Thus it appears to be an appropriate private adaptation to water scarcity. However, improved on-farm efficiency does not necessarily conserve water at the basin level, and it can actually increase consumptive use. Therefore, improved on-farm efficiency itself may not improve a larger water basin's resilience in the face of water scarcity.

CONCLUSIONS

This chapter has sought to demonstrate how a game-theoretic analysis of agricultural-urban water transfers can examine not only gains from trade, but also issues of bilateral monopoly, asymmetric capacity and bargaining power, and negative externalities of transfers. We conclude by summarizing the main policy lessons.

First, reallocation of water to urban uses through market transfers can be achieved through removing distortions in agricultural output and input markets. Second, improving irrigation efficiency itself need not encourage either rural-urban transfers or water conservation. Hence irrigation technology subsidies with these objectives may not meet intended goals. This is not to discourage efforts to improve irrigation efficiency per se. However, care must be taken to ensure that hydrologic and economic factors are such that improved efficiency truly conserves water. Third, mitigation requirements are highly sensitive to small changes in irrigation efficiency. To prevent third-party damages, large-scale water transfers will require regular environmental monitoring, recalibration of mitigation requirements, and adaptive environmental management. Fourth, collective bargaining by irrigation districts or water user associations will lead to more, not less, water transferred than if urban water buyers exercise pure monopsony power. Thus collective bargaining by itself will not necessarily remedy third-party effects of transfers. Fifth, agriculture can function to provide a reserve supply of water for cities that can be transferred in times of great urban need. Hence maintaining local agriculture serves an insurance function for urban demanders. This suggests a role for conservation easement or farmland preservation policies. Finally, in the U.S.-Mexico border region, large-scale transfers are confronting (and will continue to confront) unresolved questions about how to deal with transboundary externalities of water transfers. Unresolved legal questions remain over transboundary environmental mitigation requirements and groundwater management. We suggest that the theory of interconnected games may be of use conceptually to identify cooperative solutions to such transboundary issues.

APPENDIX 3.1: MATHEMATICAL DERIVATIONS

Irrigation District's Problem

If no water is transferred, irrigation district profits π^d are

$$\pi^d \quad = \quad Pf(h\bar{X}) - cX \tag{3.7}$$

where P is the agricultural output price, X is the district's water allotment, h is the irrigation efficiency parameter, and c is the cost of production per acre-foot of water. If the district sells water to an urban municipal provider, its profits are

$$\pi = Pf(h(X - x(1 + s))) + c(X - x(1 + s)) + wx \tag{3.8}$$

where x is the quantity of water transferred. Water required for mitigation m can be expressed as a proportion of water transferred, $m = sx$. The district's iso-payoff function in $w-x$ space shows combinations of x and w that yield a fixed profit π^0:

$$w = \frac{\pi^0 - Pf(h(X - x(1+s))) + c(X - x(1+s))}{x} \tag{3.9}$$

For a given value of x, the minimum acceptable price for the district is

$$w = \frac{Pf(hX) - Pf(h(X - x(1+s))) - cx(1+s)}{x} \tag{3.10}$$

Municipal Water Provider's Problem

Urban demand for additional water (beyond current capacity) is

$$p^m = \alpha - \beta x \tag{3.11}$$

where p^m is the inverse demand for water, and α and β are positive constants. The urban water provider is a regulated public utility whose objective is to maximize consumer benefits from obtaining new water subject to a cost-recovery constraint. Urban water consumers must pay wx, the cost of obtaining agricultural water. The urban water provider seeks to negotiate x and w to maximize this objective function, V, where

$$V = \alpha x - \frac{1}{2}\beta x^2 - wx \tag{3.12}$$

In $w-x$ space, the iso-payoff curve for the provider shows combinations of x and w that yield a fixed payoff V^0:

$$w = \frac{\alpha x - \frac{1}{2}\beta x^2 - V_0}{x} \tag{3.13}$$

For a given value of x, the provider's maximum willingness to pay is

$$w = \alpha - \frac{1}{2}\beta x \tag{3.14}$$

The Contract Curve

The contract curve, the set of efficient $x-w$ pairs, is obtained by equating the slopes of the district's and provider's iso-payoff curves (equations (3.9) and (3.13)). The efficient volume of water transferred can then be derived:

$$x^\star = \frac{(1+s)(-hPf_e + c) + \alpha}{\beta} \tag{3.15}$$

The efficient volume of water traded can be determined independently of the transfer price, w. Comparative static results can then be obtained by totally differentiating equation (3.15) to determine how changes in exogenous variables affect $x^\star$.

APPENDIX 3.2: SIMULATION MODEL PARAMETERS

This appendix describes the data sources used to specify model parameters and accounting identities. We chose the model parameters with two goals in mind: first, we wanted them to be consistent with values derived from existing studies, thus having empirical support; and second, they were chosen so that accounting relationships in the model, such as water diversions, water transfers, and transfer prices, were consistent with actual, observed values.

- *Water diversions* (X): Obtained from U.S. Bureau of Reclamation *Colorado River Accounting and Water Use Reports* for Arizona, California, and Nevada (various years). Boulder Canyon Operations Office, Boulder City, Nevada.
- *Gross farm returns* ($Pf(hX)$): Estimated from statistics for Imperial County from USDA NASS *2002 Census of Agriculture* and the Environmental Working Group Farm Subsidy Database, http://farm.ewg.org/.
- *Variable cost of productions* (c): Obtained from USDA NASS *2002 Census of Agriculture* production expenditure data for Imperial County. Excludes livestock purchases, feed purchases, and land rental payments.
- *Irrigation production function* ($f(e)$): The irrigation district production function was assumed quadratic. Parameters of the marginal value product (MVP) of water function were chosen such that the marginal willingness to pay for additional water at X was \$23.40, approximating Draper et al. (2003), and integrating the MVP curve yielded gross farm returns.
- *Irrigation efficiency* (h): Results from Rosen and Sexton (1993) suggest an h close to 0.7. Internal IID documents state that on-farm efficiency is about 0.79, but Burt (1999) suggests IID system efficiency will be less than on-farm efficiency.
- *Water transfer parameters* (x, w, m): The quantity of water transferred, price paid, and water used for Salton Sea mitigation come from the *QSA Annual Implementation Report*.
- *Return flows and consumptive use* (R, C): Data were derived based on return flows going to the Salton Sea and Colorado River. These came from U.S. Bureau of Reclamation *Colorado River Accounting and Water Use Report* and the California Environmental Protection Agency Regional Water Quality Control Board, Colorado River Basin Region.
- *Municipal demand for additional water* (α, β): Parameters were derived to obtain maximum willingness to pay for transferred water based on wholesale rates charged by SDCWA to retail municipal providers.

NOTES

1. Under bilateral monopoly, there is only one buyer and only one seller in a transaction.
2. One can think of the contract curve in terms of two people sharing a pie. Two issues need to be decided: the total size of the pie and the size of the slice each person gets. The contract curve represents the set of bargaining parameters (in this case, the water transfer quantity and price) that yields the biggest pie (i.e., the greatest total benefits to the buyer and seller). The

relative size of gains to the buyer and seller depend on the outcome of bargaining, but all else being equal, a bigger pie is preferable.

3. Draper et al.'s economic-engineering optimization model of California's water supply system (the CALVIN model; 2003) derived shadow values for constrained water supplies in different regions of the state. These shadow values may be interpreted as irrigators' willingness to pay for additional water supplies. Shadow value calculations specifically for the Imperial Irrigation District were $19 per thousand cubic meters, or roughly $23.40 per acre-foot.

REFERENCES

Bennett, L., S. Ragland, and P. Yolles. 1997. Facilitating International Agreements through an Interconnected Game Approach: The Case of River Basins. In *Decentralization and Coordination of Water Resource Management*, edited by D. Parker and Y. Tsur. Boston: Kluwer Academic Publishers, 61–85.

Binmore, K., A. Rubinstein, and A. Wolinsky. 1986. The Nash Bargaining Solution in Economic Modelling. *RAND Journal of Economics* 17:176–188.

Boggess, W., D. Zilberman, and R. Lacewell. 1993. Economics of Water Use in Agriculture. In *Agricultural and Environmental Economics*, edited by G. Carlson, D. Zilberman, and J. Miranowski. Oxford: Oxford University Press.

Brewer, J., R. Glennon, A. Ker, and G. Libecap. 2008. 2006 Presidential Address: Water Markets in the West: Prices, Trading and Contractual Forms. *Economic Inquiry* 46:91–112.

Burt, C. 1999. Irrigation Water Balance Fundamentals. Paper presented at United States Committee on Irrigation and Drainage (USCID) Conference on Benchmarking Irrigation System Performance Using Water Measurement and Water Balances. March 10, 1999, San Luis Obisbo, CA.

Caswell, M., and D. Zilberman. 1986. The Effects of Well Depth and Land Quality on the Choice of Irrigation Technology. *American Journal of Agricultural Economics* 68:798–811.

Chae, S. 1993. The N-Person Nash Bargaining Solution with Time Preference. *Economics Letters* 41:21–24.

Cortez Lara, A., M. Donovan, and S. Whiteford. 2009. The All-American Canal Lining Dispute: An American Resolution over Mexican Groundwater Rights? *Frontera Norte* 21 (41):127–150.

Dibble, S. 2005. Wetlands Become a Focus in Debate over Canal Lining. *San Diego Union-Tribune,* June 6.

———. 2006. All-American Canal Lining Halted by Court Injunction. *San Diego Union-Tribune*, August 26.

Dinar, A., and A. Subramanian. 1998. Policy Implications from Water Pricing Experiences in Various Countries. *Water Policy* 1:239–250.

Draper, A., M. Jenkins, K. Kirby, J. Lund, and R. Howitt. 2003. Economic-Engineering Optimization for California Water Management. *Journal of Water Resource Planning and Management* 129:155–164.

Folmer, H.P., P. van Mouche, and S. Ragland. 1994. Interconnected Games and International Environmental Problems. *Environmental and Resource Economics* 3:313–335.

Frisvold, G. 2009. Strategic Behaviour in Transboundary Water and Environmental Management. In *Policy and Strategic Behaviour in Water Resource Management*, edited by A. Dinar and J. Albiac. London: Earthscan, 279–300.

Frisvold, G., and M. Caswell. 1995. A Bargaining Model of Water Quality and Quantity. In *Water Quantity/Quality Disputes and Their Resolution*, edited by A. Dinar and E. Loehman. Westport, CT: Praeger Press, 399–407.

———. 2000. Transboundary Water Management: Game-Theoretic Lessons for Projects on the U.S.-Mexico Border. *Agricultural Economics* 24:101–111.

Gelt, J., J. Henderson, K. Seasholes, B. Tellman, and G. Woodard. 1998. *Water in the Tucson Area: Seeking Sustainability.* Tucson: Water Resources Research Center, College of Agriculture, University of Arizona.

Hall, R. 2004. Transboundary Groundwater Management: Opportunities under International Law for Groundwater Management in the United States-Mexico Border Region. *Arizona Journal of International and Comparative Law* 21:873–875.

Huffaker, R., and N. Whittlesey. 1995. Agricultural Conservation Legislation: Will It Save Water? *Choices* 10:24–28.

———. 2000. The Allocative Efficiency and Conservation Potential of Water Laws Encourage Investments in On-Farm Irrigation Technology. *Agricultural Economics* 24:47–60.

———. 2003. A Theoretical Analysis of Economic Incentive Policies Encouraging Agricultural Water Conservation. *International Journal of Water Resources Development* 19:37–53.

Hutson, S., N. Barber, J. Kenny, K. Linsey, D. Lumia, and M. Maupin. 2004. *Estimated Use of Water in the United States in 2000.* USGS Circular 1268. Reston. VA: U.S. Geological Survey.

IID (Imperial Irrigation District). 2004. *Quantification Settlement Agreement Annual Implementation Report.* Imperial, CA: IID.

Jacobs, K., and S. Megdal. 2004. Water Management in the Active Management Areas. In *Arizona's Water Future: Challenges and Opportunities*, edited by P. Likins. Tucson: University of Arizona, 71–94.

Kalai, E., and M. Smorodinsky. 1975. Other Solutions to Nash's Bargaining Problem. *Econometrica* 43:513–518.

Krishna, V., and R. Serrano. 1996. Multilateral Bargaining. *Review of Economic Studies* 63:61–80.

Libecap, G., R. Glennon, and A. Ker. 2005. The Implications of Legal and Political Institutions on Water Transfers in the West. *Arizona Review* 3:3–4.

Maganda, C. 2005. Collateral Damage: How the San Diego-Imperial Valley Water Agreement Affects the Mexican Side of the Border. *Journal of Environment and Development* 14 (4):486–506.

McDonald, I.M., and R.M. Solow. 1981. Wage Bargaining and Employment. *American Economic Review* 71:896–908.

Merrett, S. 2003. The Urban Market for Farmers' Water Rights. *Irrigation and Drainage* 52:319–326.

Michel, S. 2002. The Geography of Water Transfers and Urbanization in Baja and Southern California. In *Both Sides of the Border: Transboundary Environmental Management Issues Facing Mexico and the United States*, edited by L. Fernandez and R. Carson. Dordrecht, Netherlands: Kluwer Academic Publishers, 199–234.

Moore, M. 1991. The Bureau of Reclamation's New Mandate for Irrigation Water Conservation: Purposes and Policy Alternatives. *Water Resources Research* 27:145–155.

Moore, M., and A. Dinar. 1995. Water and Land as Quantity-Rationed Inputs in California Agriculture: Empirical Tests and Water Policy Implications. *Land Economics* 71:445–461.

Mumme, S.P., and D. Lybecker. 2006. The All-American Canal: Perspectives on the Possibility of Reaching a Bilateral Agreement. In *The U.S.-Mexican Border Environment: Lining the All-American Canal: Competition or Cooperation for the Water in the U.S.-Mexican Border?*, edited by V.S. Munguía. SCERP Monograph Series No. 13. San Diego: San Diego State University Press, 175–195.

Nash, J. 1953. Two-Person Cooperative Games. *Econometrica* 21:128–140.

Nelson, D.R., W.N. Adger, and K. Brown. 2007. Adaptation to Environmental Change: Contributions of a Resilience Framework. *Annual Review of Environment and Resources* 32:395–419.

Peterson, J., and Y. Ding. 2005. Economic Adjustments to Groundwater Depletion in the High Plains: Do Water-Saving Irrigation Systems Save Water? *American Journal of Agricultural Economics* 87:147–159.

Ragland, S. 1995. International Environmental Externalities and Interconnected Games. PhD diss., University of Colorado–Boulder.

Ries, N. 2008. The (Almost) All-American Canal: Consejo de Desarrollo Economico de Mexicali v. United States and the Pursuit of Environmental Justice in Transboundary Resource Management. *Ecology Law Quarterly* 35 (3):491–530.

Rosen, M.D., and R.J. Sexton. 1993. Irrigation Districts and Water Markets: An Application of Cooperative Decision-making Theory. *Land Economics* 69:39–53.

Saille, G.G., A.L. López, and J.A. Navarro Urbina. 2006. Lining the All-American Canal: Its Impact on Aquifer Water Quality and Crop Yield in Mexicali Valley. In *The U.S.-Mexican Border Environment: Lining the All-American Canal: Competition or Cooperation for the Water in the U.S.-Mexican Border?*, edited by V.S. Munguía. SCERP Monograph Series No. 13. San Diego: San Diego State University Press, 77–100.

SDCWA (San Diego County Water Authority). 2010. *Canal Lining Projects Fact Sheet.* San Diego: SDCWA.

Seager, R., M. Ting, I. Held, Y. Kushnir, J. Lu, G. Vecchi, H. Huang, N. Harnik, A. Leetmaa, N. Lau, C. Li, J. Velez, and N. Naik. 2007. Model Projections of an Imminent Transition to a More Arid Climate in Southwestern North America. *Science* 25:1181–1184.

Shiklomanov, I.A. 2000. Appraisal and Assessment of World Water Resources. *Water International* 25:11–35.

USDA NASS (United States Department of Agriculture, National Agricultural Statistics Service). 2003. *Farm and Ranch Irrigation Survey, 2003.* Washington, DC: U.S. Government Printing Office.

USDA NRCS (United States Department of Agriculture, Natural Resources Conservation Service). 2003. *EQIP Benefit Cost Analysis: Final Report.* Washington, DC: USDA NRCS.

Ward, F., and M. Pulido-Velazquez. 2008. Water Conservation in Irrigation Can Increase Water Use. *Proceedings of the National Academy of Sciences* 105 (47):18215–18220.

Woody, T. 2009. Solar Stirs Water Wars in the West. *New York Times*, Sept. 30.

Zamora Arroyo, F., P. Culp, and O. Hinojosa Huerta. 2006. Looking beyond the Border: Environmental Consequences of the All-American Canal Project in Mexico and Potential Binational Solutions. In *The U.S.-Mexican Border Environment: Lining the All-American Canal: Competition or Cooperation for Water in the U.S.-Mexican Border?*, edited by V. Sánchez Munguía. SCERP Monograph Series No. 13. San Diego State University Press, 21–58.

CHAPTER 4

Economic Tools for Climate Adaptation: Informing Water Price Negotiations

Bonnie G. Colby, Lana Jones, and Katie Adams

Water leases to improve supply reliability under climate change will typically involve moving water out of irrigation use, because agriculture accounts for the vast majority of water use in the Southwest. Low financial returns to irrigating various field crops and the relative seniority of some agricultural water allocations make transfers out of agriculture attractive to those urban and industrial users that face high costs if their supplies are curtailed. Public agencies and conservation NGOs seeking to protect and restore water-dependent habitat also may be potential lessees of irrigation water during dry periods.

This chapter outlines various mechanisms used to establish price in voluntary leases for agricultural water. As discussed in Chapters 2 and 3, water typically used to irrigate crops is being sought to improve supply reliability for cities and environmental programs, given the increased variance in interannual supplies anticipated as climate change progresses. An important component of successful temporary transfers during dry periods is the ability to negotiate mutually acceptable prices per unit of agricultural water being transferred. The residual valuation method is presented as a tool to inform negotiations for irrigation water, with an example provided for western Arizona. While this chapter focuses on negotiations involving individual agents (farms, urban water providers, habitat managers), agreements among these individual arrangements are the building blocks of regional resilience under increased temperatures and interannual water supply variability.

The temporary nature of water leases designed to improve supply reliability has several advantages. Although effective in mitigating the costs of drought-induced supply variability, dry-year leases are not suitable to provide long-term supplies for population growth. They cannot be dedicated to support growth once drought has ended, as that would make them unavailable as a buffer in future dry periods. Further, leases are less financially costly than permanent acquisitions of agricultural land and water, and they engender less heated opposition over potential third-party impacts.

Agriculturally linked rural communities and local governments, such as irrigation districts and counties, often respond to proposed water transfers out of agriculture with suspicion and alarm, concerned that movement of water away from local agricultural uses will undermine the local economy by reducing business activity and property values. Studies of transfers out of agriculture have found that local economic impacts are proportionally smaller than the amount of irrigated land that is fallowed, even when the water is moved to a new use away from the area of origin (Howe et al. 1990; Checchio 1988). Several factors account for the consistent findings of economic impacts being proportionally smaller than the percentage of irrigated acreage fallowed: farmers choose to fallow their lowest-value crops and their least productive acreage; a portion of water payments received by farmers is generally spent in the county from which the water is exported; and income from crop sales is a small portion of county income in nearly every rural county in the West.

In general, water acquirers would be astute to adopt a policy of compensating clearly documented direct impacts, both to reduce conflict and to avoid costs of economic dislocation and unemployment. Rural businesses and nearby urban areas are bound together in an interdependent regional economy that shares common labor pools and other inputs. Carefully structured compensation of third-party impacts can facilitate the ability to use temporary water transfers as a key regional climate change adaptation tool.

STRUCTURING TEMPORARY WATER ACQUISITIONS

Temporary water transfers can be structured in many different forms. In a spot market, price is often established by a bidding process, and the arrangement involves a onetime change in location or purpose of use for a specific quantity of water at a particular point in time.

Water banks are established in almost every western state. Although there are functional differences among banks, each one facilitates water transfers by taking on the role of broker, clearinghouse, or price fixer while fulfilling other administrative and technical functions. The simplest function a water bank can serve is to be an information clearinghouse, with prices determined through bilateral negotiations between lessors and lessees. The clearinghouse function reduces the transaction costs of identifying potential trading partners.

A fixed-price structure, in which the price is fixed by the bank, is another framework employed by water banks. Although the fixed-price structure may create a sense of fairness and reduce concerns about price gouging and speculation, it is unable to respond quickly to changes in market conditions. This limitation can become acute during dry years if the fixed price is too low and yields insufficient banked supplies during drought. The clearinghouse approach, by letting buyers and sellers negotiate prices themselves, allows for a more flexible price response to varying conditions.

Transaction activity in most banking programs tends to be limited to specific types of participants and water sources, and specified types of transfers (Clifford et al. 2004). The Snake River Water Bank is an example of a long-standing water bank, with periods of high trading activity during drought. It has operated for several decades to provide flexibility in Idaho's Snake River Basin.

Contingent transfer agreements, or dry-year options, are a form of temporary reallocation that may be useful to enhance dry-year supply reliability (Michelsen and Young 1993). These transfers are implemented in prespecified drought or shortage conditions. Agricultural water use during normal supply conditions is not affected. When an option is exercised, irrigators temporarily cease irrigating cropland in exchange for an exercise payment. This payment must compensate them to a level *at least* equivalent to the level of forgone profit from crop production in order for the arrangement to be attractive to irrigators. Contingent dry-year transfer contracts between senior and junior water rights holders can maintain supply reliability for juniors.

Option contracts can be onetime, short-term, or long-term arrangements. An example of the first type would be an agreement negotiated during a winter when snowpack and reservoir storage are below normal, to be exercised the following spring if the region winds up in a shortage situation that threatens urban water supplies or fish recovery flows. Long-term contracts can span many decades, giving those paying for forbearance the right to exercise an option to use irrigation water under prespecified trigger conditions (MWD 2003).

Dry-year options address two fundamental sources of uncertainty in water transactions: water supply uncertainty and uncertainty over future water values (Howitt 1998). Option contracts spread these risks between the suppliers and demanders of dry-year supplies in a different manner than spot markets and permanent acquisitions. In spot markets, supply uncertainty is borne by demanders of water, as they must rely on thinly traded markets with uncertain supply to meet their inelastic needs. In the case of permanent acquisitions, suppliers bear the uncertainty about the future value of their water and risk selling their water rights too cheaply. Dry-year option contracts spread these risks between suppliers and demanders of water during dry years. Demanders shift some of the water supply risk to suppliers by purchasing an option in advance of shortage. Likewise, suppliers are relieved of some of the risk inherent in permanent sales, as options markets facilitate the *temporary* reallocation of water, and suppliers remain in possession of their water rights, which may increase in value over time.

In addition to spreading these risks, option contracts can reduce third-party impacts of water transfers. Because options are seasonal and temporary in nature, the suppliers, typically farmers, continue normal farming operations in most years. In years when the options are exercised, the growers likely will continue participating in the local economy and spend some of the options payments locally. In comparison with spot markets, options markets also allow rural communities more time to plan for changes in local spending patterns and job opportunities that occur as a result of temporary water transfers, thus minimizing the impact of those changes.

ON-FARM ECONOMIC VALUES FOR AGRICULTURAL WATER

Information on the economic value of water in crop irrigation is important for those seeking to acquire water temporarily from agriculture and in helping parties negotiate fair prices for transactions. Accurately valuing irrigation water is an important component of structuring and negotiating dry-year transfers to improve supply reliability. Young (2005) and Gibbons (1986) described several widely employed valuation methods.

A water-crop production function models the link between the application of water and crop yield, defining a relationship between a specific level of input application (water) and its effect on output (crop production). These models are crop-, location-, and soil-specific and depend on assumptions about the levels of other crop production inputs, such as fertilizers and pest control. Crop production functions are typically estimated based on expert opinion but can also be approximated using field experiments and computer simulation. The application of this approach is limited to locations and crops for which accurate up-to-date water-crop functions are available.

Another method for valuing irrigation water is the direct observation of prices in voluntary water transactions. Complications associated with this method include inadequate transaction data and constrained market activity, including the influence of one large buyer or seller (lessor/lessee) on observed prices (Young 2005). This approach is useful only in regions with regular, voluntary transactions and access to price data, which are sometimes kept confidential by transacting parties.

The most widely applicable method for valuing water used to irrigate crops is the residual, or farm budget, method. Representative farm models are constructed to represent typical patterns of soil types, climate, labor supply and other crop production inputs, and cropping patterns. A table detailing operations and inputs for each crop is constructed based on the representative farm, with data generally obtained from farmer and extension agent interviews. This data is then used to calculate net returns to water (NRTW), above and beyond the nonwater variable costs. NRTW per acre is calculated for each crop by subtracting variable production costs (exclusive of water costs) from gross returns per acre. NRTW is a widely used indicator of the on-farm economic value of water, and it can provide an objective baseline to inform negotiations over dry-year forbearance payments. The exercise payment in a voluntary program could not lie below NRTW for a specific crop, because in that case the farmer would be better off producing crops. Exercise payments may be larger than NRTW to provide an incentive for farmers to participate in a forbearance program and to cover other costs. On the other hand, a grower weighing various risks of crop production and marketing may choose a known forbearance payment over an unknown net return to crop production.

CASE STUDY OF NET RETURNS OVER VARIABLE COSTS

This case study for Yuma County, Arizona, builds on the county crop budgets developed by the University of Arizona's Department of Agricultural and

Resource Economics (Teegerstrom et al. 2001; Teegerstrom and Knowles 1999; Teegerstrom and Tickes 1999). The most recent published farm budgets (1999–2001) were substantially updated by way of interviews and U.S. Department of Agriculture (USDA) statistics to reflect current crop yields and prices, chemical application rates, typical production operations, and fuel and labor costs (Amerigas 2006; Barkley Seed, Inc. 2006; Carr Seed Co. 2006; Ferrel Gas 2006; H&H Seed Co. 2006; Nolte 2005, 2006; Osborn 2005, 2006a, 2006b; Parker Oil 2006; Union Oil 2006; USDA NASS 2005). All figures are reported here in 2008 US$.

Federal farm support payments can substantially affect the profitability of certain crops, and thus this analysis includes loan deficiency payments (LDPs) in calculations of NRTW (CARD 2005). In Arizona, the LDP payment applies only to cotton. Other federal payments—in particular, countercyclical and direct payments—are decoupled from crop production and are not affected by temporary land fallowing due to water transfers (Reinertson 2006).

NRTW estimates the on-farm economic value of water in crop production and is calculated by subtracting variable production costs (exclusive of water costs) from gross returns per acre. In other words, the residual from the difference between the gross value of crop production and nonwater input is attributed to be the return to irrigation water in crop production (Naeser and Bennett 1998). This value is the maximum an irrigator could pay for water and still cover the variable costs of production. This is a reasonable approach to estimate the short-run value of irrigation water, as only variable costs of production are accounted for. Fixed costs are not a chief consideration in yearly production decisions (Naeser and Bennett 1998). A longer-run estimate of the on-farm value of irrigation water, however, would require the inclusion of the fixed costs of production. Moreover, attributing all of the NRTW to the water input *overstates* the on-farm economic value of irrigation water, because farm budget data do not assign a cost per acre for management expertise and farm assets such as equipment.

Tables 4.1, 4.2, and 4.3 show the NRTW calculation for three important western Arizona crops: alfalfa, cotton, and lettuce. In these calculations, costs from University of Arizona crop budgets (Teegerstrom and Knowles 1999; Teegerstrom and Tickes 1999; Teegerstrom et al. 2001) have been updated using USDA producer cost indices (USDA NASS 2008). Yields and commodity prices come

Table 4.1 *Net Returns to Water (NRTW) for Yuma Alfalfa Production (2008 US$)*

Yuma alfalfa production	*2006*	*5-yr avg.*	*High*	*Low*
Revenue per acre				
Yield/acre	9.07	9.3	10	8.62
Price/ton	$128.00	$108.20	$128.00	$89.50
Gross revenue ($/acre)	$1,160.96	$1,006.26	$1,280.00	$771.49
Total variable costs per acre	$400.79	$400.79	$400.79	$400.79
Net returns to water per acre	$760.17	$605.47	$879.21	$370.70
Acre-feet of water consumptively used per acre	5.7	5.7	5.7	5.7
Net returns to water per acre-foot of water used	$133.36	$106.22	$154.25	$65.04

from annual USDA agricultural statistics bulletins for the state (USDA NASS 2003, 2005, 2006). Crop consumptive water use was derived using Lower Colorado River Accounting System Evapotranspiration and Evaporation Calculations for annual water use (USBR 2007).

The LDP per pound are reversed in the high and low columns. In Table 4.2, the highest cotton lint price is coupled with the lowest LDP, and the lowest cotton lint price is coupled with the highest LDP. LDP are not available unless county prices fall below the loan rate set by the USDA Farm Service Agency (USDA FSA 2007). The negative NRTW of upland cotton in Yuma County suggests a potential for financial benefits to producers who participate in dry-year option arrangements instead of irrigating upland cotton.

In reviewing and comparing the NRTW for these three crops in Yuma County, the question naturally arises as to why growers do not plant more acreage in high-return specialty crops such as lettuce. Suppliers of head lettuce in the United States are almost exclusively in two states, with Arizona representing 26% and California 73% of national production in 2004 (Boriss and Brunke 2005). In Arizona, head lettuce production is carefully timed so that harvest dates coincide with the times of the year when the harvest of head lettuce in California is low, typically in November and December and again in late March and into April. These harvest dates also take advantage of the microclimates in California and Arizona that are most conducive to high-quality head lettuce production. Successfully entering the head lettuce market requires extensive planning and well-established marketing networks (Teegerstrom 2006).

Though the average annual NRTW is high, head lettuce production is extremely high-risk and capital-intensive. The market for lettuce is unstable with respect to price received by producers and is locally, not globally, driven (Kerns et al. 1999). Unlike other field crops, lettuce cannot be stored and sold when market prices are high. Its perishable nature means lettuce is sold immediately after harvest, and if market conditions are not favorable at the time of harvest, a grower has no choice but to accept the low prices or plow under the lettuce that cannot be profitably sold.

Table 4.2 *Net Returns to Water (NRTW) for Yuma Upland Cotton (2008 US$)*

Yuma upland cotton	*2006*	*5-yr avg.*	*High*	*Low*
Revenue per acre				
Yield/acre (lint)	1315	1323	1438	1213
Price/pound (lint)	$0.53	$0.52	$0.66	$0.44
LDP/pound	$0.086	$0.090	$0.034	$0.140
Yield/acre (seed)	1.03	1.05	1.24	1.02
Price/ton (seed)	$168.00	$150.00	$168.00	$134.00
Gross revenue ($/acre)	$983.77	$965.03	$1,206.00	$840.38
Total variable costs per acre	$1,150.33	$1,150.33	$1,150.33	$1,150.33
Net returns to water per acre	−$166.56	−$185.30	$55.67	−$309.95
Acre-feet of water consumptively used per acre	3.6	3.6	3.6	3.6
Net returns to water per acre-foot of water used	−$46.27	−$51.47	$15.46	−$86.10

Table 4.3 *Net Returns to Water (NRTW) for Yuma Head Lettuce (2008 US$)*

Yuma head lettuce	*2006*	*5-yr avg.*	*High*	*Low*
Revenues per acre				
Yield/acre	330	345	360	325
Price/cwt	$14.10	$19.98	$38.70	$10.30
Gross revenue ($/acre)	$4,653.00	$6,893.10	$13,932.00	$3,347.50
Total variable costs per acre	$5,082.87	$5,082.87	$5,082.87	$5,082.87
Net returns to water per acre	− $429.87	$1,810.23	$8,849.13	− $1,735.37
Acre-feet of water consumptively used per acre	1.4	1.4	1.4	1.4
Net returns to water per acre-foot of water used	− $307.05	$1,293.02	$6,320.81	− $1,239.55

The market price of lettuce has historically been volatile and is particularly sensitive to changes in supply. Flooding or disease or insect outbreaks, for instance, can dramatically reduce supply and correspondingly boost the market price. If there is surplus supply, however, the market price will plummet and only the highest-quality heads (without insect damage or contamination) will be accepted by packers (Kerns et al. 1999). And because growers cannot reliably predict what supply conditions will prevail during harvest, they must consistently incur the added costs of ensuring only the highest-quality production.

Figure 4.1 depicts the volatile nature of market prices for head lettuce. The graph shows the weekly fluctuation in prices received by Arizona growers for the first six months of 2006. Price fluctuations occur as a result of weather and other factors that affect yield and harvest dates in the numerous lettuce-growing microclimates.

In making production decisions, growers consider not only the potential profitability of a crop, but also the input costs and risks inherent in production and marketing of that crop. The variable costs of production for head lettuce and other

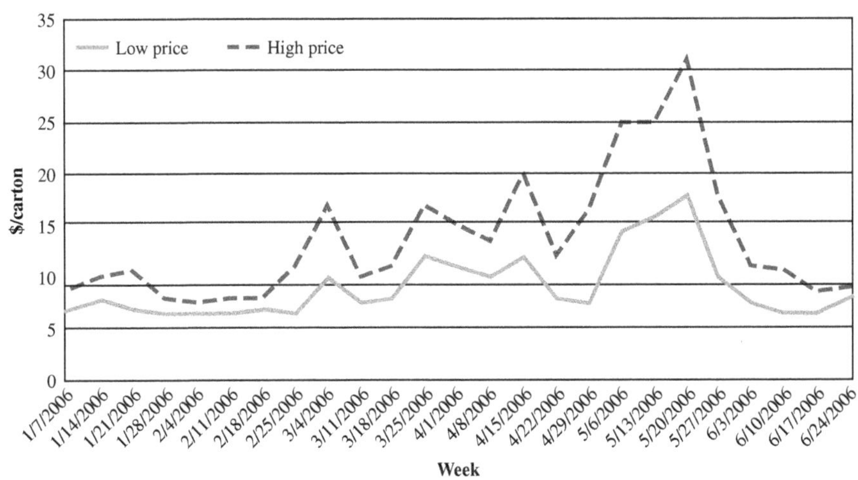

Figure 4.1 *Average Weekly Price of Head Lettuce, January–June 2006*

Source: USDA MSA (2008)

vegetables are very high compared with those of field crops. A farmer who has the necessary networks in place and decides to incur the high production costs associated with head lettuce production also is taking the risk of volatile, unpredictable market conditions. If the market price for head lettuce is low when the crop is harvested, growers must choose between accepting that price and plowing under the crop, thus averting harvest costs, and will likely incur a large financial loss for the season on their lettuce acreage.

The residual method used to calculate NRTW provides only a rough estimate of the on-farm value of irrigation water for high-value crops like lettuce. The costs that are not accounted for in NRTW are particularly high for these types of crops: special management skill, complex marketing negotiations, and a willingness to accept risk. Because monetary costs are not readily available for these factors, the NRTW overestimates water's value for high-value, high-risk crops (Young 2005).

SEASONAL CONSIDERATIONS IN STRUCTURING FORBEARANCE PROGRAMS

The time of year when growers are asked to cease irrigation has implications for compensation payments. In some instances, growers will have already produced some crops (and earned revenues) and also may have already incurred some of the variable costs of production. To accurately account for the value of forgone crop production, the timing of forbearance must be taken into account. Forbearance payments can be tied to the time of year an option is exercised. If, for example, an option is exercised before November for the upcoming season, a Yuma County cotton producer would not yet have incurred variable costs associated with cotton production, and thus the baseline calculation of NRTW for upland cotton would apply. If, however, irrigators are asked to cease irrigating late in the growing season (after July 30), they will have already incurred almost 57% of their variable costs without having produced a marketable crop, so approximately $677 arguably could be added on to minimum compensation per acre, substantially increasing the cost of the forbearance arrangement.

Unlike cotton, alfalfa production in western Arizona is a year-round process, and harvesting occurs regularly (approximately every three weeks) in April through October and once a month from November to January (Nolte 2006). As a result, NRTW in alfalfa is affected by the time of year growers are asked to cease irrigating their crop based on variable costs already incurred and also on alfalfa already harvested.

The most cost-effective acquisition programs will target lower-value field crops and avoid vegetable crops. The lower-value field crops are irrigated from winter to spring. However, to avoid unnecessary purchases by growers, fallowing notification would need to occur by early fall. Figure 4.2 illustrates the timing of planting, growing, and harvesting for major crops in Yuma based on University of Arizona crop budgets (Teegerstrom and Knowles 1999; Teegerstrom and Tickes 1999; Teegerstrom et al. 2001). The age of an alfalfa stand may also be

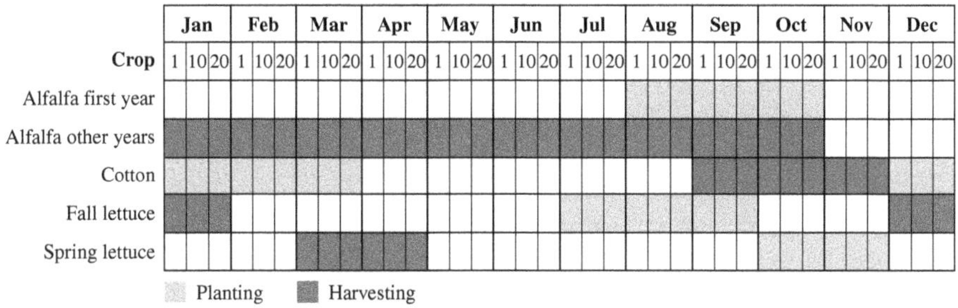

Figure 4.2 *Usual Planting and Harvesting Dates in Yuma County, Arizona*

Note: Numbers refer to days of the month
Source: USDA NASS (2003, 2005, 2006)

a determining factor in whether temporary fallowing is attractive to farmers. Alfalfa stands are most productive in their second year, followed by their first and third years; stands four or more years old can have low yields (Ottman 2008) and should probably be targeted for fallowing.

The time of year when a forbearance option is exercised has significant consequences for the on-farm net returns forgone and forbearance program costs. For the western Arizona region examined here, the most cost-effective scenario for irrigation forbearance on cropland planted in a field crop–vegetable annual crop mix is to request participating growers to forbear irrigation in the late spring and summer (forbearing production of field crops), and then provide water for production of vegetables in the winter or spring. These seasonal arrangements reduce the forgone net returns in crop production, but they may also require water storage arrangements to match the timing of irrigation cessation with seasonal need for additional water by municipalities or fish and wildlife programs.

If, however, climate and water supply prediction tools can be used to anticipate the timing and magnitude of the need for additional water in upcoming months or years, then forbearance options can be exercised for field crops, and the conserved water could be banked for use later in the year. This strategy rests on two assumptions: (1) the capacity to predict water supply shortfalls in December–February is accurate enough to inform negotiations over the structure and pricing of forbearance agreements that might be exercised the following summer (a short-term, onetime option contract), or prediction capacity is accurate enough to inform longer-term option contracts intended to ensure supply reliability for 20 to 50 years (long-term contract); and (2) adequate storage capacity is available at reasonable cost to carry the unused irrigation water into future seasons and years for use by those contracting for forbearance.

FURTHER ON-FARM FINANCIAL CONSIDERATIONS

In assessing the value of water in crop production, a few additional factors should be considered. The first is the effect of irrigation forbearance on federal price support payments. In particular, the value of forgone crop production could change if countercyclical payments (tied to the market price of cotton) and direct payments become affected by temporary irrigation forbearance. Decoupled federal payments currently would not be affected by a farmer's decision to temporarily cease irrigation (Reinertson 2005). The farmer is, however, responsible for managing noxious weeds on the fallowed lands, with some accompanying costs.

Another consideration is crop insurance. Farmers can purchase yield-based insurance, known as actual production history, which insures producers against yield losses due to natural causes such as drought, flooding, hail, wind, frost, insects, and disease (USDA RMA 2005). An individual farmer selects the percentage of average yield (50–75%) as well as the percentage of predicted price (55–100%) to be insured. Farmers can also purchase revenue-based insurance, called crop revenue coverage, which provides revenue protection based on price and yield expectations. A farmer who purchases crop insurance also receives a premium subsidy payment (USDA RMA 2006; Babcock and Hart 2005). Those who cease irrigating their cropland are not entitled to crop insurance on the acres involved in forbearance for that year (Tronstad 2006a). The forgone insurance subsidy payment, as well as potential crop insurance payments, cannot readily be factored into NRTW calculations, as they do not vary predictably with crop yield. While these payments could influence a farmer's decision to participate in a forbearance program, many Arizona growers do not purchase crop insurance because almost all of the state's crop production is irrigated, so the yield risk is lower than for dryland farming operations (Tronstad 2006b).

Risk management considerations also affect temporary water acquisition programs. Farmers routinely encounter risks through unpredictable weather, changing input cost and crop prices, the cost and availability of debt financing, tax planning, and evolving government programs and regulations (USDA ERS 2009). Diversification is a frequently suggested technique to reduce farmer risk and income variability from year to year. Farmers diversify by planting different crops, producing combinations of crops and livestock, planting different varieties of the same crop, or seeking alternate income sources. Participation in dry-year option programs could also be viewed as a source of alternate income. Dry-year option programs can even be more attractive than other forms of diversification, such as planting different crops, which can be prohibitively capital-intensive. The diversification offered by option programs can be used as a risk management tool. Because the payments are virtually risk-free, risk-averse farmers could benefit even if the per-acre payment were not as high as average—but uncertain—crop returns.

Diversification combines enterprises with varying degrees of correlation to reduce overall risk. Different crops may have highly correlated returns because of their shared reliance on the weather. Consequently, their returns will move up and down together. The payments from a fallowing program, on the other hand, are fixed and thus are likely to have low correlation with crop returns. Payment-supported fallowing can offer farmers a better way to reduce risk by diversification. Crop portfolio analyses, which include option participation as a "crop," can be used as a tool to examine how fallowing may benefit risk-averse farmers.

A final important consideration is the role of fixed costs of production. Growers may negotiate to include some fixed costs of production in payments received to refrain from irrigating. A rationale for including some fixed costs is to maintain the economic vitality of the overall farm operation, which cannot survive indefinitely if only variable costs of production are covered.

CONCLUSIONS

One anticipated effect of higher temperatures and greater variability in water supplies as a result of climate change is an increased interest in using agricultural water supplies for other purposes during extended dry periods. This chapter discussed various structures for temporary water acquisition programs and presented the residual valuation method, demonstrating its application to crop production conditions in western Arizona. Factors that influence the cost of production as well as yields and gross returns vary widely within a region, and the NRTW calculations are sensitive to input and output prices, which fluctuate over time. The NRTW results presented in this chapter should be treated as estimates for a specific area, crop, and time period. Following the basic procedures applied here, NRTW can readily be calculated to fit specific regional production and market conditions and the informational needs of a water acquisition program.

The vast majority of consumptive water use in the western United States occurs in irrigated agriculture. When drought conditions threaten to impose high costs in other sectors, voluntary irrigation forbearance can be a means to mitigate regional drought impacts. Dry-year water leases are particularly conducive to minimizing drought-induced water supply variability. This chapter examined different ways to structure temporary transfers. A critical component of negotiating dry-year transfers is determining the compensation growers receive to cease irrigating a specific amount of cropland. The residual (farm budget) approach can be useful as a baseline in negotiating payments.

Timing is important in accurately establishing a baseline for payment negotiations. If growers are asked to refrain from irrigating before field crop production begins, compensating them to the level that would leave them just as well off had they irrigated is much less costly than asking them to cease irrigation later in the season, when they have already incurred some variable costs of production. Forgone crop revenue is also substantially higher if forbearance programs interrupt high-value specialty production, so seasonal timing is necessary in structuring agreements in the most cost-effective way possible.

Temporary transfers provide important advantage over permanent acquisitions for all parties to the arrangement. Irrigators use their water for its customary purposes during years when the demand for temporary leases is low, and they can benefit financially by leasing their water in dry years. Dry-year transfers provide cities and habitat restoration programs the flexibility to lease water only when supply augmentation is necessary. Instead of incurring the economic and political costs of purchasing permanent water rights, temporary transfers allow for year-to-year decisions about leasing additional water. Further, if the original user maintains ownership of the water right, third-party impacts and transaction costs can be minimized.

In sum, addressing potential climate-related water shortages is not simple or inexpensive, but preplanning is more effective and less costly than crisis management responses. Improving supply reliability necessitates integrated participation and acceptance of reliability enhancement strategies from all parties involved. Despite the costs, complexities, and collaborative efforts required, temporary water leases and irrigation forbearance could well spare regional economies from significant economic losses in sectors lacking reliable supplies under climate-related supply variability.

ACKNOWLEDGEMENTS

The authors acknowledge funding provided by the U.S. Bureau of Reclamation and the Arizona Water Sustainability Program for the project Enhancing Water Supply Reliability through Improved Predictive Capacity and Response, which examines ways to enhance Arizona's water supply reliability from the Colorado River. This chapter has benefited significantly from conversations with public officials, research colleagues, stakeholders, and cooperative extension personnel. Report contents represent the professional expertise and research findings of the authors and do not necessarily represent the views of project sponsors.

REFERENCES

Amerigas Propane. 2006. Telephone interview, Parker, Arizona. January.

Babcock, Bruce A., and Chad E. Hart. 2005. Influence of the Premium Subsidy on Farmers' Crop Insurance Coverage Decisions. Center for Agricultural and Rural Development, Iowa State University, www.card.iastate.edu/publications/DBS/PDFFiles/05wp393.pdf (accessed September 21, 2006).

Barkley Seed, Inc. 2006. Telephone interview, Yuma, Arizona. February 6.

Boriss, Hayley, and Henrich Brunke. 2005. Commodity Profile: Lettuce. Agricultural Issues Center, University of California, http://aic.ucdavis.edu/profiles/lettuce-2005.pdf (accessed September 21, 2006).

CARD (Center for Agricultural and Rural Development at Iowa State University). 2005. http://cardsrv7.card.iastate.edu/LDPHome.aspx (accessed September 21, 2006).

Carr Seed Company. 2006. Telephone interview, Yuma, Arizona. 6 Feb.

Checchio, Elizabeth. 1988. Water Farming: The Promise and Problems of Water Transfers in Arizona. Tucson: University of Arizona, Department of Hydrology and Water Resources.

Clifford, Peggy, Clay Landry, and Andrea Larsen-Hayden. 2004. Analysis of Water Banks in the Western States. Olympia, WA, and Laramie, WY: Washington State Department of Ecology and WestWater Research.

Ferrel Gas. 2006. Telephone interview, Yuma, Arizona. January.

Gibbons, D.C. 1986. *The Economic Value of Water.* Washington, DC: Resources for the Future.

H&H Seed Company. 2006. Telephone interview, Yuma, Arizona. February 6.

Howe, Charles W., Lynne Bennett, Charles M. Brendecke, J. Ernest Flack, Robert M. Hamm, Roger Mann, Lee Rozaklis, Mark Griffin Smith, and Karl Wunderlich. 1990. Urban Water Supply Reliability: Preferences of Managers, Elected Officials and Water Users in Boulder, Colorado. Boulder, Colorado: Colorado Water Resources Research Institute.

Howitt, Richard E. 1998. Spot Prices, Option Prices, and Water Markets: An Analysis of Emerging Markets in California. In *Markets for Water: Potential and Performance*, edited by K.W. Easter, M.W. Rosegrant, and A. Dinar. New York: Kluwer Academic Publishers, 119–140.

Kerns, David, Michael Marheron, John Palumbo, Charles Sanchez, David W. Still, Barry Tickes, Kai Umeda, and Mark Wilcox. 1999. Guidelines for Head Lettuce Production in Arizona. IPM Series, No. 12, http://ag.arizona.edu/pubs/crops/az1099/ (accessed September 21, 2006).

Michelsen, A.R., and R.A. Young. 1993. Optioning Agricultural Water Rights for Urban Water Supplies during Drought. *American Journal of Agricultural Economics* 75 (November):1010–1020.

MWD (Metropolitan Water District of Southern California). 2003. Metropolitan Calls on Water Transfer Options from Sacramento Valley Rice Farmers. www.mwdh2o.com/mwdh2o/pages/news/press_releases/2003-02/Options_Exercised.htm (accessed February 12, 2005; site discontinued).

Naeser, Robert Benjamin, and Lynne Lewis Bennett. 1998. The Cost of Noncompliance: The Economic Value of Water in the Middle Arkansas River Valley. *Natural Resources Journal* 38:445–463.

Nolte, Kurt. 2005. Yuma County Agricultural Extension. E-mail communication. December 20.

———. 2006. Yuma County Agricultural Extension. Telephone interview. April 17.

Osborn, Dennis. 2005. Fertizona, Casa Grande, Arizona. E-mail communication. September 9.

———. 2006a. Fertizona, Casa Grande, Arizona. E-mail communication. February 3.

———. 2006b. Fertizona, Casa Grande, Arizona. E-mail communication. April 19.

Ottman, M. 2008. Extension Agronomist, University of Arizona. Personal communication. July.

Parker Oil. 2006. La Paz, Arizona. Telephone interview. January.

Reinertson, Jim. 2005. County Executive Director, Arizona Farm Service Agency, Yuma/La Paz County office. Telephone interview. May 1.

Teegerstrom, Trent. 2006. University of Arizona Research Specialist. Personal interview. May 8.

Teegerstrom, Trent, and Timothy Knowles. 1999. Arizona Field Crop Budgets 1999–2000: La Paz County. Cooperative Extension, University of Arizona. http://ag.arizona.edu/arec/ext/budgets/LaPaz-map.html (accessed September 21, 2006).

Teegerstrom, Trent, John Palumbo, and Mohammed Zerkoune. 2001. 2000–2001 Arizona Vegetable Crop Budgets: Western Arizona. Cooperative Extension, University of Arizona. http://ag.arizona.edu/arec/ext/budgets/western.html (accessed September 21, 2006).

Teegerstrom, Trent, and Barry Tickes. 1999. Arizona Field Crop Budgets 1999–2000: Yuma County. Cooperative Extension, University of Arizona. http://ag.arizona.edu/arec/ext/budgets/Yuma-map.html (accessed September 21, 2006).

Tronstad, Russell. 2006a. University of Arizona Extension Economist. Personal interview. May 1.

———. 2006b. University of Arizona Extension Economist. Personal interview. May 11.

Union Oil. 2006. Telephone interview, Yuma, Arizona. January.

USBR (U.S. Bureau of Reclamation). 2007. Lower Colorado River Accounting System Evapotranspiration and Evaporation Calculations. Calendar Year, Yuma Area ET Rate Table for Yuma and Parker/Palo Verde Valleys ET-Rate Table for La Paz County, 2006. Boulder City, NV: U.S. Department of the Interior, Bureau of Reclamation.

USDA ERS (U.S. Department of Agriculture, Economic Research Service). 2009. Farm Risk Management. www.ers.usda.gov/Briefing/RiskManagement/ (accessed October 10, 2010).

USDA FSA (U.S. Department of Agriculture, Farm Service Agency). 2007. Non-recourse Marketing Assistance Loan. www.fsa.usda.gov/FSA/webapp?area=home&subject=prsu&topic=col-nl (accessed March 12, 2008).

USDA MSA (U.S. Department of Agriculture, Marketing Services Agency). 2008. Weekly Truck Report. www.ams.usda.gov/mnreports/fvwtrk.pdf (accessed January 23, 2009).

USDA NASS (U.S. Department of Agriculture, National Agricultural Statistics Service). 2003. Arizona Agricultural Statistics Service 2002 Annual Statistics Bulletin. www.nass.usda.gov/Statistics_by_State/Arizona/Publications/Bulletin/02bul/main.htm (accessed April 2, 2006).

———. 2005. Arizona Field Office 2004 Annual Statistics Bulletin. www.nass.usda.gov/Statistics_by_State/Arizona/Publications/Bulletin/04bul/main.htm (accessed October 2, 2007).

———. 2006. Arizona Field Office 2005 Annual Statistics Bulletin. www.nass.usda.gov/Statistics_by_State/Arizona/Publications/Bulletin/05bul/main.htm (accessed October 2, 2007).

———. 2008. Agricultural Prices. http://usda.mannlib.cornell.edu/MannUsda/viewDocumentInfo.do?documentID=1002 (accessed March 4, 2009).

USDA RMA (U.S. Department of Agriculture, Risk Management Agency). 2005. Crop Policies and Pilots. www.rma.usda.gov/policies/ (accessed October 12, 2006).

———. 2006. Premium Subsidy Schedule. www2.rma.usda.gov/data/premium.html (accessed October 2, 2007).

Young, Robert A. 2005. *Determining the Economic Value of Water.* Washington, DC: Resources for the Future.

PART II

SECTOR IMPACTS OF CLIMATE CHANGE

CHAPTER 5

Water Shortages in the Southern Mountain States: Economic Impacts on Agriculture

George B. Frisvold and Kazim Konyar

Irrigation accounts for more than 80% of water withdrawals in the six southwestern states of Arizona, California, Colorado, Nevada, New Mexico, and Utah (Hutson et al. 2005). This region faces increasing competition for water in the wake of rapid population growth, yet new sources of water supply will be costly and difficult to develop. Historically, the region has been susceptible to severe, prolonged droughts. If such periods of severe drought recur, water available for agriculture could decline significantly.

This chapter examines the impacts of reduced water availability on agriculture in the southwestern United States. It uses an updated version of the U.S. Agricultural Resources Model (USARM) to simulate impacts of large reductions in water availability in the Colorado and Upper Rio Grande River Basins. USARM is a 32-commodity, 12-region U.S. agricultural sector model. Earlier versions of USARM have been successfully applied to examine impacts of U.S. commodity, conservation, and energy policies, as well as impacts of agricultural biotechnology adoption (Howitt 1991; Ribaudo et al. 1994; Konyar and Howitt 2000; Konyar 2001; Kim et al. 2002).

What do we mean by a "large reduction" in water availability? In the simulation, water available to agriculture in the Southern Mountain (SM) region (Arizona, New Mexico, Colorado, Utah, and Nevada) is reduced by 25%, while water available to California falls 5%, reflecting reductions in Colorado River water allocations to Southern California. The water supply shock is generic and not necessarily representative of a particular drought, water transfer, or climate change scenario. Our purpose is to examine how and how well southwestern agriculture might adapt to water shortages and identify which producer groups might be vulnerable to those shortages.

Agricultural producers can adapt in different ways and varying degrees to water shortages. First, in some areas, producers can shift from irrigated to dryland crop production. This shift is possible in Colorado and some parts of Utah or northern New Mexico, but it is not a viable option for Arizona producers. Second, producers can alter their crop mixes, shifting cultivation from more to less water-intensive crops. For example, in Arizona, producers apply, on average, 5.8 acre-feet of water per acre to alfalfa but 3.5 acre-feet per acre to wheat, nearly 40% less (USDA 2004). In principle, then, growers could significantly reduce water use simply by switching between crops. Third, producers can practice deficit irrigation, reducing the amount of water applied per acre to crops. Fourth, they can substitute other inputs for water. Fifth, within a region, water may be reallocated to areas and crops that command higher returns in the new water-scarce environment. This last adjustment assumes that growers' production is sufficiently diversified or there is a functioning water market allowing transfers from low- to high-value uses, or both.

We also conduct a simulation that increases the cost of irrigation water in the SM region and California by 25%. Price rationing is an alternative to quantity rationing in response to water shortages. U.S. Bureau of Reclamation (USBR) projects provide nearly a third of surface water deliveries for irrigation in the 17 western United States (Gollehon et al. 1997). USBR pricing of irrigation water has been controversial, with both conservative (Edwards 2007) and environmental groups (EWG 2005) calling for increases in the price irrigators pay for water. Legislation has been regularly introduced in Congress to raise the rates USBR charges irrigators (EWG 2005). Our simulation considers the impact of a 25% water price increase on agricultural water use and returns. The price increase scenario could also reflect a case where growers must resort to more expensive groundwater in response to surface water price increases or supply disruptions.

Our analysis begins with some background on the balance of water demand and supply in the Southwest, describing how changes in climate or recurrence of prolonged drought may upset that balance. Next, we provide an overview of the USARM model and discuss how the water supply shock is introduced into the model. Then we introduce a rationing model approach to estimate an upper bound of the costs of reduced supplies of water to agriculture. The rationing model assumes no scope for input substitution or altering water use intensity, relying on land fallowing as the only means of adapting to water scarcity. Under the rationing model, the water supply shock reduced SM irrigators' incomes by $65 million. One may think of this cost estimate as an upper bound, because the rationing model assumptions preclude grower responses to meet tighter water constraints at least cost. The results section reports on two simulations: a water supply shock and a water price increase. Different producer adjustments are examined in turn. These include changes in irrigated acreage, cropping patterns, water use intensity, and input substitution. These adjustments translate into changes in crop yields, production, prices, and net returns, as well as changes in consumer surplus and commodity program outlays.

The results suggest that through these adjustments, the costs of the water supply shock are substantially lower than the rationing model cost estimate. Net income to SM irrigators falls by $15.2 million. USARM assumes downward-sloping demand curves, so some of the increased production costs are passed on to consumers. If irrigators cannot pass on cost increases, regional net farm income falls by nearly $22 million.

The model does suggest sizable localized costs, however. In particular, cotton production and net returns decline by more than 40% in Arizona and New Mexico. Regional farm employment falls by 3%, with impacts felt more in Arizona than the rest of the SM region. The results also illustrate that farm price support programs do a poor job of supporting cotton income in response to water supply shocks.

The regional cost to irrigators of the water price increase is greater than the cost of the water supply shock. Costs range from $44 million to $48 million, depending on whether costs can be passed on to consumers. The price increase is more costly because it applies to all irrigators. Under the supply shock, producers can divert water from their least profitable activities. This leaves higher-valued specialty crop production largely unaffected. Under the price increase, however, costs to all irrigators increase across the board. The 25% increase in the cost of water reduces SM irrigation water use by 10.5%.

WESTERN WATER: GROWING DEMANDS, UNCERTAIN SUPPLIES

In the SM states (Arizona, Colorado, Nevada, New Mexico, and Utah) and California, agricultural irrigation accounts for 82% of freshwater withdrawals (see Table 5.1). Although it is the dominant water user, agriculture faces growing competition from other sectors. Rapid population growth in the West will increase urban water demand (domestic/public supply withdrawals).

Table 5.1 *Share of Freshwater Withdrawals by Sector, 2000*

State	*Domestic/public supply*[a]	*Irrigation*	*Other*[b]
Arizona	17%	80%	3%
California	17%	79%	4%
Colorado	8%	90%	2%
Nevada	23%	75%	2%
New Mexico	10%	88%	2%
Utah	14%	81%	5%
Regional average	15%	82%	3%

[a]Domestic/public supply includes self-supplied water for domestic use and water delivered by municipal and other public water providers to commercial and residential users

[b]Other includes direct withdrawals for industry, mining, thermoelectric power generation, livestock, and aquaculture

Source: Hutson et al. (2005)

COMPETING DEMANDS FOR WESTERN WATER

Table 5.2 shows the projected population growth from 2000 to 2030 for six western states, along with the *hypothetical* share of domestic/public supply given this growth, assuming no changes in per capita water withdrawals; share of water withdrawn for irrigation in 2030, assuming that growth in urban demand is met by reallocating water from agriculture; and implied percent reduction in agricultural water use from 2000 to 2030.

One should treat the numbers in Table 5.2 as a hypothetical scenario, not a projection. Yet they suggest two things. First, even with population growth and no improvement in urban water use efficiency, agriculture will remain the dominant water user in the West. And second, agriculture in Arizona and Nevada will face greater urban competition for water compared with Colorado and New Mexico.

While water demand grows, the scope for increasing water supplies is limited. Few potential sites for large dam projects remain, and projects face greater scrutiny over their environmental impacts. With limits on supply, growing demand for water will be met primarily by shifting water between uses (Moore 1991; Schaible et al. 1991; Vaux 1986; Howe 1985).

There are additional demands for water to maintain or restore riparian habitats and protect endangered species. The competition for water between agriculture and threatened or endangered species can grow keen during droughts. In California, the USBR is reducing water deliveries to Central Valley agriculture in attempts to protect Chinook salmon and Delta smelt (Howitt et al. 2009). Booker et al. (2005) considered the impacts of restrictions on irrigation withdrawals in the Upper Rio Grande Basin under conditions where drought combines with requirements to protect the endangered Rio Grande silvery minnow. In Arizona, some have proposed reallocating current water uses to protect riparian species in the Lower Colorado (Pitt 2001; Glennon and Culp 2002; Parrish 2003). If irrigation water supplies are restricted via regulation, this imposes direct costs on agricultural producers. If water transfers from agriculture are voluntary, via water markets, producers must still consider the opportunity costs of forgoing water.

Table 5.2 *Shares of Freshwater Withdrawals by Sector, 2030 Scenario*

		Sector's share of water withdrawals[b]			*Percent reduction in agricultural water withdrawals,*
	Population increase 2000 to 2030[a]	*Domestic/public supply*	*Irrigation*	*Other*	*2000–2030*[b]
Arizona	108.8%	34%	62%	3%	−22%
California	37.1%	23%	73%	4%	−8%
Colorado	34.7%	10%	88%	2%	−3%
Nevada	114.3%	50%	49%	2%	−35%
New Mexico	15.4%	12%	86%	2%	−2%
Utah	56.1%	21%	73%	5%	−10%

[a]Data from USDA NASS (1998)
[b]Authors' hypothetical estimates

Uncertainty of Future Water Supplies

While demands for water in the Southwest continue to grow, there are growing concerns about the reliability of water supplies. First, current surface water use is close to long-run sustainable supplies. The Colorado River Compact of 1922 apportioned 7.5 million acre-feet (ac-ft) of water to Upper Basin states (Colorado, New Mexico, Wyoming, and Utah) and 7.5 million ac-ft to Lower Basin states (Arizona, California, and Nevada). However, this calculation was made from observations during an unusually wet period in the river's history. At the time of the compact, delegates based allocations on an estimated mean flow of more than 16 million ac-ft (Gelt 1997). However, tree-ring reconstruction analyses suggest long-run annual average flows that are significantly lower. Earlier studies estimated long-run annual average flows of 13.2 million ac-ft (Hidalgo et al. 2000) and 13.5 million ac-ft (Stockton and Jacoby 1976). More recently, an analysis by Woodhouse et al. (2006) suggested higher annual flows, ranging from 14.3 million to 14.7 million ac-ft. This is still significantly below current apportioned levels. The 1944 Water Treaty between the United States and Mexico apportioned 1.5 million ac-ft more of Colorado River water per year to Mexico, for a total apportionment of 16.5 million ac-ft (Christensen et al. 2004). In 2008, the Lower Basin states and Mexico used 9.2 million ac-ft of water (USBR 2008a). Upper Basin states do not yet use their full allocation, using about 4.1 million ac-ft per year (USBR 2008b). Evaporation losses from Lakes Mead and Powell (net of miscellaneous inflows) account for an additional 1 million ac-ft in water losses to the basin (Barsugli et al. 2009). This places current consumptive use plus net losses at 14.3 million ac-ft per year, right at the lower range of Woodhouse et al.'s long-run annual average streamflow estimates of 14.3 million to 14.7 million ac-ft (2006). Reservoir storage in the Colorado River Basin is about four times average annual flows, with Lakes Powell and Mead accounting for 85% of total basin storage capacity (Christensen et al. 2004). The reservoir system buffers the basin states from large, short-run supply fluctuations. From 1906 to 2000, annual flows have ranged from 5.3 million to 24 million ac-ft. Yet average net inflows just barely exceed current demands. The present balance of regional water supply and demand is precarious.

A second concern is that the Southwest's climate is becoming warmer, with reduced surface water availability. Hoerling and Eischeid (2007) examined 42 climate change simulations conducted as part of the Fourth Assessment Report of the International Panel on Climate Change (IPCC). The models were able to replicate a realistic range of drought events when simulated over the 20th century. Further, the models predict that over 2006–2030, Colorado River streamflow will decline 25% from the 13-million-ac-ft average of 1990–2005. Although some simulations predict more rainfall, even some of these predict greater drought as higher temperatures reduce soil moisture. Hoerling and Eischeid concluded, "Indications from the simulations are that a near perpetual state of drought will materialize in the coming decades as a consequence of increasing temperature" (2007, 19).

Christensen et al. (2004) also conducted simulations of climate change on the Colorado River Basin. Their simulations predicted a 14% reduction in runoff during 2010–2039. This study focused on the effects of climate change on management of the Colorado River Basin reservoir system. It also considered the likelihood that Lake Mead elevation falls below 1,075 feet, triggering a level-one shortage declaration, and below 1,050 feet, triggering a level-two shortage. Under a level-one shortage, the Law of the River calls for reductions in deliveries of 320,000 ac-ft to the Central Arizona Project (CAP) and of 13,000 ac-ft to the Southern Nevada Water Authority (SNWA). Under a level-two shortage, supplies to CAP and SNWA are reduced by 400,000 and 17,000 ac-ft, respectively. CAP delivers about 1.5 million ac-ft to Central Arizona per year. In simulations using 1950–1999 streamflows combined with 2000 demands, a level-one shortage occurred in 60% of years. In simulations over the period 2006–2039, a level-one shortage occurred in 92% of years, and level-two shortages were declared in 77% of years. By mid-June, Lake Mead's elevation was below 1,096 feet, down from 1,214 feet at the end of 1998.

Seager et al. (2007) examined simulation results from 19 climate models used for the IPCC Fourth Assessment. Their analysis focused on the effects of increased summer evaporation combined with decreased winter and summer monsoon precipitation. Their results suggested that the combination of these effects will be significant loss in average moisture. Seager et al. concluded, "If these models are correct, the levels of aridity of the recent multiyear drought or the Dust Bowl and the 1950s droughts will become the new climatology of the American Southwest within a time frame of years to decades" (2007, 1181).

One does not have to appeal to future projections from global circulation models of climate change to consider the possibility of sharp reductions in water availability. One can look to history (Meko et al. 1995; Woodhouse et al. 2006; Hoerling and Eischeid 2007). The southwestern United States experienced multiyear droughts from 1953 to 1956 and from 2000 to 2003. In the late 1500s, the Colorado Basin was hit by a severe multidecade drought. In 1995, the Powell Consortium considered the question of how the region's hydrology and economy would respond to a drought of that magnitude, should it recur (Young 1995). Model simulations suggested that more than half of Upper Basin consumptive use demand would not be met in the worst year, and that Lake Powell's elevation would fall to the dead storage levels (Hardy 1995). Under the worst years of this drought scenario, deliveries of CAP water to Arizona decline by 60%, with reductions in CAP agricultural use falling by 75%. Agricultural water use in the Upper Basin states (Colorado, Utah, Nevada, and Wyoming) falls by more than 50% (Booker 1995). More recently, Woodhouse et al. (2006) estimated that during the late 1500s drought, the 20-year mean flows were 12.6 million ac-ft, considerably less than current uses and losses.

To recap, Arizona and the Southwest face growing demands for water for urban uses and environmental protection of aquatic species. At the same time, the delicate balance of regional supply and demand is at a tipping point, where current Colorado River inflows barely exceed current consumption. Recurrences of historic droughts combined with population growth and environmental protection mandates could create a situation where significantly less water is available for agricultural production. The remainder of this chapter develops and applies a modeling framework to examine how and how well Arizona and southwestern agriculture might adjust to severe water scarcity.

THE U.S. AGRICULTURAL RESOURCES MODEL (USARM)

The U.S. Agricultural Resources Model (USARM) is a nonlinear mathematical programming model designed to simulate farmer behavior under external market and policy shocks. A full description of the model's structure is presented in Appendix 5.1 at the end of this chapter. The model assumes that farmers try to maximize profits when they select the crop mix, acreage, and input levels. USARM includes several features that are seldom, if ever, included together in a single model. First, unlike other multiregion programming models of U.S. agriculture, it includes fruit and vegetable crops. Second, USARM explicitly includes water as an input. Third, whereas many models assume fixed-proportion, Leontief production with fixed yields, USARM allows substitution between inputs, while yields depend on input intensity. By way of comparison, the U.S. Department of Agriculture's Regional Environment and Agriculture Programming (REAP) model includes water but not (at the time of this writing) fruits and vegetables. The AGSIM model (Taylor et al. 1991) excludes water as an input, and fruits and vegetables as outputs. Programming models developed by Deepak et al. (1996) and Carpenter et al. (2000) include different fruit and vegetable crops but exclude field crops and water and assume Leontief technology.

Fourth, output prices are endogenous, with producers facing downward-sloping demand curves. Models examining impacts of changing water availability often assume that producers face fixed output prices (Sunding et al. 2002; Howitt et al. 2009). This assumption may be appropriate when analysis is confined to smaller areas where production is a small share of total output. It may be less appropriate when considering impacts on specialty crops, however. Studies often find inelastic demand for agricultural commodities (Henneberry et al. 1999; You et al. 1996; Onunkwo and Epperson 2000; Nuckton 1978). Demand elasticities will be larger for individual regions than for the national aggregate. However, for many specialty crops, just a few regions account for the bulk of production, so even regional demands may be inelastic. For example, a recent study by Russo et al. (2008) estimated that the demand for California walnuts, alfalfa, cotton, rice, and tomatoes were all inelastic. Alston et al. (1995) have demonstrated that demand elasticities have little effect on the total welfare impact of a supply shock, but they have an important effect on the distribution of impacts between producers and consumers.

Finally, USARM has been updated to account for federal commodity programs, with marketing loan and countercyclical payments explicitly modeled. Loan deficiency payments (LDPs) under marketing loan programs, when in effect, are coupled payments. That is, they encourage greater production at the margin. Countercyclical payments (CCPs) are based on the difference between the market price and a target price, but payments are based on historic, rather than current, production. CCPs do not affect marginal production decisions, but aggregate production alters the level of CCPs via its impact on market prices.

Commodity and Regional Coverage

The model accounts for 10 major field crops and 28 fruit and vegetable crops in 12 regions covering the 48 contiguous states (Table 5.3). For the field crops, the base year is 2002 and for the specialty crops, it is 2000. Additional changes (discussed below) were made to the baseline model to account for recent changes in relative prices. The model accounts for 96% of all harvested field crop acreage in the United States. Some of the vegetable crops are aggregated, so there are 22 fruit and vegetable commodity groups in the model (Table 5.3).

USARM divides the United States into 12 production regions: Appalachian (Kentucky, North Carolina, Tennessee, Virginia, West Virginia); Corn Belt (Illinois, Indiana, Iowa, Missouri, Ohio); Delta States (Arkansas, Louisiana, Mississippi); Lake States (Michigan, Minnesota, Wisconsin); Northeast (Connecticut, Delaware, Maine, Maryland, New Hampshire, New Jersey, New York, Pennsylvania, Rhode Island, Vermont); Northern Plains (Kansas, Nebraska, North Dakota, South Dakota); Southern Plains (Oklahoma, Texas); Southeast (Alabama, Florida, Georgia, South Carolina); Northern Mountain (Idaho, Montana, Wyoming), Southern Mountain (Arizona, Colorado, Nevada, New Mexico, Utah); Upper Pacific (Oregon, Washington); and Lower Pacific (California).

Despite this regional aggregation, it is possible to make inferences about economic behavior and impacts in given states within the SM region. For example, dryland wheat and barley production in the SM region is carried almost

Table 5.3 *Commodity Groupings in the USARM Model*

Field crops	
Alfalfa hay	Sorghum
Barley	Sugar beets
Corn	Sugar cane
Cotton	Soybeans
Rice	Wheat
Fruit, vegetable, and nut crops	
Almonds	Lettuce (head, leaf, romaine)
Apples	Melons (cantaloupe, honeydew, watermelon)
Asparagus	Onions, dry
Broccoli	Peaches
Cauliflower	Pears
Citrus (grapefruit, lemons, limes, oranges)	Peppers
Cucumbers (fresh, processing)	Potatoes
Grapes	Strawberries
Grapes, raisin	Tomatoes (fresh, processing)
Green beans (fresh, processing)	Walnuts
Green peas	

exclusively in Colorado and parts of Utah and New Mexico, with no such production in Arizona. Conversely, lettuce, melon, and broccoli production in the SM region is carried out almost exclusively in Arizona, and cotton production occurs only in New Mexico and Arizona.

Input Use and Substitution

The model includes seven inputs: land, labor, capital, fertilizers, agricultural chemicals, water, and energy/other inputs. USARM uses nested constant elasticity of substitution (CES) production functions that allow different degrees of substitution across inputs. There are separate production functions for irrigated and dryland crops (which exclude water as an input). In some regions, because of arid environments and high water requirements, there may be no dryland production (e.g., lettuce production in SM states). The total amounts of land and water used in each region are constrained to be no greater than region-specific, fixed amounts. Although water can be reallocated across crops within a region, water cannot be reallocated across regions.

Water Supply Shock

In the model, the total amount of land and the total amount of water that can be productively put into agricultural production are constrained at the regional level. To examine how agriculture responds to a reduction in water availability, the total amount of water available for irrigation in the SM region is reduced by 25%. The model is further shocked so that water supplies in California are reduced by 5%. This is meant to capture reductions in Colorado River water available to Southern California. The model is not disaggregated by watersheds within a region, so the shock may be thought of as an across-the-board reduction in water available in the Upper and Lower Colorado Basins, the Upper Rio Grande/Pecos Basins (above Texas), the drainage of the Great Basin that discharges into Utah and Nevada (including the Truckee-Carson watershed), and Colorado watersheds east of the Rocky Mountains.

A 25% reduction in the SM region is large, but several studies have examined similar irrigation water reductions. Booker (1995) considered reductions of 50% or more in the Upper Colorado in combination with a 75% reduction in agricultural CAP supplies. Booker et al. (2005) considered reductions of 25–50% in the Upper Rio Grande Basin. In Nevada, Elbakidze (2006) examined effects of a 27% reduction in the Truckee-Carson Irrigation District, and Seung et al. (2000) considered a 31% reduction in Churchill County.

One might question the likelihood of reductions in water availability affecting so many river basins at once. However, all states in the region face growing urban demand for water and greater regulatory constraints to increase or maintain instream flows to protect threatened or endangered fish. Moreover, climate change forecasts suggest that reductions in available runoff will span the entire region (Hoerling and Eischeid 2007; Seager et al. 2007). Historically, such widespread drought is not unknown (Hardy 1995; Meko et al. 1995). In addition, historical and paleoclimatic data suggest that streamflows for the Rio Grande and Colorado Rivers are highly correlated. Estimates from Woodhouse (2008) of tree-ring reconstruction of stream gauge data and long-term tree-ring reconstructions suggest that sharp reductions in streamflow often occur in the same years in the Colorado and Rio Grande Basins.

Water supplies to agriculture may also be reduced to maintain reservoir levels for hydropower generation. Booker (1995) identified significant trade-offs between maintaining water for hydropower and supplying water for current consumptive uses. Although the Law of the River gives precedence to (most) agricultural uses of water over hydropower generation, water management has attempted to protect hydropower production, and further, electricity sales revenue from hydropower subsidizes operation of large irrigation projects (NRLC 2005). The importance of hydropower generation is likely to grow with rising costs of operating coal-fired generation plants in the region. The Navajo Power Generation Plant near Page, Arizona, is currently facing rising costs to install equipment to comply with the Clean Air Act requirements to reduce NO_x emissions. Any future climate change mitigation legislation may further increase costs of coal-based energy generation. Thus there may well be increased future pressure to maintain reservoir levels for electricity generation.

IMPACTS OF WATER SCARCITY: A RATIONING MODEL APPROACH

This section introduces a rationing model approach to estimate costs of a regional supply shock. This approach is similar to that employed by Sunding et al. (2002) and Dale and Dixon (1998) to examine water reallocations from agriculture in California. We then discuss why this cost is likely to be an upper-bound cost estimate.

Consider a case of an irrigated crop produced under a fixed-proportion, Leontief technology with constant average and marginal costs. There is no ability to substitute between water and other inputs, the maximum amounts of water and land allocated to this crop are fixed, and these constraints are binding.[1] Now assume that all crops in the region are produced with this same Leontief technology, binding land and water constraints are imposed on each individual crop, and that output prices for all crops are fixed.

Given this highly inflexible production system, how would the region as a whole respond to a water supply shock, wherein there is a reduction in supplies? This is basically a linear programming problem. Growers adjust to the regional water supply reduction by fallowing the lowest-value crops first. Crops are ranked by net revenue per acre-foot of water applied. The rationing model assumes that growers fallow the lowest-value crops across the entire region first to meet the new, lower water supply constraint. This rationing model approach is similar to the residual method applied in Chapter 4. The main difference is that the structure is applied to production in the region as whole and not just representative farms.

In the rationing model, costs of reduced water supplies can be estimated using the net-returns-per-acre data used to calibrate the USARM model. Thus baseline returns are exactly the same as in USARM. The cost of a 25% reduction in water supplies to SM irrigators is $65 million. Although the production system as specified is highly inflexible, this result assumes that water can be reallocated freely between crops. Thus it implicitly assumes that water markets function well enough within the region so that water can be transferred between crops and producers. As Dale and Dixon explained, "For example, if Farmer A grows only fruits and nuts and Farmer B grows only low-value field crops but they both face the same proportionate cutbacks in water supply, then Farmer B will sell water to Farmer A. Farmer B will fallow his or her entire acreage before Farmer A fallows anything" (1998, 54).

Continuing with this example, if Farmer A were diversified, growing a mix of high- and low-value crops, then the supply shock might be met internally by fallowing the low-value crops while producing the high-value crops as before. Costs of the water supply shock would be higher if water could not be transferred from lower- to higher-valued uses.

One drawback of the water rationing model approach, common to linear programming, is that one gets extreme solutions. For example, production of some lower-return crops ceases completely, while production of higher-value specialty crops are completely unaffected. For example, cotton, barley, and apple production would cease in the region and alfalfa production would be reduced, but other crops would be unaffected. The positive mathematical programming approach used in USARM yields results that are less extreme and more credible.

For several reasons, the rationing model cost estimate will be an upper-bound estimate of the costs to agriculture of the water supply shock. First, the rationing model assumes that there are no substitution possibilities between water and other inputs. Growers do not have the flexibility to adjust water use intensity (applications per acre). Second, in areas where dryland production is viable, producers can shift from irrigated to dryland production instead of simply fallowing. Third, the region can respond to the water supply shock at least cost by equalizing the marginal returns per acre-foot across crops. If average and marginal production costs are not constant, but increase with the level of output (when, for example, less-productive lands are brought into production), the rationing model will not make such equi-marginal adjustments. Water will be allocated based on average rather than marginal returns, which will not be optimal. Finally, agricultural producers will be able to pass on a portion of the costs of the supply shock to consumers. Even when demand curves are elastic, producers may be able to pass on a considerable share of the costs (Taylor and Howitt 1993).

SIMULATION RESULTS

This section summarizes the results of two simulations. The first is a 25% reduction in water supply to agriculture in the SM region and a 5% reduction in California. The second is a 25% price increase for irrigation water in both the SM region and California.

Water Supply Shock Simulation

To the SM region alone, this is a reduction of 2.4 million ac-ft of water. The water shortage is modeled purely as a supply shock. Although this is a common practice, it does not capture the full impacts of drought or climate change. To fully model drought, for example, one might model changes in production functions such that water requirements for irrigated crops increase and yields of dryland crops decrease. Pure supply shocks are nevertheless important and could also represent impacts of reduced water supplies (to maintain instream flow for endangered fish) or capture the costs to agriculture of water transfers to municipalities.

Acreage Response. In response to the decline in water availability (the 25% water reduction), irrigated acreage declines by 54,700 acres, and dryland acreage increases by the same amount (Table 5.4). The biggest changes in absolute acreage are an expansion of wheat acres and a decline in cotton and irrigated alfalfa acres. Dryland wheat expands by 45,800 acres, while irrigated wheat expands by 25,100 acres. Cotton acres fall by more than 50,000 acres and irrigated alfalfa acres fall by nearly 24,000 acres. There is a 9% increase in dryland barley and more modest increases in dryland alfalfa, corn, and sorghum.

These results suggest an expansion of total acreage in Utah and Colorado and a decline in total crop acreage in Arizona and New Mexico. Cotton, which accounts for the largest acreage decline, is grown only in Arizona and New Mexico. Arizona has virtually no dryland crops, while Colorado accounts for more than 90% of dryland wheat production. So expansion of dryland wheat, barley, corn, and sorghum would be primarily in Colorado. Although dryland wheat accounts for the largest absolute increase in acreage of any crop, the change

Table 5.4 *Change in Acreage in Southern Mountain (SM) Region in Response to Water Supply Shock*

Nonirrigated crops	*Change in acreage (thousands)*	*Acres Percent*	*Irrigated crops*	*Change in acreage (thousands)*	*Acres Percent*
Alfalfa	2.0	1.1%	Alfalfa	−23.7	−1.3%
Barley	0.8	9.0%	Apples	−0.9	−9.0%
Corn	1.9	2.5%	Barley	−6.4	−4.4%
Sorghum	4.3	2.7%	Broccoli	0.0	−0.1%
Wheat	45.8	2.7%	Cauliflower	0.0	0.0%
Total	54.7		Citrus	−0.3	−1.2%
			Corn	2.8	0.3%
			Cotton	−50.1	−27.2%
			Grapes	0.0	−0.3%
			Lettuce	−0.4	−0.5%
			Melons	−0.2	−0.7%
			Onions	0.0	0.0%
			Potatoes	−0.2	−0.2%
			Sugar beets	−0.3	−0.8%
			Wheat	25.1	7.3%
			Total	−54.7	

represents only a 2.7% increase in dryland wheat acreage. Reduction in cotton acreage would occur in Arizona and New Mexico. This reduction is not made up by expansion of other irrigated crops in these states. Although some expansion of irrigated wheat acreage could occur in Arizona, this would be smaller than declines in acreage of other crops, implying a net loss in Arizona crop acreage.

Acreage of specialty crops changes very little in percentage terms. Lettuce, broccoli, melons, and cauliflower acreage decline by less than 1%. Citrus acreage declines by 1.2%. SM region acreage of these crops is concentrated almost exclusively in Arizona. Yet this represents a reduction of fewer than 1,000 acres.

It is interesting to compare how USARM predicts acreage will change in response to water shortages with actual acreage changes from 2002 to 2007. In the simulation, acres planted to apples fall by 9%. Acreage in SM apple orchards declined by 14% between agricultural census years 2002 and 2007, and by 52% between 1997 and 2007. Acreage planted to cotton decreases by 27.2% in the model, and it actually fell by 21% from 2002 and 2007. USARM predicted irrigated wheat acreage would expand by 7.3%, whereas actual irrigated wheat acres increased by 4.6%. The results suggest that a response to greater water scarcity will be a continuation of recent acreage trends already under way in the SM region.

Water Use. In the simulation, water supplies are reduced by some 2.4 million ac-ft. Most of this is achieved by a reduction of 1.36 million ac-ft in water going to alfalfa. This result is similar to that obtained by Moore and Negri (1992), who conducted a simulation experiment based on econometric modeling of land and water use, reducing USBR water supplies by 10% in 17 western states. As in our

Table 5.5 *Reduction in Irrigation Water Use in SM Region in Response to Water Supply Shock*

	Reduction in acre-feet (thousands)	*Crop's share of total regional irrigation reduction*	*Cumulative share of total regional irrigation reduction*
Alfalfa	−1,361	57%	57%
Cotton	−366	15%	72%
Wheat	−231	10%	81%
Corn	−212	9%	90%
Barley	−76	3%	93%
Lettuce	−70	3%	96%
Citrus	−22	1%	97%
Potatoes	−21	1%	98%
Sugar beets	−15	1%	98.6%
Melon	−15	1%	99.2%
Onions	−7	0%	99.4%
Apples	−7	0%	99.7%
Broccoli	−4	0%	99.9%
Cauliflower	−3	0%	100.0%
Grapes	0	0%	100.0%
All crops	−2,408		

Table 5.6 *Change in SM Region Water Use per Acre and Water Use by Crop in Response to Water Supply Reductions*

Crop	*Percent change in water use per acre*	*Percent change in total water use*
Alfalfa	−28%	−28%
Apples	−8%	−16%
Barley	−23%	−26%
Broccoli	−12%	−12%
Cauliflower	−22%	−22%
Citrus	−16%	−17%
Corn	−11%	−11%
Cotton	−35%	−53%
Grapes	−4%	−4%
Lettuce	−26%	−27%
Melons	−13%	−13%
Onions	−12%	−12%
Potatoes	−10%	−10%
Sugar beets	−28%	−28%
Wheat	−27%	−22%

study, they found that reductions in water applied to alfalfa accounted for the single largest source of water reduction. About 93% of the overall reduction is achieved by reducing water applied to alfalfa, cotton, wheat, corn, and barley (Table 5.5).

Water use per acre on cotton falls by 35%, while total water used on cotton falls by 53% (Table 5.6). To give some perspective to the size of these numbers, water applications per acre on Arizona and New Mexico cotton fell about 12% from 1998 and 2003, the years of the two most recent Farm and Ranch Irrigation Surveys. Although regional water supplies fall by 25%, crops do not experience an equi-proportional reduction in water use. Water use per acre for cotton, alfalfa, wheat, and sugar beets declines by more than 25%, while water use for many other crops declines by less than 25%. For many specialty crops, this reduction is considerably less than 25%. Water used per acre for broccoli, citrus, and melons declines by 16% or less. Irrigated wheat's water use per acre falls by 27%, more than its 22% reduction in absolute water use. This occurs because of expansion in total irrigated wheat acreage.

Fertilizer and Agricultural Chemical Use. For most crops, chemical and fertilizer input use per acre changes little (Table 5.7). Except for some dryland crops, input intensity falls. However, reductions in other inputs are much smaller than the reduction in water use per acre. Cotton is an important exception, where fertilizer and chemical use per acre fall by 24%. Input use per acre for irrigated barley falls by 9%, and for irrigated wheat, it falls by 7%. Total fertilizer and chemical use declines by 3% overall in the SM region. Total reductions are especially large for cotton (−44%) and apples and irrigated barley (−13%).

Table 5.7 *Change in SM Region Fertilizer and Chemical Use per Acre and Water Use by Crop in Response to Water Supply Reductions*

Crop	*Percent change in fertilizer and chemical use per acre*	*Percent change in total fertilizer and chemical use*
Nonirrigated production		
Alfalfa	0%	1%
Barley	1%	11%
Corn	0%	3%
Sorghum	−3%	0%
Wheat	−3%	0%
Irrigated production		
Alfalfa	−4%	−5%
Apples	−4%	−13%
Barley	−9%	−13%
Broccoli	0%	0%
Cauliflower	0%	0%
Citrus	−1%	−2%
Corn	0%	0%
Cotton	−24%	−44%
Grapes	0%	−1%
Lettuce	−1%	−1%
Melons	−1%	−1%
Onions	0%	0%
Potatoes	0%	−1%
Sugar beets	−2%	−3%
Wheat	−7%	0%
All crops		−3%

The relative change in input use means that fertilizer and chemical use *per acre-foot of water* actually increases by nearly 30% (Table 5.8). The increase is modest for apples and grapes (4%). For the remaining crops, however, fertilizer and chemical applications increase by 11% to 35% per acre-foot of water applied. Notably, input use per acre-foot increases by 34% for lettuce.

Table 5.8 *Change in SM Region Fertilizer and Chemical Use per Acre-Foot of Water Applied in Response to Water Supply Reductions*

Crop	*Percent change in fertilizer and chemical use per ac-ft of water*	*Crop*	*Percent change in fertilizer and chemical use per ac-ft of water*
Alfalfa	33%	Grapes	4%
Apples	4%	Lettuce	34%
Barley	18%	Melons	14%
Broccoli	13%	Onions	13%
Cauliflower	28%	Potatoes	11%
Citrus	18%	Sugar beets	35%
Corn	13%	Wheat	28%
Cotton	19%	All crops	30%

Knapp and Schwabe (2008), applying a dynamic, spatial production model of fertilizer water interactions, found that water conservation may significantly reduce nitrate emissions from fields. Their results suggest that reduced water applications will reduce deep percolation and hence nitrate emissions. In their quite detailed model, greater water applications leach nitrogen from the soil. Thus with reduced agricultural water supplies, there could be less runoff, drainage, and groundwater percolation from farm fields.

Yields and Production. Table 5.9 shows changes in yields, acreage, and production in the SM region, where $\%\Delta$ *output* $= \%\Delta$ *yield* $+\%\Delta$ *acreage* $+ (\%\Delta$ *yield* $\times$ $\%\Delta$ *acreage*). Cotton yields fall by nearly 24%, followed by yield losses of 9% for irrigated barley and wheat, and 4% for alfalfa. Overall, cotton production declines by 44%. This is significantly more than the 27% acreage reduction. It reflects the reduction in other inputs devoted to cotton production and subsequent reduction in cotton yields. Although irrigated wheat acreage expands, irrigated wheat production declines by 2.5%. Again, this decline results from a decline in other inputs devoted to wheat production and reduced yields. Percent reductions in output of specialty crops are modest, with declines for lettuce (1.5%), broccoli (0.5%), melons (1.4%), cauliflower (0.3%), and citrus (2.3%). Again, apples are an exception, with a production decline of 13%.

Table 5.9 *Change in Output, Yields, and Acreage in Response to Irrigation Water Supply Reduction*

	Percent change in		
	Output	*Yield*	*Acreage*
Nonirrigated crops			
Alfalfa	1.1%	0.0%	1.1%
Barley	10.1%	1.1%	9.0%
Corn	2.5%	0.0%	2.5%
Sorghum	0.7%	−2.0%	2.7%
Wheat	0.9%	−1.9%	2.7%
Irrigated crops			
Alfalfa	−5.4%	−4.1%	−1.3%
Apples	−13.0%	−4.4%	−9.0%
Barley	−13.2%	−9.2%	−4.4%
Broccoli	−0.5%	−0.5%	−0.1%
Cauliflower	−0.3%	−0.3%	0.0%
Citrus	−2.3%	−1.0%	−1.2%
Corn	−1.2%	−1.5%	0.3%
Cotton	−44.3%	−23.6%	−27.2%
Grapes	−0.8%	−0.5%	−0.3%
Lettuce	−1.5%	−1.0%	−0.5%
Melons	−1.4%	−0.7%	−0.7%
Onions	−0.3%	−0.2%	0.0%
Potatoes	−0.7%	−0.5%	−0.2%
Sugar beets	−3.2%	−2.4%	−0.8%
Wheat	−2.5%	−9.2%	7.3%

Comparing Tables 5.6 and 5.9, one can see that yields fall much less than water use per acre. This raises two questions: why is this happening, and is this response realistic? The reason yield elasticities with respect to water are low has to do with the form of our production functions. With a CES function (as with Cobb-Douglas functions), the marginal product of water (or any other input) will not fall below zero, but rather approach it asymptotically. This means that the yield curve can have a long "flat" portion where more water can have small, but still positive, effects. This also means that the marginal value product curve for water is also very flat where large amounts of water are used. If the price of water is low, then water use will be at the flat portion of the yield and marginal value product curves. Thus at low water prices and high levels of water use per acre, there is a range where large percentage reductions in water use per acre can be made without sacrificing yield or revenues.

So low water prices combined with the CES production specification account for the low yield elasticities with respect to water. How realistic is this specification? Dinar and Letey (1996) have criticized the use of irrigation production functions with this asymptotic feature for *farm-level* specifications because it does not allow for negative marginal productivity of water applications and may not accurately capture microlevel water-yield relationships. However, several empirical studies have found quite low output elasticities for water. Moore et al. estimated Cobb-Douglas and quadratic production functions with farm-level data for 13 irrigated crops in the 17 western states and found that "the output elasticities of irrigation water are highly inelastic for every crop, indicating that *reductions in water supply would have relatively small effects on crop production*" (1993, 16; emphasis added). Depending on functional form, they found yield elasticities with respect to water for alfalfa (0.138–0.145), barley (0.014–0.020), cotton (0.115–0.126), corn (0.064–0.070), sorghum (0.112–0.115), sugar beets (0.055–0.064), and wheat (0.082–0.083). These elasticities were consistently small across crops and functional form.

Estimating a sequential-decision production function for Imperial Valley wheat, Antle and Hatchett (1986) found even smaller output elasticities with respect to water. Estimating production functions for winter vegetables, Just et al. (1983) reported water output elasticities for tomatoes (0.037), bell peppers (0.046), onions (0.051), melons (0.050), and eggplant (0.079). Results from Schneider and Howell (1997) and Knapp and Schwabe (2008) also suggest that it is possible to reduce water applications significantly with relatively little yield penalty. The implied yield (or output) elasticities with respect to water in USARM are certainly in line with the estimates of these studies.

How, then, may the results be interpreted? In the model, water applications can fall over a range without much of a loss in revenue. For high-value specialty crops, water is reduced as long as it has a negligible effect on revenues. Beyond some threshold, reducing water to high-value crops would be costly. But the model stops reducing water to those crops before that threshold is reached. For crops with lower returns per acre-foot (cotton, apples, barley, wheat, and alfalfa), water applications are reduced further, however, having more significant effects on yield and overall production. Another way to think of this is that the model reduces water allocations to specialty crops *because* it can do so at little cost (up to a point). If the production functions were such that specialty crop yields fell at smaller water reductions, then the model would maintain more water for these crops and cut allocations to the lower-return crops even more.

In sum, the region adjusts to the water supply shock by preserving production of high-return crops at the expense of crops with lower returns per acre-foot. USARM assumes that yield elasticities with respect to water applications are low (at baseline levels). However, this is quite consistent with a number of estimates from farm-level production function studies. Further, the simulation accurately captures that reductions in water to high-value crops would constitute a small share of the regional water reduction, and that water use intensity would fall by a greater percentage for field crops than for specialty crops.

Labor. Total regional farm employment falls by 3%, with the greatest percentage reductions in cotton at 44%, followed by irrigated barley and apples at 13% (Table 5.10). Cotton accounts for 35% of reduction in total farm labor, while cotton and alfalfa together account for two-thirds of the total reduction, and cotton, alfalfa,

Table 5.10 *Change in Farm Labor in Response to Reduction in Regional Water Supplies*

	Percent change in farm labor	*Crop's share of regional labor reduction*[a]
Nonirrigated crops		
Alfalfa	1%	—
Barley	10%	—
Corn	3%	—
Sorghum	0%	—
Wheat	0%	—
Irrigated crops		
Alfalfa	−5%	31%
Apples	−13%	8%
Barley	−13%	7%
Broccoli	0%	—
Cauliflower	0%	—
Citrus	−2%	—
Corn	0%	—
Cotton	−44%	35%
Grapes	−1%	—
Lettuce	−1%	14%
Melons	−1%	2%
Onions	0%	—
Potatoes	−1%	—
Sugar beets	−3%	—
Wheat	0%	—
Total farm labor	−3%	

[a]A dash in this column means labor for this crop accounts for less than 1% of the total reduction in regional farm labor

and lettuce account for 80% of the total reduction. Even though lettuce experiences a small percentage reduction in labor, it is relatively labor-intensive and accounts for a significant share of regional farm labor. Arizona accounts for the bulk of cotton farm labor and virtually all of the regional lettuce labor. In addition, broccoli, cauliflower, melons, and citrus together account for about 3% of the regional decline in farm labor, and these crops are grown almost exclusively in Arizona. Therefore, these labor reductions will be felt most strongly in Arizona. The labor reductions represent farm jobs only and do not account for losses in farm-related jobs, such as in postharvest processing, transportation, or agricultural input industries.

Prices and Impacts on Consumers. Most crops in the model experience slight price increases. For 17 of 33 crops, prices rise less than 0.1%. For 24, prices rise less than 0.2%. Only three crops experience price increases of more than 0.5%. Cotton prices rise 1.1%, walnut prices rise 0.6%, and sugar beet prices rise 1.1%.

Because of reduced production and higher commodity prices, total consumer surplus declines by $129.7 million. About 55% of this loss is borne by first purchasers of cotton and alfalfa. Cotton consumer surplus falls $35.6 million, while alfalfa consumer surplus falls $36.2 million. Cotton gins and dairy producers will feel these losses as first purchasers of cotton and alfalfa. Small grain (wheat and barley) consumer surplus falls by $5.1 million, and lettuce, citrus, broccoli, melons, cauliflower, onions, and walnuts together account for $10.8 million in consumer losses. Somewhat surprisingly, consumer surplus losses for corn make up 14% of the consumer losses, at $18.1 million. This occurs even though percent changes in national corn production (−0.01%) and price (< 0.05%) are minuscule. Why does this happen? Sales from corn are so large that base consumer surplus is more than $66 billion. So the $18.1 million consumer loss represents a reduction of less than 0.03% in corn consumer surplus.

Net Farm Income. Net farm income from growing nonirrigated crops (in Colorado, Utah, and New Mexico) increases by $3.9 million, while net income from growing irrigated crops falls by over $15 million (Table 5.11). The largest decrease in income is in irrigated alfalfa production (nearly $10.5 million) and cotton production (nearly $2.5 million). Irrigated alfalfa accounts for 69% of the total loss in regional net farm income, while cotton accounts for 16%. Cotton faces the largest percentage reduction in net farm income, however, a loss of 44%. Net income from specialty crop production falls little in percentage terms. In fact, income for some specialty crops actually increases.

Why does such a large water supply shock appear to have such a small effect on the net incomes of SM producers? There are a number of reasons. First, producers respond to water shortages by cutting back water use and production for the least profitable activities. This will be production on the least productive land, which has the highest marginal and average production costs. This generates a pivotal shift in supply curves, such that marginal costs at low production levels change very little, so the percent change in net income can be small (see Figure 5.1 in Appendix 5.2 at the end of this chapter).

Table 5.11 *Change in SM Region Net Income in Response to Water Supply Shock*

	Change in net farm income ($ thousands)	Percent change in net farm income
Nonirrigated crops		
Alfalfa	$581	2%
Barley	$31	10%
Corn	$175	3%
Sorghum	$259	2%
Wheat	$2,899	3%
Subtotal	$3,944	
Irrigated crops		
Alfalfa	−$10,455	−2%
Apples	−$130	−9%
Barley	−$811	−6%
Broccoli	$108	1%
Cauliflower	$34	0%
Citrus	−$406	−2%
Corn	−$423	0%
Cotton	−$2,481	−44%
Grapes	−$5	0%
Lettuce	$595	6%
Melons	−$307	−1%
Onions	$138	0%
Potatoes	−$332	0%
Sugar beets	$72	1%
Wheat	−$775	−3%
Subtotal	−$15,178	
Total (irrigated + nonirrigated)	−$11,234	

A second reason has to do with the price elasticity of demand for specialty crops. Demands for these commodities tend to be highly inelastic. California and Arizona account for a large share of national production, so regional water supply shocks have more of an impact on national production and output prices. Rising prices create a "natural hedge" against the water supply shock. As the supply curve shifts back from S to S', output rises from P_m to P'_m (Figure 5.1). Even though producers lose from rising marginal costs and reduced output, these losses are offset somewhat by higher market prices as part of the cost of the shock is passed on to consumers.

With pivotal supply curve shifts and downward-sloping demand curves, producers can gain from a negative shock, even if demand is elastic (Taylor and Howitt 1993). In studies of labor supply shocks in Imperial Valley lettuce production, Carter et al. (1981, 1987) found that union labor strikes were ineffective because they actually increased grower profits. In these studies, the input shock was to labor instead of water, but the same principle applies. In a programming model simulation of wetlands regulations, Claassen et al. (1998) found that relaxing restrictions on planting on wetlands (and increasing national acreage planted) reduced net farm income. In their case, a pivotal positive supply shock (bringing marginal land into production) reduced farm income.

We can use the aggregate consumer surplus estimates to approximate how much of the cost of the water supply shock SM irrigators pass on to consumers. Assume that consumer surplus losses from sales of irrigated crops in a region are proportional to a region's share of total sales (regional irrigated crop sales/aggregate crop sales). If so, then lower production and higher prices for SM crops generated approximately $6.5 million in consumer surplus losses. Of this, $4 million in losses were to first purchasers of alfalfa, $1 million to purchasers of lettuce, and $1.4 million to purchasers of other fruit and vegetable crops.[2]

This $6.5 million figure is approximately what SM irrigators were able to pass on to consumers. So if producers were not able to pass on these costs (for example, if one assumed that demand curves were completely flat), then losses to SM irrigators from the supply shock would be closer to $22 million. Dairy producers in the SM region would feel much of the $4 million in losses as first purchasers of alfalfa from this region. If lettuce producers were not able to pass on the $1 million in costs to first purchasers, lettuce producer returns would decrease rather than increase as they do in Table 5.11. In fact, all producer groups would see their net incomes fall.

Cotton Program Payments and Water Shocks. Cotton LDP payments do not shelter producers from water supply shocks (see Figure 5.2 in Appendix 5.2). In fact, regional LDP payments fall by $2.4 million, roughly the same amount as the overall loss in regional net income from cotton. Thus nearly all the loss to cotton producers is loss of government payments. This is consistent with results from a study of a positive supply shock to cotton: adoption of Bt cotton (Frisvold et al. 2006). That study estimated that 82% of the producer gain from Bt cotton adoption came from increased program payments. Results are also consistent with the general finding of Lichtenberg and Zilberman (1986) that price support programs increase losses to producers from a negative supply shock.

Welfare Impacts. Turning to overall welfare effects, the water supply shock redistributes income from consumers to agricultural producers outside the SM region. As production in the SM region and California contracts, producers elsewhere gain from the higher prices this generates. Producers in other regions are also able to increase their sales volumes slightly. Thus producer surplus in the SM region falls by $11 million (from $15 million in irrigator losses and a $4 million gain in nonirrigated net income), but producer surplus in the rest of the United States increases by $101 million (Table 5.12). Consumer surplus falls by nearly $130 million, and government payments decrease by about $30 million. Almost all of this reduction came from reductions in national cotton program payments. Changes in other output prices were not large enough to trigger large changes in other payments. The change is welfare, ΔW, calculated as

$$\Delta W = \Delta CS + \Delta PS - \Delta GP$$

Table 5.12 *Welfare Impacts of the Water Supply Shock*

Welfare measure	*Change from water supply shock ($ million)*
Consumer surplus	−$130
Other producer surplus	$101
SM producer surplus	−$11
Government payments	−$30
Net welfare	−$10

where ΔCS = change in consumer surplus, ΔPS = change in producer surplus, and ΔGP = change in government payments. Thus the water supply shock reduces net welfare by $10 million.

25% Water Price Increase Simulation

We next consider the effects of a 25% increase in the cost of water in the SM region and California. Again, this is a generic simulation, not intended to represent a particular policy. Nevertheless, it does inform policy debates about the pricing of water. USBR projects provide nearly a third of surface water deliveries for irrigation in the 17 western states (Gollehon et al. 1997). Under the National Reclamation Act, the USBR charges irrigators based on their ability to pay for water and allows projects to charge other water and energy users higher rates to make up for the difference between capital costs of project construction and irrigation water delivery costs and what irrigators pay. In addition, irrigators are exempt from interest charges on USBR project capital costs. USBR pricing of irrigation water has been controversial (Wahl 1989; CBO 2006; Holland and Moore, 2003; GAO 1981, 1989, 1990, 1996, 1997, 2007). Contract prices charged to irrigators in SM states range from $6 to $35 per acre-foot below full-cost prices (Moore 1991).

Both conservative (Edwards 2007) and environmental groups (EWG 2005) have called for increases in rates irrigators pay. Legislation has been introduced regularly to raise rates irrigators are charged for USBR deliveries (EWG 2005). Moore (1991) and Gardner (1996) note, however, that the values of past irrigation subsidies have been capitalized into land values captured by original landowners. Moore argues that attempts to recover full costs from past subsidies would be politically and economically infeasible, because growers simply could not afford to pay such high rates. Irrigators would be more able (if not willing) to pay rates closer to current costs of supplying water.

Howitt (1991) examined the effects of a 25% increase in the price of water across the Northern Mountain states as well as the Southern Mountain states and California. In programming model simulations, the 25% price increase reduced production of program crops (corn, cotton, rice, sorghum, and wheat) by 33%, and wheat acreage in the Mountain region shifted from irrigated to dryland acreage. In a study simulating elimination of USBR subsidies in central California,

Ulibarri et al. (1998) found that cotton and sugar beets experienced the largest decrease in per-acre profits.

USBR projects supply only a portion of the total irrigation water used in the SM region and California. In USARM, the water price is a weighted average price of surface water and groundwater. Thus the 25% price increase captures increased surface water prices and increased groundwater pumping costs. The 25% increase in the cost of water reduces SM irrigation water use by 10.5% and California water use by 4.7%. Another way to think of this experiment is a 10% (4%) reduction in SM (California) water availability, with a 25% increase in surface water costs and a 25% increase in groundwater pumping costs as growers shift from surface water supplies to more costly groundwater. For example, central Arizona irrigators receive CAP water at concessional rates in exchange for forgoing pumping groundwater. CAP irrigators have junior water rights among Lower Colorado River water users, however, and would be the first to forgo deliveries if a shortage were declared on the Colorado River. If a shortage were declared, growers may resort to pumping groundwater to make up for the reduction in CAP deliveries.

Table 5.13 *Impacts on SM Region Acreage, Yield, and Output from the 25% Water Price Increase Simulation*

	Change in acres (thousands)	*Percent change in*		
		Acreage	*Yield*	*Output*
Nonirrigated crops				
Alfalfa	1.4	1%	0%	1%
Barley	0.7	7%	1%	8%
Corn	1.6	2%	0%	2%
Sorghum	3.6	2%	−2%	1%
Wheat	38.3	2%	−2%	1%
Total	45.6			
Irrigated crops				
Alfalfa	−3.5	0%	−1%	−1%
Apples	−2.6	−25%	−10%	−33%
Barley	−7.5	−5%	−8%	−12%
Broccoli	0.2	2%	1%	3%
Cauliflower	0.0	1%	0%	1%
Citrus	0.1	0%	0%	0%
Corn	1.6	0%	−1%	0%
Cotton	−46.6	−25%	−19%	−40%
Grapes	0.0	−1%	−1%	−2%
Lettuce	1.1	2%	0%	2%
Melons	0.0	0%	0%	0%
Onions	0.1	0%	0%	0%
Potatoes	−0.3	0%	0%	−1%
Sugar beets	0.1	0%	−1%	0%
Wheat	11.6	3%	−3%	0%
Total	−45.6			

The regional water quantity limits would not be binding constraints at a 25% increase in the price of water.

Acreage, Yield, and Output Effects. As with the pure water supply shock, dryland acreage increases while irrigated acreage declines (Table 5.13). Wheat accounts for the bulk of the dryland acreage increase. Yet irrigated wheat acreage also increases. This differs from Howitt's earlier simulation results (1991), where dryland wheat increased but irrigated wheat declined. One reason irrigated wheat acreage increases is that wheat has a higher relative price (calibrated at 2007 prices) than it did in Howitt's simulation. Cotton acreage and production are again hard hit, as they were in the water supply shock scenario, with output falling by 40% and acreage falling by more than 46%. This is also attributable to relative changes in output prices. Cotton prices have fallen relative to other crops, leaving net returns per acre-foot of water relatively low. Apple and barley production also decline more than most crops, as in the supply shock scenario. Again, specialty production changes little.

Water Use. The 25% increase in the cost of water reduces SM irrigation water use by 10.5%, for an arc elasticity of − 0.5 (Table 5.14) Water use falls by more than 1 million ac-ft. Water use for alfalfa and cotton falls by 446,000 and 326,000 ac-ft (Table 5.14). Reductions in water use for these two crops alone account for more than three-quarters of the total reduction in irrigation water use. Although alfalfa accounts for a large share of total water reduction, water use per acre does not decline as much as for cotton, barley, and apples. Alfalfa's water reductions per acre are intermediate between these and the specialty crops.

Table 5.14 *Impact of 25% Water Price Increase on SM Region Water Use*

	Change in water use (thousand ac-ft)	*Percent change in water use per acre*	*Percent change in water use*
Alfalfa	−446	−9%	−9%
Apples	−16	−18%	−38%
Barley	−68	−19%	−24%
Broccoli	−2	−7%	−5%
Cauliflower	0	−2%	−1%
Citrus	−3	−3%	−2%
Corn	−93	−5%	−5%
Cotton	−326	−29%	−47%
Grapes	−1	−9%	−10%
Lettuce	7	1%	3%
Melons	−5	−5%	−5%
Onions	−4	−8%	−7%
Potatoes	−20	−9%	−10%
Sugar beets	−4	−8%	−8%
Wheat	−38	−7%	−4%
Total	−1,019		−11%

Farm Labor. Under the price increase, total farm employment falls by 1% (compared to 3% under the supply shock). Employment falls by 44%, 33%, and 12% for cotton, apple, and barley production, respectively. Employment increases, however, for specialty crop production.

Net Farm Income. The regional cost to irrigators of the water price increase is $44 million (Table 5.15). Net returns to nonirrigated production (in Colorado, Utah, and New Mexico) increase by $3 million. One reason the price increase is more costly is that it applies to all irrigators. Under the supply shock, producers can divert water from the least profitable activities, leaving higher-valued activities relatively unaffected. Under the price increase, however, all irrigators' costs increase for all activities. Cotton production is the most vulnerable to water price increases, with net returns falling by 86%.

Returns to broccoli, cauliflower, and onion production increase, but this is because their producers are able to pass cost increases on to consumers. SM irrigators are able to pass on about $4 million in losses to consumers. About $1.4 million is passed on to first purchasers of alfalfa. Amounts passed on to broccoli purchasers ($0.64 million), onion purchasers ($0.66 million), and cauliflower

Table 5.15 *Change in SM Region Net Income in Response to Water Price Increase*

	Net income change ($ thousands)	*Percent change in net income*
Nonirrigated crops		
Alfalfa	$355	1%
Barley	$25	8%
Corn	$144	2%
Sorghum	$218	2%
Wheat	$2,376	2%
Subtotal	$3,118	2%
Irrigated crops		
Alfalfa	−$18,627	−3%
Apples	−$864	−61%
Barley	−$2,576	−19%
Broccoli	$678	5%
Cauliflower	$137	2%
Citrus	−$21	0%
Corn	−$11,956	−4%
Cotton	−$4,903	−86%
Grapes	−$146	−4%
Lettuce	−$100	−1%
Melons	−$472	−1%
Onions	$558	1%
Potatoes	−$2,069	−1%
Sugar beets	$142	1%
Wheat	−$3,839	−16%
Subtotal	−$44,059	−3%
Total	−$40,941	−3%

purchasers ($0.1 million) allow these producers to have higher total net returns after the water price increase.

CONCLUSIONS

This chapter examined how large changes in irrigation water supplies or prices might affect agriculture in the Southern Mountain (SM) region of the United States. Agriculture can adjust in different ways to water scarcity by fallowing land, switching to less water-intensive crops, reducing water applications per acre, or substituting other inputs for water. U.S. Agricultural Resources Model (USARM) simulations quantified the importance of such adjustments for reducing the costs of water scarcity. Estimates from a rationing model suggested that a 25% reduction in irrigation water supplies to the SM region would reduce irrigator income by $65 million. The rationing model only permitted fallowing as a response to water shortages. When USARM simulations allowed growers to adjust more flexibly, irrigator losses fell to $15 million to $22 million, depending on whether growers could pass costs on to consumers. Whereas models examining costs of water shortages usually assume that output prices are fixed, our results demonstrate the importance of output price effects and impacts on consumers.

While SM agriculture, as a whole, was quite resilient to the water supply shock, particular crops were vulnerable. Among these were cotton, barley, alfalfa, and apples. Cotton program payments, designed to support income under low prices, did not protect cotton producers from water supply shocks.

Regional farm employment fell by 3%. Impacts would be felt relatively more in Arizona, because labor demand fell more for crops grown primarily in Arizona than for other crops grown in the SM region. For similar reasons, regional acreage reductions were modest overall but would be proportionally greater in Arizona. The water supply shock could have the unintended benefit of reducing fertilizer and agricultural chemical pollution. Greater water applications can leach fertilizers and chemicals from the soil. Thus with reduced agricultural water supplies, there may also be less runoff, drainage, and groundwater percolation from farm fields.

Under a 25% water price increase simulation, total irrigation water use in the SM region fell by 10.5%. Regional irrigator losses were greater for the price shock, ranging from $44 million to $48 million, depending on whether costs could be passed on to consumers. Under the supply shock, producers could divert water from their least profitable activities. The price increase was more costly overall because it raised costs for all irrigated production.

As a holder of junior water rights to Colorado River deliveries, the Central Arizona Project and its irrigators would be among the first affected by severe regional water shortages. Results suggest that field and forage crops would be the first to contract in the face of water shortages, while producers maintained production of higher-valued specialty crops. This implies that field and forage crop production in central Arizona would be relatively vulnerable to severe water shortages in the Southwest. Dairies in central Arizona would also be negatively affected from reduced supplies and higher prices of alfalfa. However, western

Arizona, along the Colorado River mainstem, would continue to be a national center of specialty crop production, even in the face of large regional water shortages.

APPENDIX 5.1: THE U.S. AGRICULTURAL RESOURCES MODEL (USARM)

This appendix describes the basic structure of USARM. It describes the model equations and their functional form. It also describes the model calibration approach used and data sources used to derive model parameters.

Objective Function

The objective function represents the aggregate consumer (domestic and foreign) and producer welfare for all regions and activities:

$$\Pi = \sum_i \left[\alpha_i \sum_w \sum_r q_{iwr} + 0.5\delta_i \left(\sum_w \sum_r q_{iwr} \right)^2 \right] - \sum_i \sum_w \sum_r \tau_{iwr} q_{iwr} - \sum_r \left[v_{r1} \sum_i \sum_w x_{iwr1} \right.$$
$$\left. + 0.5\omega_{r1} \left(\sum_i \sum_w x_{iwr1} \right)^2 \right] - \sum_i \sum_w \sum_r \eta_{iwr1} x_{iwr1} - \sum_i \sum_w \sum_r \sum_j \left(\rho_{iwrj} x_{iwrj} + \varphi_{iwrj} x^2{}_{iwrj} \right) \quad (5.1)$$

This formulation ensures a competitive market equilibrium solution. The first expression in brackets measures the area under the crop-specific linear market quantity-dependent demand equations for each crop, i. The variable q_{iwr} represents the output of crop i, produced under cultivation condition w (1 = dry and 2 = irrigated), in region r. The coefficient τ_{iwr} accounts for marketing and transportation costs. The third expression allows the land rents to be endogenous at the regional level, where v_{r1} and ω_{r1} are the intercept and slope of regional linear land supply equations. The coefficient η_{iwr1} accounts for the difference between the regional average land rent and the crop activity specific land rents in that region.

The last term in the objective function is a quadratic cost function, where ρ and φ are the coefficients, and the variable x_{iwrj} is the amount of input j ($j = 1, \ldots, 7$ with 1 = land input) used in the cropping activity i, w, r. This function is quadratic in the land input and linear in the others. The quadratic cost function captures the fact that as more land is allocated to a specific crop, the marginal cost increases as marginal lands with lower yield potential come into production. It also allows for the exact calibration of the model solutions to the base year levels of crop acreage following Howitt's Positive Mathematical Programming (PMP) method (1995a, 1995b). The PMP approach eliminates the need to use upper and lower bound constraints on the activity levels when simulating policy scenarios.

Nested-CES Production Functions

The regional total output from each cropping activity q_{iwr} (indices are dropped for brevity) is defined by the following nested-CES production function with seven categories of inputs:

$$Q = C\left\{\beta_F\left[C_F\left(\sum_{j=1}^{2}\beta_j x_j^{\gamma_F}\right)^{1/\gamma_F}\right]^{\gamma} + \beta_V\left[C_V\left(\sum_{j=3}^{7}\beta_j x_j^{\gamma_V}\right)^{1/\gamma_V}\right]^{\gamma}\right\}^{1/\gamma} \quad (5.2)$$

The function consists of two nests. The first nest, in the first set of brackets, includes the allocable inputs, land and water. The second nest, in the second set of brackets, is for the remaining five variable inputs: agricultural chemicals, fertilizers, labor, capital, and energy/other inputs. Each nest is in itself a CES function. A nested-CES function is more flexible than a regular CES function, because there can be more than one elasticity of substitution between inputs. In agricultural crop production, the ability to substitute inputs varies significantly (Debertin et al. 1990; Hertel et al. 1989; Rendleman 1993; Ray 1982).

In the nested-CES production function, the nests can be thought of as hierarchies. Equation (5.2) has the higher nest parameters on the outside. The scalar C is the top-nest scale parameter, and β_F and β_V are the top-nest share parameters for allocable and variable inputs. Moving to the lower nests, C_F and C_V are scale parameters for allocable and variable input nests, respectively. The quantity of input j allocated to each cropping activity is indicated by x_j, where the j values of 1 and 2 correspond to allocable inputs of land and water, and the remaining values of j (from 3 to 7) correspond to variable inputs. The parameter β_j is the share parameter of the jth input. In addition to index j, input quantity x is indexed over i, w, and r, which are dropped here for brevity. The coefficient $\gamma = (s - 1)/s$, where s is the top-nest elasticity of substitution coefficient. Finally, $\gamma_F = (s_F - 1)/s_F$ and $\gamma_V = (s_V - 1)/s_V$, where s_F and s_V are the elasticity of substitution between the allocable inputs and the elasticity of substitution between the variable inputs, respectively. For the dryland cropping activities, the allocable input nest has only the land input as its argument.

Resource Constraints

In the following discussion, indices on x are reintroduced: i (crop), w (irrigation condition), r (region), and j (input). Regional irrigation water constraints limit the total irrigation water used by all irrigated crops ($w = 2$) in a region to the actual total irrigation water ($j = 2$) used in the region in the base year, $\underline{X}_{2r2}$. This constraint is specified as

$$\sum_i x_{i2r2} \leq \sum_i \underline{X}_{i2r2} \quad (5.3)$$

The regional irrigation land constraint restricts the total land allocated to irrigated cultivation to the total actual base year irrigated acreage:

$$\sum_i x_{i2r1} \leq \sum_i \underline{X}_{i2r1} \tag{5.4}$$

Empirical Specification

The coefficients of the demand equations, α_i and δ_i, are derived by solving crop-specific linear demand equations, where a crop's aggregate demand elasticity and the base year aggregate market price and quantities are given. The coefficient τ_{iwr} is measured as the difference between the national market and the regional price of each crop in the base year. Parameter τ reflects the deviation of a region from the average transportation and marketing costs. In the policy runs, the national-level market price for each crop is endogenously determined. During the solution, τ maps the national crop prices to their regional levels, maintaining the deviation of regional crop price from the national average.

The coefficients of the regional land supply equation, ν_{r1} and $\omega_{r1,}$ are calculated by solving a region-specific linear equation, where the regional average land rents, land in crop production, and land supply elasticity are given. The supply elasticity is derived from a special run of a model presented in Lewandrowski et al. (1999). The coefficient η_{iwr1} is calculated as the difference between the crop-specific land rent in each region and the average land rent for that region. The method of calculating the cost function coefficients ρ and φ (for land only) is based on Howitt's PMP approach (1995a, 1995b).

The elasticity of input substitution coefficients of the nested-CES function (equation (5.2)) are derived from several empirical sources. The values of the substitution parameter between land and water are obtained from Hatchett (1997). In this study, separate elasticities for truck crops, row crops, and alfalfa for California are estimated. Because of the lack of empirical estimates for the other states, and the fact that the range of irrigation technologies is similar across regions, the California elasticities for irrigated activities are used in all regions.

The values of the input substitution parameters for the top nest—in other words, the allocable and variable inputs nest and the substitution parameter for the inputs in the variable input, lower nest 2—are based on Hertel et al. (1989) and Rendleman (1993). Rendleman estimates own-price and cross-price elasticities of derived demand for several inputs used in the production of food grains, feed grains, oilseed crops, and cotton. His input categories, with the exception of "energy/other" inputs, closely resemble ours. For the "energy/other" category, we rely on estimates from Hertel et al. (1989). From the Allen partial elasticities of substitution found in Hertel et al. and Rendleman (1993), we calculate Morishima elasticities of substitution. Morishima elasticities are relevant for the type of CES function we use here (Blackorby and Russell 1989). The pairwise Morishima elasticities are weighted by each input's cost share in the variable input nest to come up with a single crop- and region-specific elasticity of substitution for nest 2.

The elasticities of substitution for the top nest (i.e., between allocable and variable nests) are calculated in a similar fashion, using original estimates from the three empirical sources.

The values of the remaining parameters of equation (5.2), C, β_F, β_V, C_F, C_V, and β_j, are obtained by solving the first-order conditions of a profit maximization problem defined by equations (5.1)–(5.4). The known quantities are elasticity of input substitution, base year levels of acreage, output, input use under each activity, and regional crop prices, along with shadow values of the regional land and irrigation water constraints.

To summarize, in the first stage, a linear programming problem is solved. The shadow value estimates from the solution, along with input substitution coefficients and the base year levels of the variables, are used to calculate coefficients ρ and φ in equation (5.1) and coefficients C, β_F, β_V, C_F, C_V, and β_j in equation (5.2). In the second stage, the model defined by equations (5.1)–(5.4) is solved. The crop prices are endogenous in the second model. Both models are written in GAMS software and solved sequentially with the MINOS optimization algorithm on a PC (Brooke et al. 1992). The base year solution of the second model exactly replicates regional base year acreage and input use in each cropping activity and output price.

USARM was originally calibrated to match acreage and price data for field crops in 2002 and specialty crops in 2000. Recent relative price changes, however, have significantly altered the crop mix in the region. Output prices for major field crops were changed to reflect changes in relative prices in 2007. For example, corn and wheat prices were increased, and cotton prices were reduced. Thus relative prices and acreage planted to major crops have adjusted to reflect the new environment, with more acreage planted to wheat and corn and less to cotton.

Data

The primary budget data on the field crops, except alfalfa hay and sugar cane, have been extracted from the results of the most recently available cost of production surveys, which are conducted by the U.S. Department of Agriculture (USDA) agencies National Agricultural Statistics Service (NASS) and Economic Research Service (ERS) every five to eight years for each commodity. Occasionally, the surveys leave out states with significant acreage. This also occurs for alfalfa and sugar cane. In these instances, the data have been obtained from the crop budgets published by various state cooperative extension service offices. State-level acreage, yield, and market prices were obtained from NASS, Historical Data, Crops County Data Files.

A total of 396 detailed cost of production budget estimates from 35 states, covering 28 vegetable and fruit crops, were collected. The crops were aggregated into 22 groups. Most of the 396 budgets are authored by state cooperative extension economists or crop specialists. In a few cases, state growers' association experts provided budgets. In some states, most notably, Arizona, California, Florida, Idaho, Oregon, Texas, and Washington, multiple budgets for the same crop were collected to capture the diversity in the states' growing conditions.

The model captures 97.8% of the actual year 2000 acreage of the included vegetable and fruit crops.

Domestic and export elasticities for the field crops are obtained from Green and Price (1987), Hertel et al. (1989), and Konyar and Knapp (1990). For the vegetable and fruit crops, usually only aggregate demand elasticities were available (Alston et al. 1993; Henneberry et al. 1999; You et al. 1996; Hatchett 1997; Onunkwo and Epperson 2000; Nuckton 1978). Elasticities were disaggregated into domestic and export elasticities using the following procedure. Total output, domestic consumption, export quantity, market prices, and aggregate demand elasticities are given. Using these data and employing a maximum entropy approach, domestic and export demand elasticities were solved optimally such that the resulting elasticities guarantee market clearance in the two markets, as well as that the individual elasticities sum up to the aggregate elasticity.

Data on program acres and yield by state were obtained from USDA ERS (2009) data files. Available data include base acres designated under the 2002 Farm Act, by commodity, including base acres updated to 1998–2001 plantings under the act, as well as state-level program yields used to determine direct and CCP rates for each program commodity. Young et al. (2005) described the data and programs in detail.

APPENDIX 5.2: SUPPLY SHOCKS, COMMODITY PROGRAMS, AND PRODUCER RETURNS

The reduction in available water supplies to SM agriculture appears to have relatively minor effects on farm income. This may seem counterintuitive. This is not the result of any unusual properties of USARM. Rather, it results from the relationship between the type of supply curve shift and producer returns. This relationship can be best explained using a simple, graphic supply-and-demand approach (Figure 5.1). The supply-demand approach can also be used to illustrate how price support policies do not protect producers from losses from water shortages (Figure 5.2).

Why Does Net Income Fall so Little?

We have reason to expect that the water supply shock causes a divergent (or pivotal) shift in the supply curve. In other words, marginal costs increase more for higher-cost producers and areas (marginal areas with lower land quality) than for lower-cost producers and areas. The supply curve thus "pivots upward," with little cost change near the intercept. Under an initial market solution, with supply curve S, equilibrium market price is P_m and producer surplus is $g + a$. A negative supply shock (such as reduced water supplies) shifts the supply curve to S'. Market price rises to P'_m, and the change in producer surplus is $f - a$. Growers lose a because their marginal costs rise, but gain f because market price rises. The more elastic the demand curve is, the smaller area f will be. However, the more *inelastic* the demand curve is, the greater f will be. Also, for smaller supply shocks, f will be larger relative

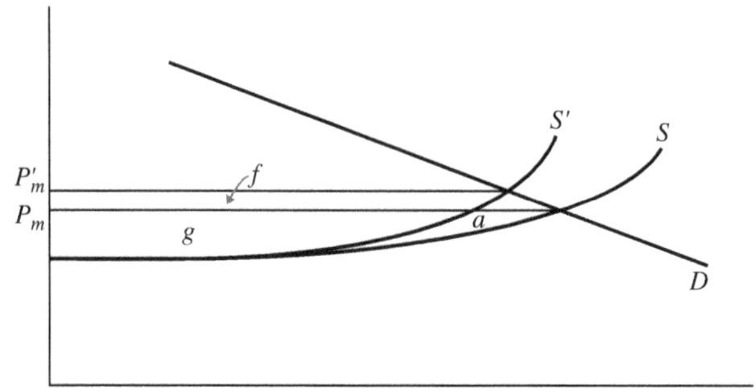

Figure 5.1 *Change in Producer Returns with a Pivotal Supply Curve Shift*

to *a*. So for specialty crops, which experience smaller shocks and have inelastic demand, it is not surprising that the change in producer surplus is small and in some cases positive. Even if output price did not increase (for example, if growers faced a flat demand curve at P_m), the *percentage* reduction in net income, which is $a/(g + a)$ can be small for small supply shocks.

Water Supply Shocks and Loan Deficiency Payments

Under an initial market solution, with regional supply curve *S*, equilibrium national market price is P_m and regional producer surplus is $f + a$. A negative supply shock (such as reduced water supplies) shifts the supply curve to S'. If national market price rises to P'_m, producer surplus is now $e + f$, and the change in producer surplus is $e - a$. Growers lose *a* because their marginal costs rise, but gain *e* because market price rises. Now consider a commodity whose price is supported with coupled loan deficiency payments. With the initial supply curve, *S*, producer surplus is $a + b + c + d + e + f$. Water shortages that shift supply back to S' change producer surplus by $-(a + b + c)$, back to $d + e + f$. So, somewhat paradoxically, loan deficiency payments (LDPs) lead to greater producer surplus

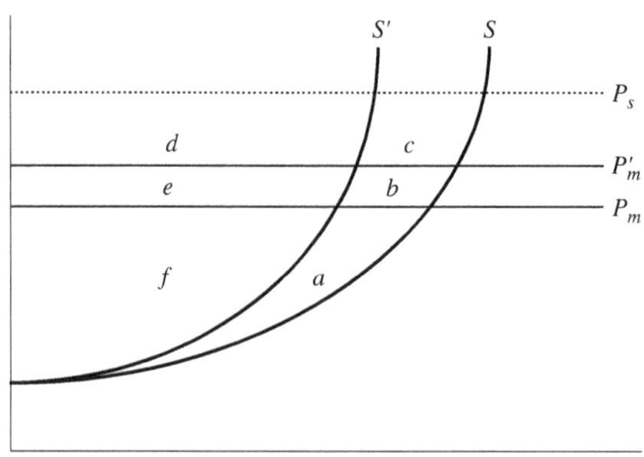

Figure 5.2 *Price Supports, Supply Shifts, and Producer Returns*

than the market solution with abundant water (S) and water shortages (S'), yet the losses to producers from water shortages under LDPs, $-(a + b + c)$, are larger than losses under the market solution, $e - a$.

NOTES

1. This generates a reverse L-shaped supply curve for each crop. Sunding et al. (2002) assumed a more generalized specification in their rationing model, where average and marginal costs increase as more land is put into production for each crop. Thus they have upward-sloping supply curves.
2. Exact consumer surplus losses would be slightly different than costs producers avoid by passing them on, and also slightly different than the figures quoted here. Figures presented here are approximations, but close ones.

REFERENCES

Alston, J.M., J. Christian, J.R. Murua, and R.J. Sexton. 1993. Restricting Flow of Almonds to Export Markets May Raise Profits. *California Agriculture* 47:7–10.

Alston, J.M., G.W. Norton, and P.G. Pardey. 1995. *Science under Scarcity: Principles and Practice for Agricultural Research Evaluation and Priority Setting*. Ithaca, NY: Cornell University Press.

Antle, J.M., and S. Hatchett. 1986. Dynamic Input Decisions in Econometric Production Models. *American Journal of Agricultural Economics* 68:939–949.

Barsugli, J.J., K. Nowak, B. Rajagopalan, J.R. Prairie, and B. Harding. 2009. Comment on "When Will Lake Mead Go Dry?" by T.P. Barnett and D.W. Pierce. *Water Resources Research* 45:W09601, doi:10.1029/2008WR007627.

Blackorby, C., and R.R. Russell. 1989. Will the Real Elasticity of Substitution Please Stand Up? (A Comparison of Allen/Uzawa and Morishima Elasticities). *American Economic Review* 79:882–888.

Booker, J.F. 1995. Hydrologic and Economic Impacts of Drought under Alternative Policy Responses. *Water Resources Bulletin* 31:889–906.

Booker, J.F., A.M. Michelsen, and F.A. Ward. 2005. Economic Impact of Alternative Policy Responses to Prolonged and Severe Drought in the Rio Grande Basin. *Water Resources Research* 41:W02026, doi:10.1029/2004WR003486.

Brooke, A., D. Kendrick, and A. Meeraus. 1992. *GAMS: A User's Guide*. San Francisco: Scientific Press.

Carpenter, J., L. Gianessi, and L. Lynch. 2000. *The Economic Impact of the Scheduled U.S. Phaseout of Methyl Bromide*. Washington, DC: National Center for Food and Agricultural Policy.

Carter, C., D. Hueth, J. Mamer, and A. Schmitz. 1981. Labor Strikes and the Price of Lettuce. *Western Journal of Agricultural Economics* 6:1–14.

———. 1987. Agricultural Labor Strikes and Farmers' Income. *Economic Inquiry* 25:121–133.

CBO (Congressional Budget Office). 2006. *How Federal Policies Affect the Allocation of Water*. Washington, DC: CBO.

Christensen, N.S., A.W. Wood, N. Voisin, D.P. Lettenmaier, and R.N. Palmer. 2004. The Effects of Climate Change on the Hydrology and Water Resources of the Colorado River Basin. *Climatic Change* 62:337–363.

Claassen, R., R.E. Heimlich, R.M. House, and K.D. Wiebe. 1998. Estimating the Effects of Relaxing Agricultural Land Use Restrictions: Wetland Delineation in the Swampbuster Program. *Review of Agricultural Economics* 20:390–405.

Dale, L., and L. Dixon. 1998. *The Impact of Water Supply Reductions on San Joaquin Valley Agriculture during the 1986–1992 Drought.* Monograph Report. Santa Monica, CA: RAND Corporation.

Debertin, D.L., A. Pagoulatos, and A. Aoun. 1990. Impacts of Technological Change on Factor Substitution between Energy and Other Inputs with U.S. Agriculture, 1950-79. *Energy Economics* 12:2–10.

Deepak, M.S., T.H. Spreen, and J.J. VanSickle. 1996. An Analysis of the Impact of a Ban of Methyl Bromide on the US Winter Fresh Vegetable Market. *Journal of Agricultural and Applied Economics* 28:433–443.

Dinar, A., and J. Letey. 1996. *Modeling Economic Management and Policy Issues of Water in Irrigated Agriculture.* Westport, CT: Greenwood Publishing Group.

Edwards, C. 2007. *Agricultural Subsidies.* Washington, DC: Cato Institute.

Elbakidze, L. 2006. Potential Economic Impacts of Changes in Water Availability on Agriculture in the Truckee and Carson River Basins, Nevada, USA. *Journal of the American Water Resources Association* 42:841–849.

EWG (Environmental Working Group). 2005. *Double Dippers: How Big Ag Taps into Taxpayers' Pockets—Twice.* Washington, DC: EWG.

Frisvold, G., J.M. Reeves, and R. Tronstad. 2006. Bt Cotton Adoption in the United States and China: International Trade and Welfare Effects. *AgBioForum* 9:69–78.

GAO (U.S. General Accounting Office/Government Accountability Office). 1981. *Reforming Interest Provisions in Federal Water Laws Could Save Millions.* CED-82-3. Washington, DC: GAO.

———. 1989. *Water Subsidies: Basic Changes Needed to Avoid Abuse of the 960-Acre Limit.* GAO-RCED-90-6. Washington, DC: GAO.

———. 1990. *Water Subsidies: The Westhaven Trust Reinforces the Need to Change Reclamation Law.* GAO-RCED-90-198. Washington, DC: GAO.

———. 1996. *Bureau of Reclamation: Information on Allocation and Repayment of Costs of Constructing Water Projects.* GAO-RCED-96-109. Washington, DC: GAO.

———. 1997. *Bureau of Reclamation: Reclamation Law and the Allocation of Construction Costs for Federal Water Projects. Testimony before the Subcommittee on Water and Power Resources, Committee on Resources, House of Representatives.* GAO/T-RECED-97-150. Washington, DC: GAO.

———. 2007. *Bureau of Reclamation: Reimbursement of California's Central Valley Project Capital Construction Costs by San Luis Unit Irrigation Water Districts.* GAO-080307R. Washington, DC: GAO.

Gardner, B.D. 1996. Irrigation Subsidies: Some Efficiency, Equity, and Environmental Implications. In *The Economics of Agriculture, vol. 2, Papers in Honor of D. Gale Johnson*, edited by J. Antle and D. Sumner. Chicago: University of Chicago Press, 300–319.

Gelt, J. 1997. Sharing Colorado River Water: History, Public Policy and the Colorado River Compact. *Arroyo* 10:1–11.

Glennon, R.J., and P.W. Culp. 2002. The Last Green Lagoon: How and Why the Bush Administration Should Save the Colorado River Delta. *Ecology Law Quarterly* 28:903–992.

Gollehon, N., W. Quinby, and M. Aillery. 1997. Water Use and Pricing. In *Agricultural Resources and Environmental Indicators, 1996–1997*, edited by M. Anderson and R. Magleby. Agriculture Handbook No. AH712. Washington, DC: U.S. Department of Agriculture, Economic Research Service, 67–82.

Green, R.C., and J.M. Price. 1987. Estimates of Short-Run Price Elasticities for Major U.S. Field Crop and Livestock Commodities Using FAPSIM. Unpublished manuscript Washington, DC: U.S. Department of Agriculture, Economic Research Service.

Hardy, T.B. 1995. Assessing Environmental Effects of Severe Sustained Drought. *Water Resources Bulletin* 31:867–875.

Hatchett, S. 1997. Description of CVPM. In *Central Valley Project Improvement Act Programmatic Environmental Impact Statement.* Sacramento, CA: U.S. Bureau of Reclamation, II-1–II-23.

Henneberry, S., P. Kullapapruk, and H. Qiang. 1999. Consumer Food Safety Concerns and Fresh Produce Consumption. *Journal of Agricultural and Resource Economics* 24:98–113.

Hertel, T.W., V.E. Ball, K.S. Huang, and M.E. Tsigas. 1989. *Computing Farm Level General Equilibrium Demand Elasticities for Agricultural Commodities.* Bulletin No. 988. West Lafayette, IN: Purdue University Agricultural Experiment Station.

Hidalgo, H.G., T.C. Piechota, and J.A. Dracup. 2000. Alternative Principal Components Regression Procedures for Dendrohydrologic Reconstructions. *Water Resources Research* 36:3241–3249.

Hoerling, M., and J. Eischeid. 2007. Past Peak Water in the Southwest. *Southwest Hydrology* 6:18–19.

Holland, Stephen, and Michael R. Moore. 2003. Cadillac Desert Revisited: Property Rights, Public Policy, and Water-Resource Depletion. *Journal of Environmental Economics and Management* 46:131–155.

Howe, C.W. 1985. Economic, Legal, and Hydrologic Dimensions of Potential Interstate Water Markets. *American Journal of Agricultural Economics* 67:1226–1230.

Howitt, R.E. 1991. Water Policy Effects on Crop Production and Vice Versa: An Empirical Approach. In *Commodity and Resource Policies in Agricultural Systems*, edited by R.E. Just and N. Bockstael. New York: Springer-Verlag, 234-253.

———. 1995a. A Calibration Method for Agricultural Economic Production Models. *Journal of Agricultural Economics* 46:147–159.

———. 1995b. Positive Mathematical Programming. *American Journal of Agricultural Economics* 77:329–342.

Howitt, R.E., D. MacEwan, and J. Medellín-Azuara. 2009. Economic Impacts of Reductions in Delta Exports on Central Valley Agriculture. Giannini Foundation of Agricultural Economics, University of California. *Agricultural Resource Economics Update* 12 (3):1–4.

Hutson, S.S., N.L. Barber, J.F. Kenny, and K.S. Linsey. 2005. *Estimated Use of Water in the United States in 2000.* Rev. ed. U.S. Geological Survey Circular 1268. Washington, DC: U.S. Geological Survey.

Just, R.E., D. Zilberman, and E. Hochman. 1983. Estimation of Multicrop Production Functions. *American Journal of Agricultural Economics* 65:770–780.

Kim, H., K. Konyar, and K. Sargent. 2002. Economic Viability of Bt-Corn in the U.S. Paper presented at the American Agricultural Economics Association Annual Meeting. July 28–31, 2002, Long Beach, CA.

Knapp, K., and K. Schwabe. 2008. Spatial Dynamics of Water and Nitrogen Management in Irrigated Agriculture. *American Journal of Agricultural Economics* 90:524–539.

Konyar, K. 2001. Assessing the Role of US Agriculture in Reducing Greenhouse Gas Emissions and Generating Additional Environmental Benefits. *Ecological Economics* 38:85–103.

Konyar, K., and R.E. Howitt. 2000. The Cost of the Kyoto Protocol to U.S. Crop Production: Measuring Crop Price, Regional Acreage, Welfare, and Input Substitution Effects. *Journal of Agricultural and Resource Economics* 25:347–367.

Konyar, K., and K. Knapp. 1990. Dynamic Regional Analysis of the California Alfalfa Market with Government Policy Impacts. *Western Journal of Agricultural Economics* 15:22–32.

Lewandrowski, J., R.F. Darwin, M. Tsigas, and A. Raneses. 1999. Estimating Costs of Protecting Global Ecosystem Diversity. *Ecological Economics* 29:111–125.

Lichtenberg, E., and D. Zilberman. 1986. The Welfare Economics of Price Supports in U.S. Agriculture. *American Economic Review* 76:1135–1141.

Meko, D., C.W. Stockton, and W.R. Boggess. 1995. The Tree-Ring Record of Severe Sustained Drought. *Water Resources Bulletin* 31:789–801.

Moore, M.R. 1991. The Bureau of Reclamation's New Mandate for Irrigation Water Conservation: Purposes and Policy Alternatives. *Water Resources Research* 27:145–155.

Moore, M.R., N.R. Gollehon, and D.H. Negri. 1993. Alternative Forms for Production Functions of Irrigated Crops. *Journal of Agricultural Economics Research* 44:16–32.

Moore, M.R., and D.H. Negri. 1992. A Multicrop Production Model of Irrigated Agriculture, Applied to Water Allocation Policy of the Bureau of Reclamation. *Journal of Agricultural and Resource Economics* 17:29–43.

NRLC (Natural Resources Law Center, University of Colorado). 2005. Conference Primer (Draft 3.0): Hard Times on the Colorado River: Drought, Growth, and the Future of the Compact. 26th Annual Conference at the Natural Resources Law Center. June 8–10, 2005, Boulder, CO.

Nuckton, C. 1978. *Demand Relationship for California Tree Fruits, Grapes, and Nuts: A Review of Past Studies*. Special Report. Davis, CA: Giannini Foundation of Agricultural Economics, Davis, CA.

Onunkwo, I.M., and J.E. Epperson. 2000. Export Demand for U.S. Walnuts: Impacts of U.S. Export Promotion Programs. *Journal of Food Distribution Research* 31:21–27.

Parrish, D. 2003. Where Has All the Water Gone? Water Marketing and the Colorado River Delta. *Transnational Law and Contemporary Problems* 13:369–396.

Pitt, J. 2001. Can We Restore the Colorado River Delta? *Journal of Arid Environments* 49:211–220.

Ray, S.C. 1982. A Translog Cost Function Analysis of U.S. Agriculture, 1939–77. *American Journal of Agricultural Economics* 64:490–498.

Rendleman, M. 1993. *Estimation of Aggregate U.S. Demands for Fertilizer, Pesticides, and Other Inputs*. Technical Bulletin No. 1813. Washington, DC: U.S. Department of Agriculture, Economic Research Service, Resources and Technology Division.

Ribaudo, M., C. Osborn, and K. Konyar. 1994. Land Retirement as a Tool for Reducing Agricultural Nonpoint Source Pollution. *Land Economics* 70:77–87.

Russo, C., R. Green, and R. Howitt. 2008. *Estimation of Supply and Demand Elasticities of California Commodities*. Working Paper No. 08-001. Davis: University of California, Department of Agricultural and Resource Economics.

Schaible, G.D., C.S. Kim, and N.K. Whittlesey. 1991. Water Conservation Potential from Irrigation Technology Transitions in the Pacific Northwest. *Western Journal of Agricultural Economics* 16:194–206.

Schneider, A.D., and T.A. Howell. 1997. Methods, Amounts and Timing of Sprinkler Irrigation for Winter Wheat. *Transactions of the American Society of Agricultural Engineers* 40:137–142.

Seager, R., M. Ting, I. Held, Y. Kushnir, J. Lu, G. Vecchi, H. Huang, N. Harnik, A. Leetmaa, N. Lau, C. Li, J. Velez, and N. Naik. 2007. Model Projections of an Imminent Transition to a More Arid Climate in Southwestern North America. *Science* 25:1181–1184.

Seung, C.K., T.R. Harris, J.E. Englin, and N. Netusil. 2000. Impacts of Water Reallocation: A Combined Computable General Equilibrium and Recreation Demand Model Approach. *Annals of Regional Science* 34:473–487.

Stockton, C.W., and G.C. Jacoby. 1976. *Long-Term Surface Water Supply and Streamflow Levels in the Upper Colorado River Basin*. Lake Powell Research Project Bulletin 18. Los Angeles: University of California, Institute of Geophysics and Planetary Physics.

Sunding, D., D. Zilberman, R. Howitt, A. Dinar, and N. McDougall. 2002. Measuring the Costs of Reallocating Water from Agriculture: A Multi-model Approach. *Natural Resource Modeling* 15:201–225.

Taylor, C.R., and R. Howitt. 1993. Aggregate Evaluation Concepts and Models. In *Agricultural and Environmental Resource Economics*, edited by G.A. Carlson, D. Zilberman, and J.A. Miranowski. Oxford: Oxford University Press, 142–174.

Taylor, C.R., J.B. Penson, E.G. Smith Jr., and R.D. Knutson. 1991. Economic Impacts of Chemical Use Reduction on the South. *Southern Journal of Agricultural Economics* 23:15–24.

Ulibarri, C.A., H.S. Seely, and D.B. Wills. 1998. Farm Profitability and BuRec Water Subsidies: An LP Look at a Region. *Contemporary Economic Policy* 16:442–451.

USBR (U.S. Department of the Interior, Bureau of Reclamation). 2008a. *Colorado River Accounting and Water Use Report Arizona, California, and Nevada*. Boulder City, NV: USBR, Lower Colorado Region Boulder Canyon Operations Office.

———. 2008b. *Upper Colorado River Basin Consumptive Uses and Losses Report: 2006–2010 (Provisional)*. Salt Lake City: U.S. Department of the Interior.

USDA (U.S. Department of Agriculture). 2004. *Farm and Ranch Irrigation Survey: 2003*, vol. 3, *Special Studies, Part 1 of the 2002 Census of Agriculture*. AC-02-SS-1. Washington, DC: USDA, National Agricultural Statistics Service.

USDA ERS (U.S. Department of Agriculture, Economic Research Service). 2009. Data Sets: Farm Program Acres. http://ers.usda.gov/Data/BaseAcres/Download.aspx (accessed September 26, 2010).

USDA NASS (U.S. Department of Agriculture, National Agricultural Statistics Service). 1998. *Census of Agriculture: Farm and Ranch Irrigation Survey*. www.agcensus.usda.gov/Publications/1997/Farm_and_Ranch_Irrigation_Survey/index.asp (accessed September 26, 2010).

Vaux, H.J. Jr. 1986. Economic Factors Shaping Western Water Allocation. *American Journal of Agricultural Economics* 68:1135–1142.

Wahl, Richard W. 1989. *Markets for Federal Water: Subsidies, Property Rights, and the Bureau of Reclamation*. Washington, DC: Resources for the Future.

Woodhouse, C. 2008. Paleo and Historical Variability. Paper presented at the Border-Area Climate Change Impacts and Water Sector Adaptation Workshop. April 1–2, 2008, Tucson, AZ.

Woodhouse, C.A., S.T. Gray, and D.M. Meko. 2006. Updated Streamflow Reconstructions for the Upper Colorado River Basin. *Water Resources Research* 42:W05415.

You, Z., J. Epperson, and C. Huang. 1996. A Composite System Demand Analysis for Fresh Fruits and Vegetables in the U.S. *Journal of Food Distribution Research* 27:11–22.

Young, C.E., D.W. Skully, P.C. Westcott, and L. Hoffman. 2005. *Economic Analysis of Base Acre and Payment Yield Designations under the 2002 U.S. Farm Act*. Economic Research Report No. ERR-12. Washington, DC: U.S. Department of Agriculture, Economic Research Service.

Young, R.A. 1995. Coping with Severe Sustained Drought on the Colorado River: Introduction and Overview. *Water Resources Bulletin* 31:779–788.

CHAPTER 6

Climate, Water Availability, Energy Costs, and National Park Visitation

George B. Frisvold, Xudong Ma, and Srinivasa Ponnaluru

National parks in the U.S Southwest are areas of spectacular beauty, rich in unique biological, cultural, scientific, and historical significance. In this chapter, we use the term "national park" generically for any site in the National Park System, including national parks, monuments, recreation areas, historic sites, and so on. National parks make important contributions to local rural economies. In 2007, southwestern national parks received more than 28 million recreation visits. Visitors spent roughly $1.5 billion in local gateway communities, supporting more than 32,000 jobs (calculated from Stynes 2008).

Many are increasingly concerned, however, that rapid population growth, drought, climate change, and other environmental stressors may place park resources at risk (Wall 1998; Burns et al. 2003; Saunders et al. 2006; Mitchell 2006; NPCA 2007; Uscher 2008). Surprisingly, relatively little research has been done to examine the potential impacts of climate change on outdoor recreation in general (Cline 1992; Mendelsohn and Markowski 1999; Loomis and Crespi 1999; Morris and Walls 2009) and national park visitation in particular (Weiler et al. 2002; Richardson and Loomis 2004; Loomis and Richardson 2006; Jones and Scott 2006; Scott et al. 2007). Researchers have done extensive work considering the economic impacts of national park visitation on local economies (e.g., Stynes et al. 2000; Stynes 2008). With a few exceptions (Johnson and Suits 1983; Wicks et al. 1994; Hanink and White 1999; Hanink and Stutts 2002; Weiler and Seidl 2004; Weiler 2006; Pergams and Zaradic 2006), however, there has been little estimation of factors affecting demand for park visitation in the first place.

This chapter uses multivariate regression to estimate factors affecting annual visitation to southwestern national parks—those in New Mexico, Arizona, and southern parts of California, Nevada, and Utah. We model demand for park visits as a function of spatial demand factors, park attributes, climate, and water availability. We then use input-output analysis to translate changes in visits into changes in visitor spending, then into changes in local sales, personal income, and employment. Our study extends previous literature on demand for national park visitation and adds to the literature on the impacts of climate change on outdoor recreation. It also considers the impact of climate mitigation policy on park visitation and the local economies of park gateway communities. Our estimated elasticity of visits with respect to gasoline price is -0.18. Thus we examine how mitigation policies might affect visits via changes in gasoline prices.

Previous research has suggested that climate change would have a net positive effect on outdoor recreation (Mendelsohn and Markowski 1999; Loomis and Crespi 1999) and park visitation (Weiler et al. 2002; Richardson and Loomis 2004; Loomis and Richardson 2006; Jones and Scott 2006; Scott et al. 2007). This research has involved using visitation regression results to conduct comparative static exercises, simulating response to some hypothesized change in climate; conducting visitor surveys to elicit contingent responses to climate change; or combining and comparing both methods. The pure effect of higher temperatures appears to be positive (at least, if temperature increases are not too extreme; see Richardson and Loomis 2004). Reduced snowfall discourages skiing (Cline 1992; Mendelsohn and Markowski 1999; Chapter 7 of this volume), but warmer temperatures appear to contribute to a net increase in total recreation and park visitation.

Changes in temperature or precipitation will not be the only impacts of climate change. The ecology and resources of parks also will change, as will the number and types of plant and animal life forms. Streamflows and lake levels will change as well, affecting fishing and water-based recreation. Visitor response to park ecology and hydrology is more difficult to measure than responses to climate variables, and changes in ecology and hydrology are harder to predict.

A few studies have attempted to consider both temperature and precipitation changes and resource quality changes. Even accounting for changes in resource quality, the effects of climate change appear to be positive, up to a point (Scott et al. 2007; Richardson and Loomis 2004). Scott et al. found that visitor responses to ecological changes were slightly positive for small changes associated with near-term climate change, but they grew increasingly negative with extreme changes associated with more distant climate change. Examining demand for reservoir recreation, Loomis and Crespi (1999) considered the positive effect of higher temperature together with the negative effect of reduced lake surface areas and found that the net effect on reservoir recreation was positive.

In our study, we conducted comparative static exercises examining the impacts of long-term average temperature increases of 1 degree F for July and for January. These changes were considered alone and, where relevant, in combination with reductions in lake surface areas. The effects of warmer temperatures depend on when and where they occur. Winter warming tends to increase visits, except in warmer, low-elevation, lower-latitude parks (low desert parks). These low desert parks, such as Saguaro National Park in the Sonoran Desert or Joshua Tree National Park in the Mojave Desert, tend to lose visitors with temperature increases. Although cooler parks may benefit from extended seasons—warmer winters, earlier springs, later falls—low desert parks are already quite warm in winter. Beyond a threshold, greater heat only discourages visits. These findings are consistent with those of Richardson and Loomis (2004), who found negative visitor responses to an "extreme heat" scenario. Summer warming tends to decrease visits, except in some higher-elevation and higher-latitude parks. The combined effects of summer and winter warming decreased total regional visits but actually led to an increase in visitor spending. This occurs because spending patterns vary by park. Visitor spending increases in some parks outweighed spending declines in others. The projected increase in annual visitor spending was a modest \$7.8 million, from a baseline of \$1.2 billion.

As a result of recent droughts, the surface area of Lake Powell decreased by 29% and that of Lake Mead by 24% from 1998 to 2003. Declining reservoir levels reduced visits and accounted for a loss of more than 700 jobs in addition to \$29 million in lost sales and \$10 million in lost personal income. Assessing the effects of climate change on reservoir recreation is complicated, because there is a high degree of uncertainty about climate change impacts on water runoff and reservoir inflows, and reservoirs are managed systems that depend on human reactions to water shortages. Because visits to Lakes Powell and Mead account for a large share of regional park system visits (36%), results suggest that falling lake levels could substantially reduce total visits to the region.

MODEL ESTIMATION

Data in this study come from 42 southwestern national parks over 25 years, from 1979 to 2003. This provides 1,050 observations of time-series–cross-section (TSCS) data. The multivariate regression equation specification is

$$Y_{it} = \alpha + \beta X_{it} + \gamma W_t + \delta Z_i + \varepsilon_{it} \tag{6.1}$$

where Y_{it} is (the log of) annual visits to park i in year t. X_{it} is a vector of explanatory variables that vary over both space and time. W_t represents variables that vary over time, but not across parks. This includes variables such as regional population or gasoline prices. Z_i represents variables that vary across parks but are time-invariant. This includes fixed park attributes, such as the year the park was established, local topography, or distance from other parks. To correct for biases introduced by panel heteroskedasticity, contemporaneous correlation of the errors, and serially correlated errors, we estimate equation (6.1) using panel-corrected standard errors (PCSEs) and correction for first-order autocorrelation (Beck and Katz 1995; Beck 2001).[1]

DATA AND VARIABLES

Data on annual park visitation come from the National Park Service Public Use Statistics Office (NPS n.d.). The units in the southwestern states of our study include 24 national monuments (NMs), 10 national parks (NPs), 3 national historical parks (NHPs), 2 national historic sites (NHSs), 2 national recreation areas (NRAs), and 1 national memorial (NMem), as listed in Table 6.1. The top 10 parks accounted for 80% of annual recreation visits in 2003, while the top 5 accounted for

Table 6.1 *Recreation Visits to Southwestern National Parks in Sample, 2003*

Park	*Recreation visits*
Lake Mead NRA	7,915,581
Grand Canyon NP	4,124,900
Zion NP	2,458,792
Glen Canyon NRA	1,876,984
Joshua Tree NP	1,283,346
Bryce Canyon NP	903,760
Canyon de Chelly NM	866,498
Arches NP	757,781
Saguaro NP	643,697
Montezuma Castle NM	637,024
Petrified Forest NP	586,245
Cedar Breaks NM	569,455
Capitol Reef NP	535,441
White Sands NM	492,008
Carlsbad Caverns NP	457,631
Canyonlands NP	386,986
Bandelier NM	287,096
Organ Pipe Cactus NM	277,566
Wupatki NM	266,159
Hubbell Trading Post NHS	163,883
Sunset Crater Volcano NM	159,073
Tuzigoot NM	115,216
Walnut Canyon NM	111,465
Natural Bridges NM	98,874
Rainbow Bridge NM	98,865
Coronado NMem	89,309
Casa Grande Ruins NM	87,687
Chaco Culture NHP	84,621
Navajo NM	76,620
Capulin Volcano NM	61,373
Tonto NM	59,216
El Morro NM	57,889
Pipe Spring NM	56,341
Tumacacori NHP	52,393
Gila Cliff Dwellings NM	47,869
Chiricahua NM	44,976
Aztec Ruins NM	42,493
Pecos NHP	38,326
Salinas Pueblo Missions NM	33,827
Hovenweep, NM	29,737
Fort Union NM	12,944
Fort Bowie NHS	8,445
Total	26,958,392

Source: NPS (n.d.)

nearly two-thirds of visits. The two sites with lakes—Lake Mead National Recreation Area (NRA) and Glen Canyon NRA (home of Lake Powell)—accounted for 36% of visits, illustrating the importance of water-based recreation.

We can group explanatory variables used in the regression analysis into three categories: demand factors, park attributes, and climate/environmental variables. Table 6.2 reports descriptive statistics for the model's explanatory variables. Demand factors include variables for the regional price of gasoline, the exchange rate, and regional population, as well as a Competing Destination Index and a Market Potential Index. The variable *lnGas* is the log of the U.S. average price of gasoline taken from the *Short Term Energy Outlook* (EIA various dates). Prices were deflated using the Consumer Price Index for urban consumers in the Western Region (BLS various dates). Previous research has found fuel prices to be significantly and negatively associated with national park visits (Johnson and Suits 1983; Pergams and Zaradic 2006). The variable *lnXRate* is the log of the real exchange rate weighted by airline passenger fares, published by the Board of Governors of the Federal Reserve System (Goldberg 2004). This variable is meant to capture relative price effects on foreign versus domestic travel. Tourism demand studies usually include some measure of exchange rates between source and competing target destinations (Witt and Witt 1991). Exchange rate fluctuations affect the costs of traveling to the United States. For example, a weak dollar may

Table 6.2 *Descriptive Statistics for Variables Used in Park Visitation Regression*

		Mean	*Standard deviation*	*Minimum*	*Maximum*
lnVisit	ln (annual visits)	12.163	1.525	8.50	16.10
lnGas	ln (real gasoline price)	5.119	0.243	4.78	5.66
lnXRate	ln (real exchange rate)	4.693	0.115	4.54	4.93
lnPop	ln (regional population)	17.451	0.143	17.20	17.68
Post911	Post 9/11 dummy variable	0.092	0.274	0	1
Petrified	Change in counting methods	0.006	0.075	0	1
Sunset	Change in counting methods	0.008	0.087	0	1
Casa	Change in counting methods	0.002	0.044	0	1
Dry	Year with SPI < -2	0.01	0.10	0	1
Cerro	Dummy for fire year	0.00	0.03	0	1
Tonto	Dummy for road closures	0.00	0.03	0	1
CDIndex	Competing Destination Index	0.160	0.040	0.09	0.28
MPIndex	Market Potential Index (000)	1,276	2,731	306	25,529
Estab	Year NPS site established	1929	19	1906	1966
TopoVar	Topographic variation index	0.477	0.656	−0.44	1.84
Lodging	Concessional lodging dummy	0.119	0.324	0	1
Area	Park area	152,369	349,155	40	1,495,664
NP	National park	0.208	0.406	0	1
NHP	National historical park	0.049	0.216	0	1
NHS	National historical site	0.059	0.235	0	1
JanTemp	Average January temperature	35.7	8.3	21.7	54.7
JulTemp	Average July temperature	76.8	7.7	60.1	94.2
lnPowell	ln (Lake Powell surface area)	0.28	1.80	0.00	11.94
lnLMNRA	ln (Lake Mead NRA surface area)	0.28	1.82	0.00	11.97

reduce the costs of a U.S. vacation for European or Japanese visitors. If the dollar depreciates, this means that domestic travel (e.g., to U.S. national parks) is relatively less expensive for U.S. citizens. Conversely, a strong dollar relative to the peso, for example, may encourage trips to Mexico. An exchange rate weighted by aggregate trade may not accurately reflect differences in travel-related exchange rates, because such an aggregate weight would emphasize manufacturing as opposed to travel purchases. This index is a real index, adjusted for inflation in the United States and among trading partners (Goldberg 2004).

Hanink and White (1999) and Hanink and Stutts (2002) examined spatial aspects of demand for national park visits. They applied two types of indices: a Competing Destination Index (*CDIndex*) and a Market Potential Index (*MPIndex*). The *CDIndex* measures a park's spatial relationship to other parks, expressed as

$$CDIndex_i = \sum_{j=1}^{n} 1/D_{ij} \tag{6.2}$$

where D_{ij} is the distance of park i from park j. The closer a park is to other parks in the area considered, the greater the *CDIndex* will be. Isolated parks will have a low *CDIndex*. Both Hanink and White (1999) and Hanink and Stutts (2002) argued that if visits to parks were complementary, the regression coefficient for *CDIndex* would be positive. A negative coefficient would imply that visits to individual parks substituted for each other. Hanink and both associates found negative coefficients for *CDIndex*, implying that parks substitute for each other. Hanink and Stutts confined their study to Civil War national battlefield parks, whereas Hanink and White focused on major, high-volume parks with facilities for overnight stays. They did not examine possible complementarities between these high-volume parks and other, nearby parks. The *CDIndex* was constructed by Ponnaluru (2005), who calculated the index using distances between all southwestern parks in the study from online driving distance calculations provided by the web tool of the American Automobile Association (AAA).

The Market Potential Index (*MPIndex*) measures a park's location relative to areas of potential market demand:

$$MPIndex_i = \sum_{j=1}^{n} Y_j/D_{ij} \tag{6.3}$$

where Y_j is real personal income in a metropolitan or micropolitan statistical area and where D_{ij} is the distance from that area and park i. Thus *MPIndex* measures a park's location in relationship to sources of demand for visits. Hanink and both associates used population instead of income to capture demand. In our study, however, personal income is used to capture both the effects of population and income on demand for visits. We would expect both to be important. Wicks et al. (1994) found personal income to be important in explaining visits to Everglades National Park. Again, we rely on Ponnaluru (2005), who constructed an *MPIndex* for southwestern parks using the AAA driving distance data combined with data from the Regional Economic Information System of the Bureau of Economic Analysis on personal income for all metropolitan and micropolitan regions of the Southwest.

We also account for the effects of population growth separately by including the log of regional population (*lnPop*) as an explanatory variable. Data come from the U.S. Bureau of Census and include populations of Arizona, New Mexico, as well as southern counties in Utah, Nevada, and California.

Different attributes of parks will also affect visitor demand. Previous studies have found park area (in square miles) to be significantly and positively associated with visits (Hanink and White 1999; Weiler and Seidl 2004; Weiler 2006). Data on park area are available from the National Park Service (NPS various dates). Both park area (*Area*) and its squared term (*AreaSq*) were used in the regressions.

Hanink and White (1999) and Hanink and Stutts (2002) have used park age as an indicator of the quality or uniqueness of park attributes. They argued that many of the most spectacular parks in the United States (e.g., Yosemite, Grand Canyon, and Yellowstone) were established first. They included park age as an explanatory variable, and it had a positive and significant effect in both studies. Rather than use age, we designate the year a park was first established. If we follow the argument that year of establishment relative to other parks is a signal of quality, then we might expect that it is the relative age of parks that matters, not their absolute age. Data on year of park establishment were obtained from the National Park Service (NPS 2003).

A variable (*Post911*) was included to account for the effects of the aftermath of the 9/11 attacks in the United States. Weiler (2006) notes that this could have altered visitation, and our own conversations with National Park System managers suggested a post-9/11 effect. The variable took on a value of 0.306 for the year 2001 (to account for the number of days from September 11 to December 31, 2001), 1 for the year 2002, and 0 otherwise.

The variable *TopoVar* is an index measuring topographical variation. Developed by McGranahan (1999), the variable is an index measuring diversity of landforms in each U.S. county. Each park was assigned a *TopoVar* value based on the county where it is located (or principally located). McGranahan found that topographical diversity helped explain in-migration to rural counties, arguing that greater diversity was a desirable natural feature. We include it here as a potentially important, positive park attribute. Puustinen et al. (2009) found topographical features to be important predictors of visits to national parks in Finland.

The variable *Lodging* is a dummy variable indicating whether a park has concessional lodging within its borders. Lodging within a park is likely an attractive feature. It may also be a signal of quality. The National Park Service (NPS) and concessionaires may desire to build lodging in attractive parks with high visitor demand. This raises a question about whether the variable might be endogenous. However, construction of concessional lodging predates our survey data, often by several decades, so it is strongly predetermined. Because this variable is a proxy for quality attributes and not just the presence of lodging amenities, one should take care in interpreting the regression coefficient for this variable.

Weiler and Seidl (2004) and Weiler (2006) have examined the effect of National Park System designation on annual visitation. The NPS has 16 different designations for sites, such as national monument, national park, national historical park, and so forth. The researchers estimated time-series–cross-section regressions for eight parks that changed designations between 1979 and 2000, treating redesignation from national monument to national park as a natural experiment. National park designation, they argue, may be an important signal of quality to potential visitors, particularly out-of-state or foreign visitors. Their results are consistent with this argument, finding that, controlling for other factors, national park designation significantly increased the number of visits.

In our analysis, we include dummy variables for national park (*NP*), national historical park (*NHP*) and national historic site (*NHS*) status to distinguish sites from national monuments and the one national memorial in the sample. Although some sites change designation, most do not. So the impact of a change in designation is mixed with any long-term differences that site designation might imply. The NPS designates system sites with lakes as national recreation areas. Instead of a single static variable, we include measures of lake surface area (discussed below).

The last group of dummy variables accounts for park anomalies. *Tonto* indicates the year 1990, when construction led to driving delays and closures of roads leading to the Tonto National Monument. Visits during that year were 25% lower than 1989 visits and 27% lower than 1991 visits. *Cerro* indicates the year when the Cerro Grande fire took place in Bandelier National Monument. The variables *Petrified*, *Sunset*, and *Casa* mark years when structural changes in visitor counting methods and protocols were put in place at Petrified Forest National Park, Sunset Crater Volcano National Monument, and Casa Grande Ruins National Monument.

Climate/environmental variables were the third category used in the regression. Following Mendelsohn and Markowski (1999) and Jones and Scott (2006), we used long-term winter and summer temperature averages to capture long-run adjustments to climate. These included average January temperature and temperature squared (*JanTemp*, *JanTempSq*) and July temperature and temperature squared (*JulTemp*, *JulTempSq*). Data for these variables come from the Western Regional Climate Center (WRCC), which reports 1971–2000 average monthly temperatures for its sites. NPS sites usually have WRCC weather-monitoring sites within their boundaries. When this was not the case, we chose the nearest site to the park visitor center.

We also experimented with different variables to capture the possible effects of drought or dryness on park restrictions to curb fire risk. None of the variables tried were statistically significant. One variable, *Dry*, was included in the final regression. This was a variable based on the December 12-month Standardized Precipitation Index (SPI) reported by the National Oceanic and Atmospheric Administration (NOAA). The SPI measures precipitation in an area in relation to historical averages and is standardized to range -4 to 4, with 0 as a normal year. When SPI < -2, the year is designated as "extremely dry." *Dry* equals 1 if the SPI < -2 and 0 otherwise.

Lake Mead NRA and Glen Canyon NRA, home of Lake Powell, along with Grand Canyon NP and Zion NP, are the four most visited National Park Service sites in the Southwest (Table 6.1). Following Loomis and Crespi (1999) and Ward et al. (1995), we specified visits to these lakes as a function of lake surface area. With declining surface area, marinas and launches may be closed. There may be more hazards to boaters and skiers, and water-based recreation can become more congested. Conversely, lower lake levels can expose previously submerged attractions, such as unique archeological or geological features. Thus lower lake levels could be an attraction to some. Separate variables *lnPowell* and *lnLMNRA* were included in the regressions, where *lnPowell* is the log of Lake Powell's surface area, and *lnLMNRA* is the log of the combined surface areas of Lake Mead and Lake Mohave, both of which are part of the Lake Mead NRA. When both lakes are at capacity, Lake Mohave's surface area is about 18% of that of Lake Mead. In times of drought, however, it can rise to about 28% of Lake Mead's surface area because of Bureau of Reclamation management of reservoir levels. The hypothesis of a single surface area coefficient for Glen Canyon and Lake Mead NRAs was rejected using a log-likelihood test, so results with separate coefficients are reported. We also experimented with a specification that allowed the temperature variables to interact with dummy variables indicating the presence of a lake. Based on a log-likelihood test, we could not reject the null hypothesis that temperature affected sites with lakes no differently than it did other sites.

REGRESSION RESULTS

The overall model fit of the regression is good ($R^2 > 0.96$) (Table 6.3). The hypothesis of panel homoskedasticity was rejected using a Lagrange multiplier test (Greene 2003); the hypothesis of no contemporaneous correlation of the errors was rejected using a Breusch-Pagan Lagrange multiplier statistic (Breusch and Pagan 1980); and the hypothesis of serially uncorrelated errors was rejected based on a Durbin-Watson test. Thus the model was estimated using panel-corrected standard errors and correction for first-order autocorrelation (AR(1)) (Beck and Katz 1995).

The coefficient for *lnGas*, the elasticity of visits with respect to the price of gasoline, is -0.18 (Table 6.3), implying that a 10% increase in the price of gas reduces visits 1.8%. The negative coefficient for *lnXRate* implies that a real appreciation of the dollar has a negative impact on park visits, as expected. The coefficient for *lnPop* is close to 1, meaning visits increase at the same rate as regional population. This is an important result, given that the population of the study area is projected to grow by more than 45% from 2005 to 2030 and more than 75% from 2005 to 2050. It suggests that the effect of population growth would likely swamp other economic or environmental effects.

The Competing Destination Index, *CDIndex*, coefficient is positive, suggesting complementarities among parks. This result differs from Hanink and White (1999), who found this coefficient to be negative. One possible reason for the difference is that their study included only major, high-volume parks and not

Table 6.3 *Annual Recreation Visits: Regression Estimates with Panel-Corrected Standard Errors and AR(1) Correction*

Dependent variable: ln (annual recreation visits)

Variable	*Definition*	*Coefficient*	*Standard deviation*	*Significance*
lnGas	ln (real gasoline price)	−0.18387	0.0595898	0.2%
lnXRate	ln (real exchange rate)	−0.276949	0.1012295	0.6%
lnPop	ln (regional population)	0.9964413	0.2239983	0.0%
Post911	Post 9/11 dummy variable	−0.1274745	0.0454946	0.5%
CDIndex	Competing Destination Index	10.60369	0.8020792	0.0%
MPIndex	Market Potential Index	4.02E-08	1.83E-08	2.8%
Estab	Year NPS site established	−0.0136317	0.0017216	0.0%
TopoVar	Topographic variation index	0.3262711	0.0691121	0.0%
Lodging	Concessional lodging dummy	1.910868	0.1229902	0.0%
Area	Park area	8.54E-06	6.41E-07	0.0%
AreaSq	Park area squared	−6.46E-12	5.21E-13	0.0%
Petrified	Change in counting methods	−0.03677	0.0752489	62.5%
Sunset	Change in counting methods	−0.8013567	0.1131472	0.0%
Casa	Change in counting methods	−0.3171525	0.1051311	0.3%
NP	National park	0.7198789	0.1168318	0.0%
NHP	National historical park	−0.4543525	0.1200482	0.0%
NHS	National historical site	−0.5737084	0.1577949	0.0%
JanTemp	Average January temperature	0.1690673	0.0532566	0.2%
JanTempSq	Avg. January temperature squared	−0.0018679	0.0007252	1.0%
JulTemp	Average July temperature	0.2211722	0.0878374	1.2%
JulTempSq	Avg. July temperature squared	−0.0015441	0.0005926	0.9%
lnPowell	ln (Lake Powell surface area)	0.1056788	0.0176344	0.0%
lnLMNRA	ln (Lake Mead NRA surface area)	0.4451444	0.0264338	0.0%
Dry	Year with SPI < -2	−0.0203223	0.0304931	50.5%
Cerro	Dummy for fire year	−0.2612433	0.1246288	3.6%
Tonto	Dummy for road closures	−0.3022662	0.082419	0.0%
Intercept		9.229082	6.478796	15.4%

Notes: number of observations = 1,050; adjusted $R^2 = 0.9627$

clusters of parks situated more closely together. The Market Potential Index (*MPIndex*) coefficient was positive. This suggests that proximity to sources of market demand encourages visits. This index is personal income of metropolitan and micropolitan areas in the Southwest, weighted by their distance from a park. This index increases as personal incomes in the area increase. Thus a positive coefficient appears to differ from Weiler (2006), who found park visits to be an inferior good, declining with rising national- and state-level income, and Wicks et al. (1994), who found it to be decreasing in regional income. The *MPIndex* shows small temporal variation relative to spatial variation, so it might be better thought of as Hanink and White (1999) considered it, a measure of proximity to sources of demand. Topographical variation of the park and its environs (*TopoVar*) has a positive effect on visits, as does having concessional lodging. The coefficient on concessional lodging (*Lodging*) is quite large, but one must take care interpreting it. As noted earlier, decisions were made years ago to build lodging within parks with more desirable attributes. For this reason, we interpret *Lodging* as capturing both the effects of facilities for overnight stays and unobserved aspects of quality. Thus, although the coefficient for *Lodging* is 1.9, we would not expect that constructing new lodging at a site where none previously existed would more than double visits.

National Park System designation also matters as a signal of quality. Weiler and Seidl (2004) and Weiler (2006) found a national park designation has a strong positive effect on visits. We also found that compared with national monuments (the default), visits were less at national historical sites or national historical parks.

Visits first increase, then decrease in January temperature (based on *JanTemp* and *JanTempSq*). Warmer Januaries appear to encourage visits up 45 degrees F, then have negative effects. Warmer July temperatures also appear to encourage visits up 72 degrees F, then have negative effects.

The elasticity of visits to Glen Canyon NRA with respect to Lake Powell surface area was about 0.11, while for Lake Mead NRA, the elasticity with respect to the combined surface areas of Lakes Mead and Mohave was about 0.45. Loomis and Crespi (1999) estimated monthly visits to nine U.S. Army Corps of Engineers reservoirs in California and found an elasticity of reservoir recreation visits with respect to lake surface area of 0.39. In a study estimating visits to 115 Corps of Engineers sites, Ward et al. (1995) estimated reservoir-visit elasticity with respect to surface area of 0.47.

The coefficient for *Dry*, the variable indicating a year with a SPI < -2, was negative, but statistically insignificant. We experimented with different discrete and continuous variables based on either the Standard Precipitation Index (SPI) or the Palmer Drought Severity Index. Although the coefficients of these variables were negative, indicating a negative association between drought and visits, none proved statistically significant.

The dummy variable *Cerro* accounts for the Cerro Grande fire, which burned many acres within the Bandelier National Monument and surrounding areas in 2000, while the dummy variable *Tonto* accounts for road construction delays and closures decreasing visits to Tonto National Monument in 1990. Regression results suggest that controlling for other factors, the Cerro Grande fire accounted for a 23% drop in visits in 2000. Since 1994, visits have been trending downward at Bandelier. Actual 2000 visits were 25% below 1999 levels and 21% below 2001 levels. The coefficient for *Tonto*, intended to capture effects of road closures, was also negative and significant.

COMPARATIVE STATICS

Next, we use regression coefficients to translate changes in January and July temperatures to percent changes in visits to each park. We then convert visits to party days, visitor spending, and local economic impacts using the Money Generation Model 2 (MGM2), developed by Stynes et al. (2000) for the NPS. MGM2 is an input-output model that translates visits into party days and spending for different visitor segments. Segments include local and nonlocal visitors on day trips, campers, and visitors using hotels. Different types of visitors have different spending patterns, so the share of visitors in each segment affects overall spending. Converting visits to party days is important, because the latter measures the number of days (and nights) visitors are in the area, as opposed to the number of times parties enter a park. Visitor spending depends on the number of days a party spends in the area and the size of the party more than the number of times individuals enter a park. Visitor spending per party day estimates were derived directly from surveys of visitors and park managers or based on spending patterns from similar parks.

Visitor spending overstates the amount of money that visits inject into the local economy. Local sales include only local retail markups (or wholesale markups, where appropriate). Multipliers are derived from input-output models such as IMPLAN or RIMS II (Rickman and Schwer 1995).[2] Visitor spending generates local sales, which in turn support local jobs and provide personal income. The jobs measure includes all jobs, making no distinction between full-time and part-time or seasonal jobs. Personal income includes wages, salaries, and benefits, as well as proprietor's income. MGM2 uses the more conservative IMPLAN Type SAM multipliers instead of Type II multipliers (Stynes et al. 2000; Alward and Lindall 1996).

MGM2 reports both direct effects of visitor spending and multiplier effects. Multiplier effects may be divided into indirect effects (changes in sales, income, and jobs in backward-linked industries) and induced effects (changes in sales, income, and jobs from changes in household income). Indirect effects include demand generated for inputs used to deliver goods and services to visitors. Induced effects are the multiplier effects from increased income from visitor-related wages and profits. For example, more visitors can increase demand for hotel workers, who in turn spend part of their incomes locally.

CLIMATE CHANGE SCENARIO

We conduct a comparative static exercise, simulating the effects of a 1 degree F increase in July and January temperatures on the percent change in annual visits. These temperature increases are projected onto the observations at a 2003 baseline. Studies suggest warming of this magnitude over the next 20 years in the Southwest (Hayhoe et al. 2004; NMOSE 2006; Christensen and Hewitson 2007; NMBGMR 2007; Ray et al. 2008). Previous research has found warmer temperatures to have a net positive impact on outdoor recreation (Mendelsohn and Markowski 1999; Loomis and Crespi 1999) and park visitation (Weiler et al. 2002; Richardson and Loomis 2004; Loomis and Richardson 2006; Jones and Scott 2006; Scott et al. 2007).

Our simulation results are similar in some ways to these previous findings, with some important exceptions. The simulated temperature increases had a positive effect on visits to some parks, but negative effects on others. While 27 parks experienced increased visitation, 15 had reduced visitation. Whereas visits to higher-elevation and higher-latitude parks increased, visits to low-elevation low desert parks declined with the temperature increases (Table 6.4). For

Table 6.4 *Estimated Change in Visits and Visitor Spending with a 1°F July and a 1°F January Temperature Increase*

		Change in		
Park	*Percent change in visits*	*Visits*	*Party days*	*Visitor spending ($ million)*
Arches NP	1.9%	14,763	7,235	$1.2
Aztec Ruins NM	3.9%	1,667	580	$0.0
Bandelier NM	7.7%	22,056	9,761	$0.9
Bryce Canyon NP	11.9%	107,864	63,015	$4.9
Canyon de Chelly NM	2.9%	25,383	13,062	$1.2
Canyonlands NP	4.3%	16,585	9,968	$0.7
Capitol Reef NP	4.7%	25,056	14,324	$1.1
Capulin Volcano NM	5.6%	3,412	1,267	$0.1
Carlsbad Caverns NP	− 1.8%	− 8,385	− 3,990	− $0.3
Casa Grande Ruins NM	− 7.8%	− 6,883	− 2,753	− $0.2
Cedar Breaks NM	6.4%	36,367	18,218	$1.1
Chaco Culture NHP	5.9%	4,994	2,071	$0.1
Chiricahua NM	− 0.9%	− 383	− 212	$0.0
Coronado Nmem	− 1.5%	− 1,352	− 676	$0.0
El Morro NM	7.2%	4,150	1,721	$0.1
Fort Bowie NHS	− 4.0%	− 338	− 169	$0.0
Fort Union NM	9.4%	1,218	517	$0.0
Gila Cliff Dwellings NM	3.6%	1,707	698	$0.0
Glen Canyon NRA	− 0.7%	− 12,806	− 9,660	− $0.7
Grand Canyon NP	6.9%	282,970	110,104	$23.2
Hovenweep NM	4.9%	1,455	741	$0.1
Hubbell Trading Post NHS	3.8%	6,252	3,126	$0.3
Joshua Tree NP	− 7.0%	− 90,298	− 42,260	− $3.4
Lake Mead NRA	− 7.7%	− 611,579	− 186,624	− $18.6
Montezuma Castle NM	− 2.6%	− 16,342	− 8,172	− $0.7
Natural Bridges NM	5.7%	5,628	2,885	$0.2
Navajo NM	5.5%	4,207	2,130	$0.2
Organ Pipe Cactus NM	− 8.9%	− 24,647	− 12,785	− $1.1
Pecos NHP	5.9%	2,263	779	$0.0
Petrified Forest NP	2.2%	12,887	7,220	$0.8
Pipe Spring NM	2.0%	1,136	568	$0.0
Rainbow Bridge NM	1.5%	1,510	755	$0.1
Saguaro NP	− 7.1%	− 45,581	− 25,545	− $2.8
Salinas Pueblo Missions NM	1.9%	659	275	$0.0
Sunset Crater Volcano NM	8.4%	13,402	6,702	$0.6
Tonto NM	1.9%	1,154	577	$0.1
Tumacacori NHP	− 4.3%	− 2,277	− 1,139	− $0.1
Tuzigoot NM	− 3.8%	− 4,369	− 2,185	− $0.2
Walnut Canyon NM	7.4%	8,291	4,146	$0.4
White Sands NM	− 0.6%	− 2,787	− 1,039	− $0.1
Wupatki NM	0.6%	1,627	814	$0.1
Zion NP	− 1.7%	− 42,993	− 16,496	− $1.4
Total	− 1.0%	− 262,357	− 30,449	$7.8

high-elevation, high-latitude parks, warmer winters had a positive effect on visits, but they had a negative effect on low desert parks. Although the total number of

Table 6.5 *Local Economic Impacts of a 1°F July and a 1°F January Temperature Increase*

	Direct effects			Total effects		
Park	Sales ($ million)	Jobs	Income ($ million)	Sales ($ million)	Jobs	Income ($ million)
Grand Canyon NP	20.5	413	8.4	29.8	536	11.6
Bryce Canyon NP	4.0	114	1.4	5.3	134	1.8
Arches NP	1.1	29	0.4	1.4	34	0.5
Canyon de Chelly NM	1.0	24	0.3	1.4	30	0.5
Cedar Breaks NM	0.9	26	0.3	1.2	31	0.4
Capitol Reef NP	0.9	25	0.3	1.2	30	0.4
Bandelier NM	0.7	17	0.3	1.0	22	0.4
Petrified Forest NP	0.7	16	0.2	1.0	21	0.4
Canyonlands NP	0.6	16	0.2	0.7	19	0.2
Sunset Crater Volcano NM	0.5	12	0.2	0.7	15	0.3
Walnut Canyon NM	0.3	8	0.1	0.4	9	0.2
Hubbell Trading Post NHS	0.2	6	0.1	0.3	7	0.1
Natural Bridges NM	0.2	5	0.1	0.3	6	0.1
Navajo NM	0.2	4	0.1	0.2	5	0.1
El Morro NM	0.1	3	0.0	0.1	3	0.0
Chaco Culture NHP	0.1	2	0.0	0.1	3	0.0
Wupatki NM	0.1	1	0.0	0.1	2	0.0
Capulin Volcano NM	0.1	2	0.0	0.1	2	0.0
Rainbow Bridge NM	0.1	1	0.0	0.1	2	0.0
Hovenweep NM	0.1	1	0.0	0.1	2	0.0
Tonto NM	0.0	1	0.0	0.1	1	0.0
Pipe Spring NM	0.0	1	0.0	0.1	1	0.0
Aztec Ruins NM	0.0	1	0.0	0.1	1	0.0
Pecos NHP	0.0	1	0.0	0.0	1	0.0
Gila Cliff Dwellings NM	0.0	1	0.0	0.0	1	0.0
Fort Union NM	0.0	1	0.0	0.0	1	0.0
Salinas Pueblo Missions NM	0.0	0	0.0	0.0	1	0.0
Fort Bowie NHS	0.0	0	0.0	0.0	0	0.0
Chiricahua NM	0.0	0	0.0	0.0	0	0.0
Coronado Nmem	0.0	− 1	0.0	0.0	− 1	0.0
Tumacacori NHP	− 0.1	− 2	0.0	− 0.1	− 2	0.0
White Sands NM	− 0.1	− 3	0.0	− 0.1	− 3	0.0
Casa Grande Ruins NM	− 0.1	− 4	0.0	− 0.2	− 5	− 0.1
Tuzigoot NM	− 0.2	− 4	− 0.1	− 0.2	− 5	− 0.1
Carlsbad Caverns NP	− 0.3	− 7	− 0.1	− 0.4	− 9	− 0.1
Glen Canyon NRA	− 0.6	− 14	− 0.2	− 0.9	− 18	− 0.3
Montezuma Castle NM	− 0.6	− 15	− 0.2	− 0.9	− 19	− 0.3
Organ Pipe Cactus NM	− 0.9	− 22	− 0.3	− 1.3	− 28	− 0.5
Zion NP	− 1.1	− 28	− 0.4	− 1.6	− 35	− 0.6
Saguaro NP	− 2.4	− 58	− 0.8	− 3.5	− 74	− 1.2
Joshua Tree NP	− 2.8	− 62	− 1.0	− 4.3	− 81	− 1.6
Lake Mead NRA	− 13.7	− 402	− 4.9	− 18.0	− 468	− 6.3
Total	9.4	110	4.3	14.4	171	6.0

regional visits declined by 1%, total visitor spending actually increased. This occurred because average spending at high-elevation, high-latitude parks (especially the Grand Canyon) tends to be higher than in low desert parks.

Regional visitor spending rose by $7.8 million, from a baseline of $1.2 billion. Accounting for multiplier effects, this increased sales by $14.4 million, employment by 171 jobs, and personal income by $6 million (Table 6.5). Four sites (Grand Canyon NP, Lake Mead NRA, Bryce Canyon NP, and Joshua Tree NP) account for 74% of the absolute value of changes in visitor spending. Twelve parks account for 90% of this change. Future research on climate change and park-based recreation could capture much of the economic impact by focusing on a smaller subset of parks.

IMPACTS OF CHANGES IN LAKE SURFACE AREA

One of the most important ways drought and climate change may affect recreation visits and local economies is by reducing lake levels at Lake Powell and Lake Mead. Results above suggest that visits to Lake Mead National Recreation Area and Glen Canyon National Recreation Area (home of Lake Powell) are sensitive to lake surface area. In this section, we develop estimates of the local economic impacts of recent drought-induced reductions in lake levels. Projections of future regional climate suggest that lake levels may fall even more in the future. We discuss these projections and estimate possible economic impacts in terms of visitor spending and implications for local economies.

Recent Drought Effects

Because of recent droughts, the surface area of Lake Powell fell 29% and the surface area of Lake Mead fell 24% from 1998 to 2003. Lake Mohave's surface area fell only 0.4%. Table 6.6 reports estimates of what visits to Glen Canyon NRA and Lake Mead NRA would have been in 2003 had lake surface areas been at 1998 levels. It thus captures some economic losses from declining lake levels. The total effect of the decline in reservoir levels was $29 million in sales, $10 million in personal income, and a loss of more than 700 jobs. From 2003 to 2008, the surface area of Lake Mead declined further, for a 35% reduction from 1998 to 2008. The other lakes remained essentially unchanged. Extending the analysis to 2008, the total effects of declining lake levels from 1998 to 2008 were a loss of roughly $42 million in sales, $15 million in personal income, and more than 1,000 jobs (Table 6.6).

Future Drought Effects

Levels at Lakes Powell and Mead have fallen farther than previously anticipated by the Bureau of Reclamation, which manages the reservoirs (Garrick et al. 2008), and studies suggest the possibility of significant reduction in lake levels in the Colorado River Basin (Harding et al. 1995; Christensen et al. 2004; Christensen and Lettenmaier 2006; Barnett and Pierce 2009).

Table 6.6 *Economic Impacts of Declining Reservoir Levels: Losses in 2003 and 2008 from Lake Surface Area below 1998 Levels*

	Glen Canyon NRA	*Lake Mead NRA*	*Total*
2003 reductions in			
Visits	69,634	827,662	897,296
Party days	52,530	252,562	305,092
Visitor spending ($ million)	$4.0	$25.1	$29.1
Direct effects (reductions)			
Sales ($ million)	$3.2	$18.5	$21.7
Jobs	79	543	622
Income ($ million)	$1.2	$6.6	$7.8
Total effects (reductions)			
Sales ($ million)	$4.7	$24.3	$29.0
Jobs	99	633	732
Income ($ million)	$1.6	$8.6	$10.2
2008 reductions in			
Visits	67,566	1,273,815	1,341,381
Party days	50,970	388,707	439,677
Visitor spending ($ million)	$3.9	$38.7	$42.6
Direct effects (reductions)			
Sales ($ million)	$3.1	$28.5	$31.6
Jobs	77	836	913
Income ($ million)	$1.1	$10.2	$11.3
Total effects (reductions)			
Sales ($ million)	$4.5	$37.4	$41.9
Jobs	96	974	1,070
Income ($ million)	$1.6	$13.2	$14.8

In the late 1500s, the Colorado Basin was hit by a severe multidecade drought. Harding et al. (1995) examined how the region's hydrology would respond to the recurrence of a drought of that magnitude. Model simulations suggested that in the worst years of the drought, the level of Lake Mead would fall to 1,075 feet, while Lake Powell would fall to the dead pool level, 3,370 feet. The dead pool level is the level below which water can no longer be extracted from a reservoir via gravity. Christensen et al. (2004) conducted simulations of climate change on management of the Colorado River Basin reservoir system. They considered the likelihood that the level of Lake Mead would fall below 1,075 feet, triggering a level-one shortage declaration, and below 1,050 feet, triggering a level-two shortage. Under a level-one shortage, the Law of the River calls for reductions in deliveries of 320,000 acre-feet to the Central Arizona Project (CAP) and 13,000 to the Southern Nevada Water Authority (SNWA). Under a level-two shortage, supplies to the CAP and SNWA are reduced by 400,000 and 17,000 acre-feet. In simulations using 1950–1999 streamflows combined with 2000 demands, a level-one shortage occurred in 60% of years. In simulations over the period 2006–2039, a level-one shortage occurred in 92% of years and a level-two shortage in 77% of years. Using updated General Circulation Model (GCM) results, Christensen and Lettenmaier (2006) obtained more optimistic results. Nevertheless, simulations project that level-one shortages would occur in 21% of years from 2010 to 2039, and level-three shortages would occur in 10 – 11% of years. A level-three shortage could be declared if Lake Mead fell below 1,025 feet.

Barnett and Pierce (2008), using probabilistic simulation modeling, reported a 50% chance that Lakes Mead and Powell would fall below their minimum power pool levels by 2017. These levels are the minimum needed to generate hydroelectric power: 1,050 feet for Lake Mead and 3,490 for Lake Powell. Assuming a climate change–induced 20% reduction in runoff, the researchers found a 50% probability that both lakes would reach their dead pool levels (3,370 for Powell, 900 for Mead) by 2028. This article, titled "When Will Lake Mead Go Dry?" received considerable press coverage. It has also received criticism from scientists (Barsugli et al. 2009; Rajagopalan et al. 2009) and water managers, in part because of hydrologic modeling assumptions, but more importantly because of assumptions about how the Bureau of Reclamation (USBR), states, and other water management entities will respond to reduced water supplies. For example, the study did not include the USBR's *Interim Guidelines* for dealing with shortages on the Colorado River (USBR 2007). Subsequent work by Rajagopalan et al. (2009) suggests that the probability of the lakes falling below dead pool levels are less than 10% before 2026 and highlights the importance of proactive management to reduce risks of shortages. After 2026, however, the authors found that the risk of lake "drying" could rise dramatically, depending on how climate change might affect runoff levels. More recently, Barnett and Pierce (2009) attempted to account for criticisms of their earlier work and considered reductions in deliveries needed to keep Lake Mead above 1,000 feet, the level needed to maintain water deliveries to Las Vegas. They still estimated that given 20% reductions in runoff, shortage declarations (and reduced deliveries) would be common occurrences by 2030. They also predicted that deliveries would have to be curtailed sharply to maintain Lake Mead at 1,000 feet, even if Lake Powell were reduced to its dead pool level.

Table 6.7. *Decline in Lake Visitation in Response to Declining Reservoir Levels*

	Lake level (ft.)	*Percent reduction in visits*
Lake Mead NRA	1,145 (baseline)	
	1,075	– 10.5%
	1,050	– 14.1%
	1,000	– 21.2%
Glen Canyon NRA (Lake Powell)	3,608 (baseline)	
	3,560	– 2.6%
	3,490	– 6.8%
	3,370	– 16.2%

Table 6.8 *Economic Impacts of Lake Reduction Scenarios*

	Scenario 1: Level-One Shortage			*Scenario 2: Power Pool*			*Scenario 3: Powell Dead Pool*		
	Glen Canyon	*Lake Mead*	*Total*	*Glen Canyon*	*Lake Mead*	*Total*	*Glen Canyon*	*Lake Mead*	*Total*
Recreation visits	−48,342	−830,705	−879,047	−128,203	−1,118,904	−1,247,107	−303,432	−1,679,373	−1,982,805
Party days	−36,467	−253,491	−289,958	−96,713	−341,435	−438,148	−228,899	−512,463	−741,362
Visitor spending ($ million)	−$2.8	−$25.2	−$28.0	−$7.4	−$34.0	−$41.4	−$17.5	−$51.0	−$68.5
Direct effects									
Sales ($ million)	−$2.2	−$18.6	−$20.8	−$5.8	−$25.0	−$30.8	−$13.9	−$37.5	−$51.4
Jobs	−55	−545	−600	−145	−735	−880	−342	−1103	−1,445
Income ($ million)	−$0.8	−$6.6	−$7.4	−$2.1	−$8.9	−$11.0	−$5.0	−$13.4	−$18.4
Total effects									
Sales ($ million)	−$3.2	−$24.4	−$27.6	−$8.5	−$32.9	−$41.4	−$20.2	−$49.4	−$69.6
Jobs	−68	−635	−703	−183	−855	−1,038	−431	−1284	−1,715
Income ($ million)	−$1.2	−$8.6	−$9.8	−$3.1	−$11.6	−$14.7	−$7.3	−$17.4	−$24.7

A key point about this literature is that the levels of Lakes Mead and Powell are not just functions of climate and hydrology. Rather, human management of the Colorado River and reservoir system is quite critical. Indeed, the USBR's *Interim Guidelines* illustrate how reservoir levels can vary depending on management options chosen. The final environmental impact statement (EIS) for the guidelines reports estimates of the probabilities of lakes falling below certain levels that would create boating hazards or make marinas or launches inaccessible (USBR 2007). The EIS does not attempt, however, to calculate the effects of these reduced lake levels on visits or subsequent economic impacts. The results reported here could be used in future analyses of the economic consequences of different reservoir management options.

Lake Level and Climate Change Scenarios

Table 6.7 reports estimates of how changing lake elevations affect visitation to Lakes Powell and Mead. We use these results to construct three water shortage scenarios in Table 6.8. In the Level-One Shortage scenario, Lake Powell drops from its 2003 level of 3,608 feet to 3,560 feet, while Lake Mead's level drops from 1,145 feet to 1,075 feet. If Lake Mead reaches a level of 1,075 feet, this would trigger a level-one shortage declaration on the Colorado River. At 3,560 feet for Lake Powell, several launch facilities could not be operated. In the Power Pool scenario, lake elevations fall farther, from 2003 levels to their minimum power pool levels. The Powell Dead Pool scenario matches the case examined in Barnett and Pierce (2009), where Lake Powell falls to its dead pool level, while Lake Mead falls to 1,000 feet, the minimum level needed to continue to transfer water to Las Vegas. A key question for future research is how climate change affects the probability of occurrence of the scenarios examined here. Christensen et al.'s (2004) research suggests that lake levels of the Level-One Shortage scenario would become the norm, while conditions corresponding to the Power Pool scenario would become frequent.

CLIMATE CHANGE MITIGATION POLICIES

We next consider the effects of carbon taxes on gasoline prices and subsequent impacts on park visitation. Nordhaus and Boyer (2000) and Nordhaus (2008) estimate an optimal global price for CO_2, rising from \$9.5/ton of carbon in 2015 to \$23/ton by 2050. Estimates of the marginal social costs of carbon emissions range widely, with large outliers (Tol 2005, 2007). Most studies report estimates below \$75/ton of carbon (\$18/ton of CO_2,) with most estimates from peer-reviewed studies falling below \$50/ton of carbon (\$12/ton of CO_2) (Tol 2007). Clarke et al. (2007) simulated carbon price trajectories needed to stabilize atmospheric CO_2 concentrations at 450, 550, and 650 parts per million (ppm). They estimate that by 2025, the price of CO_2 emissions would need to rise to \$40–\$95 per ton of CO_2 to reach the 450 ppm emission target, \$5–\$30 to reach the 550 ppm target, and \$1–\$10 to reach the 650 ppm target. After 2025, prices would have to continue to rise by 3–5%/year in real terms.

Table 6.9. *Impacts of Carbon Taxes on Park Visits and Local Economies*

Scenario parameters		
$/ton CO_2	$5.45	$10
$/ton carbon	$20	$36.67
Gasoline price increase (cents/gallon)	5	8.8
Percent increase from base	3.1%	5.5%
Changes in		
Recreation visits (percent)	−0.6%	−1.0%
Recreation visits	−153,679	−272,656
Party days	−66,888	−118,672
Visitor spending ($ million)	−$7.2	−$12.7
Direct effects		
Sales ($ million)	−$5.9	−$10.4
Jobs	−145	−257
Income ($ million)	−$2.2	−$3.9
Total effects		
Sales ($ million)	−$8.3	−$14.7
Jobs	−179	−318
Income ($ million)	−$3.0	−$5.4

Table 6.9 estimates the impacts of two carbon tax levels on the price of gasoline and the subsequent implications for park visits and local economic impacts. Conversion factors translating taxes rates on CO_2 to rates on carbon and gasoline prices are based on Aldy et al. (2008). The $10 per ton of CO_2 tax increases the price of gasoline by 8.8 cents/gallon. At 2003 prices, this represents a 5.5% price increase. A $10 per ton of CO_2 tax reduces the number of annual regional park visits by more than 270,000. Accounting for multiplier effects, sales fall by $14.7 million, local personal income falls by $5.4 million, and employment falls by 318 jobs. Thus while gateway communities near low desert parks and national recreation areas may suffer from the effects of warmer temperatures or falling reservoir levels from climate change, they also will be negatively affected by policies to mitigate climate change.

One way to counteract the negative employment effects of a carbon tax would be to increase the level of park service employment in the region. Stynes (2007) estimates that for every two NPS jobs, one additional non-NPS job is generated. The Public Lands Service Corps Act of 2009 (HR 1612) introduced in Congress would increase construction and employment in national parks and other public lands through the creation of a service corps. This would be something similar to the old Civilian Conservation Corps (CCC) instituted as part of the New Deal. Results reported in Table 6.9 suggest that a relatively small increase in jobs under this proposed program (a few hundred) would suffice to counteract the negative effects of a $10/ton CO_2 tax.

CONCLUSIONS

To summarize, this study estimated time-series–cross-section data for 42 National Park System sites in the U.S. Southwest over 25 years to examine how climate, energy price, and other variables affected annual park visits. Econometric results were then used as inputs for input-output analysis to examine impacts of climate change, water shortages, and carbon taxes on sales, personal income, and employment in gateway communities to national parks.

Many of our regression results were consistent with earlier empirical analyses of park visitation. Explanatory variables previously used by geographers (Hanink and White 1999; Hanink and Stutts 2002), market potential and competing destination indices, were significant predictors of visits. Results suggest complementarities between southwestern parks such that proximity to other parks encouraged visits. Results were also similar to those of other studies that found topographical features (Puustinen et al. 2009), gasoline price (Johnson and Suits 1983; Pergams and Zaradic 2006), and national park designation status (Weiler and Seidl 2004; Weiler 2006) to be important explanatory variables. For sites with reservoirs, we estimated elasticities of visits with respect to lake surface areas of 0.11 for Glen Canyon NRA (Lake Powell) and 0.45 for Lake Mead NRA (Lake Mead and Lake Mohave combined areas). Loomis and Crespi (1999) estimated a surface area elasticity of 0.39 for reservoirs in California, whereas Ward et al. (1995) estimated an elasticity of 0.47 in a study of 115 reservoirs.

Previous research has projected that warmer temperatures from climate change may have a net positive impact on outdoor recreation (Mendelsohn and Markowski 1999; Loomis and Crespi 1999) and park visitation (Weiler et al. 2002; Richardson and Loomis 2004; Loomis and Richardson 2006; Jones and Scott 2006; Scott et al. 2007). Our findings regarding climate warming and park visits are mixed, suggesting that warming will affect parks differently depending on when seasonal warming occurs and the baseline climate of a park. Warming had positive effects on visitation at high-elevation, higher-latitude parks, but negative effects at low desert parks. This should not be too surprising, given the high temperatures (especially in summer) at the region's low desert parks. Model results suggest that a 1 degree F increase in January and July temperatures would reduce regional park visits by 1%. Total visitor spending increases slightly (less than 0.7%). This occurs because spending patterns vary by park, with spending gains in higher-altitude and higher-latitude parks outweighing losses in low desert parks.

Combining regression analysis with input-output modeling is a powerful tool for examining the effects of environmental or economic changes. We estimate that the total effects of declining lake levels from 1998 to 2008 at Lakes Powell and Mead led to a substantial reduction in visits, resulting in economic losses of $42 million in sales, $15 million in personal income, and more than 1,000 jobs. Future climate change–induced reductions in reservoir levels could have significant economic impacts. While the Bureau of Reclamation's *Interim Guidelines* for managing the Lower Colorado River Basin estimates probabilities of lakes falling below certain levels, its final EIS did not attempt to calculate the economic impacts on water-based recreation and communities dependent on recreation spending. The results and methods reported here could be used in future analyses of the economic consequences of different reservoir management options.

Carbon taxes would have modest negative impacts on local employment. The elasticity of park visits with respect to gas price was about −0.18. Recent proposed legislation to increase public works and employment in national parks could counteract these negative impacts.

Finally, our estimated elasticity of visits with respect to regional population was 1.0. This is important, given that the population of the study area is projected to grow by more than 75% from 2005 to 2050. It suggests that the effects of population growth would likely swamp other economic or environmental effects.

NOTES

1. Beck and Katz (1995) recommend using panel-corrected standard errors (PCSEs) to address heteroskedasticity and contemporaneous correlation of errors and an AR(1) correction for autocorrelation. They also recommend using a single autocorrelation (ρ) parameter instead of attempting to estimate separate parameters for each unit, because it is difficult to estimate multiple ρ parameters efficiently and the single parameter performs well in general.
2. IMPLAN Pro Version 2.0 is social accounting and impact analysis software from Minnesota IMPLAN Group, Inc., Stillwater, MN (1999).

REFERENCES

Aldy, J.E., E. Ley, and I.W.H. Parry. 2008. A Tax-Based Approach to Slowing Global Climate Change. Discussion paper DP 08-26. Washington, DC: Resources for the Future.

Alward, G., and S. Lindall. 1996. Deriving SAM Multiplier Models Using IMPLAN. In *Deriving SAM Multiplier Models Using IMPLAN*. Paper presented at the IMPLAN Users Conference. August 15–16, 1996, Minneapolis.

Barnett, T.P., and D.W. Pierce. 2008. When Will Lake Mead Go Dry? *Water Resources Research* 44:W03201–W03300.

———. 2009. Sustainable Water Deliveries from the Colorado River in a Changing Climate. *Proceedings of the National Academy of Sciences* 106:7334–7338.

Barsugli, J., K. Nowak, B. Rajagopalan, J. Prairie, and B. Harding. 2009. Comment on "When Will Lake Mead Go Dry?". *Water Resources Research* 45:W09601, doi:10.1029/2008WR007627.

Beck, N. 2001. Time-Series-Cross-Section Data: What Have We Learned in the Past Few Years? *Annual Review of Political Science* 4:271–293.

Beck, N., and J. Katz. 1995. What to Do (and Not to Do) with Time-Series Cross-Section Data. *American Political Science Review* 89:634–647.

BLS (U.S. Bureau of Labor Statistics). Various dates. Consumer Price Index. www.bls.gov/cpi/home.htm (accessed October 4, 2010).

Breusch, T.S., and A.R. Pagan. 1980. The Lagrange Multiplier Test and Its Applications to Model Specification in Econometrics. *Review of Economic Studies* 47:239–253.

Burns, C.E., K.M. Johnston, and O.J. Schmitz. 2003. Global Climate Change and Mammalian Species Diversity in U.S. National Parks. *Proceedings of the National Academy of Sciences* 100:11474–11477.

Christensen, J.H., and B. Hewitson. 2007. Regional Climate Projections. In *Climate Change 2007: The Physical Science Basis*, edited by S. Solomon, D. Qin, M. Manning, Z. Chen, M. Marquis, K.B. Averyt, M. Tignor, and H.L. Miller. New York: Cambridge University Press, 847–940.

Christensen, N., and D.P. Lettenmaier. 2006. A Multimodel Ensemble Approach to Assessment of Climate Change Impacts on the Hydrology and Water Resources of the Colorado River Basin. *Hydrology and Earth System Sciences Discussions* 3:3727–3770.

Christensen, N.S., A.W. Wood, N. Voisin, D.P. Lettenmaier, and R.N. Palmer. 2004. The Effects of Climate Change on the Hydrology and Water Resources of the Colorado River Basin. *Climatic Change* 62:337–363.

Clarke, L.E., J.A. Edmonds, H. Jacoby, H.M. Pitcher, J.M. Reilly, and R.G. Richels. 2007. *Scenarios of Greenhouse Gas Emissions and Atmospheric Concentrations*. Washington, DC: U.S. Climate Change Science Program.

Cline, W. 1992. *The Economics of Global Warming*. Washington, DC: Institute for Institutional Economics.

EIA (Energy Information Administration, U.S. Department of Energy). Various dates. Short Term Energy Outlook. www.eia.gov/emeu/steo/pub/contents.html (accessed October 4, 2010).

Garrick, D., K. Jacobs, and G. Garfin. 2008. Models, Assumptions, and Stakeholders: Planning for Water Supply Variability in the Colorado River Basin. *Journal of the American Water Resources Association* 44:381–398.

Goldberg, L.S. 2004. Industry-Specific Exchange Rates for the United States. *Federal Reserve Bank of New York Economic Policy Review* 10:1–16.

Greene, William H. 2003. Econometric Analysis. 5th ed. Upper Saddle River, NJ: Prentice Hall.

Hanink, D., and M. Stutts. 2002. Spatial Demand for National Battlefield Parks. *Annals of Tourism Research* 29:707–719.

Hanink, D., and K. White. 1999. Distance Effects in the Demand for Wildland Recreational Services: The Case of National Parks in the United States. *Environment and Planning A* 31:477–492.

Harding, B.L., T.B. Sangoyomi, and E.A. Payton. 1995. Impacts of a Severe Sustained Drought on Colorado River Water Resources. *Water Resources Bulletin* 31:815–824.

Hayhoe, K., D. Cayan, C. Field, P. Frumhoff, E. Maurer, N. Miller, S. Moser, S. Schneider, K. Cahill, E. Cleland, L. Dale, R. Drapek, R. Hanemann, L. Kalkstein, J. Lenihan, C. Lunch, R. Neilson, S. Sheridan, and J. Verville. 2004. Emissions Pathways, Climate Change, and Impacts on California. *Proceedings of the National Academy of Sciences* 101:12422–12427.

Johnson, R., and D. Suits. 1983. A Statistical Analysis of the Demand for Visits to U.S. National Parks: Travel Costs and Seasonality. *Journal of Travel Research* 22:21–23.

Jones, B., and D. Scott. 2006. Climate Change, Seasonality and Visitation to Canada's National Parks. *Journal of Park and Recreation Administration* 24:42–62.

Loomis, J., and J. Crespi. 1999. Estimated Effects of Climate Change on Selected Outdoor Recreation Activities in the United States. In *The Impact of Climate Change on the United States Economy*, edited by R. Mendelsohn and J.F. Neumann. Cambridge, UK: Cambridge University Press, 289–314.

Loomis, J.B., and R.B. Richardson. 2006. An External Validity Test of Intended Behavior: Comparing Revealed Preference and Intended Visitation in Response to Climate Change. *Journal of Environmental Planning and Management* 49:621–630.

McGranahan, D.A. 1999. *Natural Amenities Drive Rural Population Change*. Agricultural Economic Report No. 781. Washington, DC: U.S. Department of Agriculture, Economic Research Service.

Mendelsohn, R., and M. Markowski. 1999. The Impact of Climate Change on Outdoor Recreation. In *The Impact of Climate Change on the United States Economy*, edited by R. Mendelsohn and J.F. Neumann. Cambridge, UK: Cambridge University Press, 267–288.

Mitchell, J.G. 2006. National Parks in Peril. *National Geographic* 210:68–87.

Morris, D., and M. Walls. 2009. Climate Change and Outdoor Recreation Resources. Backgrounder. Washington, DC: Resources for the Future.

NMBGMR (New Mexico Bureau of Geology and Mineral Resources). 2007. *Climate Change and Water Resources in New Mexico*. Socorro: New Mexico Bureau of Geology and Mineral Resources, New Mexico Institute of Mining and Technology.

NMOSE (New Mexico Office of the State Engineer). 2006. *The Impact of Climate Change on New Mexico's Water Supply and Ability to Manage Water Resources*. Albuquerque: NMOSE.

Nordhaus, W.D. 2008. *A Question of Balance: Weighing the Options on Global Warming Policies*. New Haven, CT: Yale University Press.

Nordhaus, W.D., and J. Boyer. 2000. *Warming the World: Economic Models of Global Warming*. Cambridge, MA: MIT Press.

NPCA (National Parks Conservation Association). 2007. *Unnatural Disaster: Global Warming and Our National Parks*. Washington, DC: NPCA.

NPS (National Park Service, U.S. Department of the Interior). 2003. History E-Library: National Park System Birthdays, www.nps.gov/history/history/hisnps/NPSHistory/birthdays.htm (accessed September 30, 2010).

———. n.d. National Park Service Public Use Statistics Office: NPS Stats. www.nature.nps.gov/stats/index.cfm (accessed September 30, 2010).

———. Various dates. National Park Service Public Use Statistics Office: Acreage Reports, www.nature.nps.gov/stats/acreagemenu.cfm (accessed September 30, 2010).

Pergams, O., and P. Zaradic. 2006. Is Love of Nature in the US Becoming Love of Electronic Media? 16-Year Downtrend in National Park Visits Explained by Watching Movies, Playing Video Games, Internet Use and Oil Prices. *Journal of Environmental Management* 80:387–393.

Ponnaluru, S. 2005. Visitation to the National Parks in the Southwest: the Influence of Economic and Climate Variables. Master's thesis. University of Arizona.

Puustinen, J., E. Pouta, M. Neuvonen, and T. Sievänen. 2009. Visits to National Parks and the Provision of Natural and Man-made Recreation and Tourism Resources. *Journal of Ecotourism* 8:18–31.

Rajagopalan, B., K. Nowak, J. Prairie, M. Hoerling, B. Harding, J. Barsugli, and B. Udall. 2009. Water Supply Risk on the Colorado River: Can Management Mitigate? *Water Resources Research* 45:W08201, doi:10.1029/2008WR007652.

Ray, A.J., J.J. Barsugli, and K.B. Avery. 2008. *Climate Change in Colorado: A Synthesis to Support Water Resources Management and Adaptation*. Boulder: University of Colorado, Western Water Assessment.

Richardson, R., and J. Loomis. 2004. Adaptive Recreation Planning and Climate Change: A Contingent Visitation Approach. *Ecological Economics* 50:83–99.

Rickman, D.S., and R.K. Schwer. 1995. A Comparison of the Multipliers of IMPLAN, REMI, and RIMS II: Benchmarking Ready-made Models for Comparison. *Annals of Regional Science* 29:363–374.

Saunders, S., T. Easley, J.A. Logan, and T. Spencer. 2006. *Losing Ground: Western National Parks Endangered by Climate Disruption*. Louisville, CO, and Washington, DC: Rocky Mountain Climate Organization and the Natural Resources Defense Council.

Scott, D., B. Jones, and J. Konopek. 2007. Implications of Climate and Environmental Change for Nature-Based Tourism in the Canadian Rocky Mountains: A Case Study of Waterton Lakes National Park. *Tourism Management* 28:570–579.

Stynes, D. 2008. *National Park Visitor Spending and Payroll Impacts, 2007*. Report prepared for the National Park Service, Social Science Program. East Lansing: Michigan State University, Department of Park, Recreation and Tourism Resources.

Stynes, D.J., D.B. Propst, W.H. Chang, and Y. Sun. 2000. *Estimating Regional Economic Impacts of Park Visitor Spending: Money Generation Model Version 2 (MGM2)*. East Lansing: Michigan State University, Department of Park, Recreation and Tourism Resources.

Tol, R.S.J. 2005. The Marginal Damage Costs of Carbon Dioxide Emissions: An Assessment of the Uncertainties. *Energy Policy* 33:2064–2074.

———. 2007. *The Social Cost of Carbon: Trends, Outliers and Catastrophes*. Economics Discussion Paper No. 2007-44. Kiel, Germany: Kiel Institute for the World Economy.

USBR (U.S. Department of the Interior, Bureau of Reclamation). 2007. *Colorado River Interim Guidelines for Lower Basin Shortages and Coordinated Operations for Lake Powell and Lake Mead: Final Environmental Impact Statement, vol. 1, Upper and Lower Colorado Regions*. Boulder City, NV: USBR.

Uscher, J. 2008. Visit Them Now: National Parks Threatened by Climate Change. AARP, www.aarp.org/travel/destinations/info-11-2008/national_parks.html (accessed September 21, 2010).

Wall, G. 1998. Implications of Global Climate Change for Tourism and Recreation in Wetland Areas. *Climatic Change* 40:371–389.

Ward, F., B. Roach, J. Loomis, R. Ready, and J. Henderson. 1995. *Regional Recreation Demand Models for Large Reservoirs*. Final report to U.S. Army Corps of Engineers. Las Cruces, NM: New Mexico State University, Department of Agricultural Economics.

Weiler, S. 2006. A Park by Any Other Name: National Park Designation as a Natural Experiment in Signaling. *Journal of Urban Economics* 60:96–106.

Weiler, S., J. Loomis, R. Richardson, and S. Shwiff. 2002. Driving Regional Economic Models with a Statistical Model: Hypothesis Testing for Economic Impact Analysis. *Review of Regional Studies* 32:97–111.

Weiler, S., and A. Seidl. 2004. What's in a Name? Extracting Econometric Drivers to Assess the Impact of National Park Designation. *Journal of Regional Science* 44:245–262.

Wicks, B., M. Uysal, and S. Kim. 1994. The Effect of Lodging Prices on Visitors' Demand: Everglades National Park. *Journal of Hospitality & Tourism Research* 17:51–61.

Witt, S.F., and C.A. Witt. 1991. Modeling and Forecasting Demand in Tourism. London: Academic Press.

WRCC (Western Regional Climate Center). Various dates. Historical Climate Information. www.wrcc.dri.edu/CLIMATEDATA.html (accessed October 4, 2010).

CHAPTER 7

Climate, Changing Snowpack, and the Future of Winter Recreation

Rosalind H. Bark and Bonnie G. Colby

Along with water-based recreation (discussed in Chapter 6), winter snow-based recreation is potentially vulnerable to the higher temperatures and increased variability in precipitation anticipated with climate change. Ski resorts rely on good natural snow conditions or the capability to manufacture snow. Many top winter recreation resorts rely on a combination of both, enhancing their ability to adapt to variable conditions. Snowmaking technology enables resorts to open earlier in the autumn, remain open later in the spring, and replenish snowpack after a warm or wet period. In the U.S. Southwest, ski resorts often experience high interannual variability in snowpack conditions. Much of the variability on an interannual timescale is explained by the El Niño Southern Oscillation phenomenon. In general, El Niño conditions are associated with wetter winters, whereas La Niña conditions often portend dry winters. This variability in natural snow conditions can affect the year-to-year financial viability of ski resorts. In response, some resort managers invest in snowmaking to ensure that resort infrastructure is used and personnel are employed even in poor natural snowpack years.

Overlying interannual snowpack variability is longer-term climate change. General Circulation Models (GCMs) predict that the winter season in much of the southwestern United States later in this century will be both warmer and drier. A worrisome outcome of these predictions for ski resort managers is that the ski season is likely to become shorter as a result of poor snowpack—a consequence of less overall winter precipitation, precipitation falling as rain instead of snow early and late in the ski season, and overnight temperatures that are too warm for snow to be manufactured. Furthermore, the water and energy requirements for snowmaking make this adaptive activity particularly susceptible to uncertainties in water and energy costs and availability, which are expected to become more problematic as a result of climate change. Paradoxically, increased capitalization in ski resorts has likely made many resorts and their surrounding communities more dependent on winter recreation revenues and therefore more financially vulnerable to the negative impacts of climate change.

This chapter summarizes climate prediction results for snowpack in Arizona and illustrates the linkage of climate prediction science with local economic activity, specifically snow-based recreation. Climate variability in the region and climate change adaptation strategies for the ski industry and nearby communities are explored, including investing in snowmaking and diversifying the activities and services offered. We demonstrate the application of several economic methodologies to climate change concerns, using local economic impacts modeling and assessing the financial feasibility of investment in adaptation technology in the form of snowmaking. Snowmaking is both water- and energy-intensive. As an adaptation strategy, it can have significant regional implications for both of these resources and is vulnerable to changes in energy and water supplies and costs.

The ski industry has long been recognized as vulnerable to climate change impacts. This vulnerability has been assessed by researchers in many countries, including Australia (Galloway 1988; König 1998), Austria (Breiling et al. 1997), Canada (McBoyle and Wall 1992; Lamothe and Périard Consultants 1988), Scotland (Harrison et al. 1999), Switzerland (König and Abegg 1997; Elsasser and Bürki 2002), and the United States (Bloomfield and Hamburg 1997). All of these studies predict that negative impacts from climate change will be most problematic for low-elevation ski resorts. A key mechanism is the prediction that climate change will reduce snow cover at lower elevations, with precipitation falling as rain instead of snow, and thus fewer days in the season will meet minimum operational snow base thresholds for winter sports. These thresholds vary with the sport: 10 centimeters (3.9 inches) for cross-country skiing, 30 centimeters (11.8 inches) for downhill skiing and snowboarding, and greater than 30 centimeters for sports taking place above the treeline in rocky terrain (Breiling and Charamza 1999; Scott et al. 2003). Without the use of adaptation strategies to compensate for the lack of snow, ski resorts would have to make an adequate level of profit during a shorter season, and financially marginal resorts would likely close.

Whereas the previous research focused on low-elevation ski resorts, our study investigates the likely impact of climate change on low-latitude, *high*-elevation ski resorts. Our conclusions based on our area of study in Arizona are also generally applicable to other low-latitude, high-elevation ski resorts in South America and New Zealand.

ARIZONA'S SKI RESORTS

Arizona is famous for its saguaro cacti and the Grand Canyon, not for its ski industry, but the state has four ski resorts (see Figure 7.1) that cater primarily to in-state skiers, with a small number also coming from nearby states and Mexico. Two of the resorts are small and marginal because seasonal snowpack is unreliable. The Mount Lemmon Ski Valley resort, located in the mountains north of Tucson, is the southernmost resort in the United States. The Elk Ridge Ski Area, west of Flagstaff, is the lowest-elevation resort in the state. Elevation ranges from 2,286 to 2,484 meters (7,500 to 8,150 feet) at Elk Ridge and 2,499 to 2,789 meters (8,199 to 9,150 feet) at Mount Lemmon. The two other, larger resorts are at higher elevations and thus have more reliable snowpack. The largest ski resort in the state, with 324 skiable hectares, is Sunrise Park Resort, located in the rural White Mountains in the eastern-central part of the state, ranging from 2,836 to 3,354 meters (9,304 to 11,004 feet) in elevation. This ski resort, located on the White Mountain Apache Reservation, is also the only tribally owned resort in the state. The Arizona Snowbowl, currently with 51 skiable hectares (expansion is planned to 414 skiable hectares), is the highest-elevation resort in the state, at 2,805 to 3,506 meters (9,203 to 11,503 feet). It is also the northernmost and is conveniently located on the outskirts of Flagstaff, with easy road access to the largest metropolitan area in the state, Phoenix. Sunrise and Snowbowl cater to an average 250,000 skiers per season and are the only resorts in the state that can afford to invest in snowmaking.

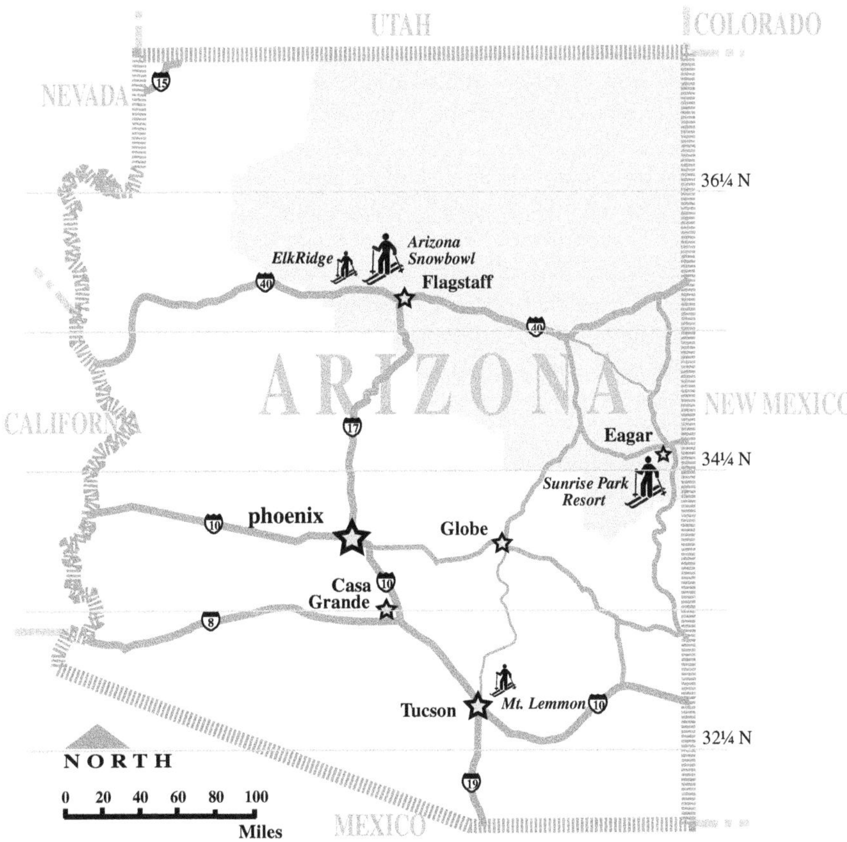

Figure 7.1 *Arizona's Ski Resorts*

Different landownership and water rights also influence the relative competitiveness of the four main players in Arizona's ski industry. As Sunrise is located on reservation land, it is not restricted by USDA Forest Service (USFS) regulations, and it is the only Arizona ski resort offering on-site accommodations and night skiing. The other three resorts are constrained by their location on USFS land, and any plans to expand facilities must comply with the National Environmental Policy Act (NEPA), which "requires federal agencies to integrate environmental values into their decision making processes by considering the environmental impacts of their proposed actions and reasonable alternatives to those actions" (EPA 2010). When Snowbowl made plans for facility expansion, it faced litigation with local tribes and environmental groups but eventually was allowed to proceed.[1] Water rights are a major restraint on future development of skiing in Arizona and were the focus of this legal action. Snowbowl has no water rights, and therefore its snowmaking plan was dependent on approval of a proposed water lease with the city of Flagstaff for treated effluent, or potable supplies, to use in snow manufacture at the resort. If the city approves the water lease, Snowbowl's snowmaking operation will require 5,678 cubic meters (m^3) of reclaimed water per day during the ski season, which will be delivered by a 22.5-kilometer (14-mile) pipeline from the wastewater treatment plant. New snowmaking infrastructure will cover 95% of the skiable terrain. This investment will put Snowbowl on a par with climate change–vulnerable low-elevation, high-latitude ski resorts in central Ontario, Canada, that have snowmaking capabilities for 100% of skiable terrain (Scott et al. 2003).

In the case of Sunrise, water rights settlement legislation with the Tribe (Public Law No. 111–291, enacted December 8, 2010) requires the federal government to plan, design, and construct snowmaking capacity and infrastructure for the Tribe at its Sunrise resort. Currently only Sunrise Peak has snowmaking facilities, and as such, it is the peak that opens first when natural snowpack is poor and stays open latest in the season when demand for skiing wanes (Scott et al. 2003; Murray 2006). To fund this snowmaking investment, characterized as tribal water-related economic development, the new 2010 legislation includes an appropriation.

The future of the ski industry in Arizona appears clear: The two large resorts will adapt by investing in snowmaking for most of their skiable terrain, enabling them to open more consistently in the short term and remain open as temperatures are projected to warm over the long term. The comparative advantage of Arizona's ski resorts relative to substitute (competitor) resorts in nearby New Mexico will depend on the adaptive responses of such competitors, the response of consumers to changing relative conditions and costs, and natural interannual variability in snowpack among rival resorts. Meanwhile, the smaller resorts will continue to rely on natural snowpack and probably will eventually be forced to close as climate change impacts worsen. Researchers have predicted such a pattern of closures worldwide, with marginal low-elevation and low-latitude resorts being forced to close as the snowline migrates to higher elevations. The bottom line for resorts is that "location can be crucial to success" (National Assessment Synthesis Team 2000; see also OECD 2006).

INTERANNUAL SNOWPACK VARIABILITY

Arizona's ski resorts are characterized by a high degree of interannual variability in snowpack, skiable days, and visitor numbers. Winter precipitation in the Southwest is influenced by the location of the winter storm tracks. There is evidence that storm tracks across the continental United States are shifted by sea-surface temperature (SST) anomalies in the Pacific Ocean. The El Niño Southern Oscillation (ENSO) phenomenon occurs in a two- to seven-year cycle. In the warm phase, El Niño, winter storm tracks shift south, bringing generally wetter conditions to the Southwest (Arizona and New Mexico). In the cold phase, La Niña, storms track over the Northwest (Gutzler et al. 2002; Brown and Comrie 2004). ENSO also has a neutral phase, which is neither particularly dry nor particularly wet. Additionally, a decadal-scale cycle called the Pacific Decadal Oscillation (PDO), which has a cold phase and a warm phase, may modulate ENSO conditions. The cold phase of the PDO intensifies dry conditions in both the neutral and La Niña phases of ENSO, whereas wetter winters are generally associated with neutral and El Niño ENSO conditions during the warm phase of the PDO (Gutzler et al. 2002; Brown and Comrie 2004; Goodrich 2004). These oscillations provide some predictability to winter precipitation conditions in the Southwest that might be useful for skiers, as well as ski resort and water managers (Weiher 2000; Patten et al. 2003).

Figure 7.2 plots snowfall data from Snowbowl and the ENSO Index for the 1981–1982 through 2005–2006 seasons. In general, neutral and El Niño conditions (positive ENSO) are correlated with higher average seasonal snowpack, whereas La Niña conditions (negative ENSO) are associated with less snowpack. Snowpack is a necessary condition for skiing at Snowbowl, because during the period illustrated in the figure, the resort did not have snowmaking capability. A regression modeling skier visitation against a binary variable set equal to 1 for El Niño conditions and 0 otherwise finds evidence that the El Niño phase is a significant (at the 5% level) and positive predictor of skier visits at Snowbowl. This means El Niño seasons tend to attract more skiers and thus are more profitable ski seasons. Furthermore, just two variables, total seasonal snowfall and days open, explain 77% of the variation in skier visitation over the 25-season record.[2]

Table 7.1 shows the ENSO phase for each season from 1981–1982 through 2005–2006, along with seasonal snowfall, days open, and ski visitor data for Snowbowl; ski visitor data for Arizona; and an estimate of ski visitor data for Sunrise.[3] During the data period, Sunrise visitor numbers were steadier than those at Snowbowl, perhaps indicating that natural snowfall combined with limited snowmaking capacity enabled the resort to open more consistently. Nevertheless, the data also show that although Sunrise had six times more skiable hectares than Snowbowl during this period, it attracted on average only 50,000 more skiers per season. This may reflect the superior location of Snowbowl. The table also reports the "reliability score" (Bürki et al. 2003), defined as the number of seasons in the last 10 that met or exceeded 100 skiable days, in this case at Snowbowl. Bürki et al. considered "reliability" to be 7 out of 10 good winter seasons in their study area, Switzerland. Snowbowl's reliability data indicate a downward trend that is clearly

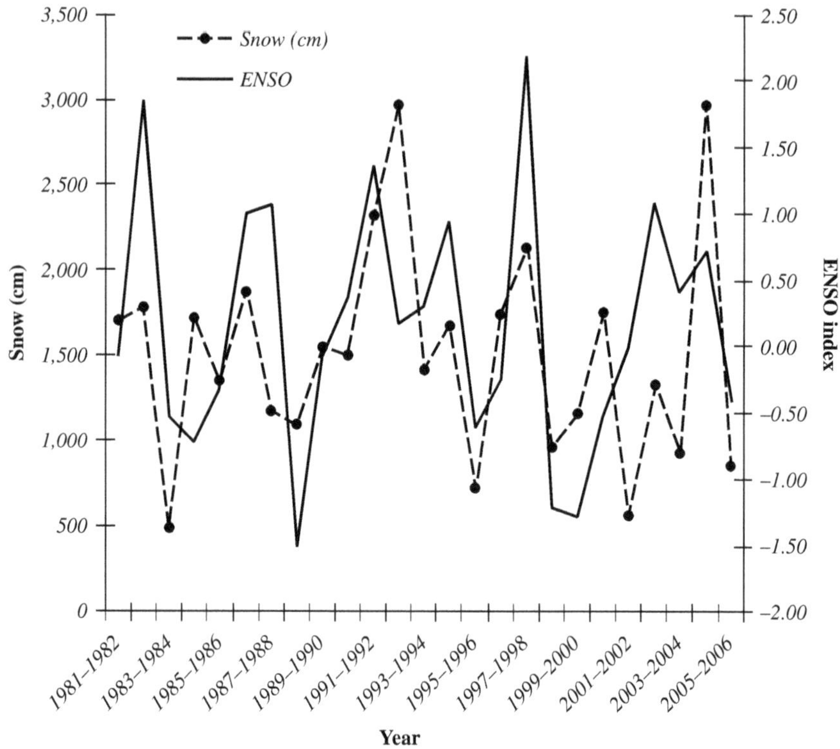

Figure 7.2 *Snowbowl Snowfall Correlated with the ENSO Index, 1981–1982 to 2005–2006*

Note: 1 centimeter (cm) = 0.39 inches
Source: NOAA (2010)

worrisome for Snowbowl's manager and contributed to the decision to invest in snowmaking. In fact, as noted in the recent Intergovernmental Panel on Climate Change Fourth Assessment Report (IPCC AR4), "adaptation measures are seldom undertaken in response to climate change alone" (Adger et al. 2007). Snowmaking is an adaptation not only to climate change, but also to more general interannual climate variability. The cost-effectiveness of this adaption is improved by its having this dual role.

Another aspect of interannual natural snowpack variability is the timing of snowfall, which can be crucial to the viability of a ski resort. In an Austrian study by Breiling and Charamza (1999), adequate snowpack over the Christmas and New Year holidays and in February, the most important skiing month in Austria, proved essential for a good season. In Arizona, timing is also crucial and can sometimes be more important than abundant snow. For instance, the 2000–2001 ski season was good at Sunrise from a revenue perspective, even though snowfall was not the highest on record, because snow accumulated in October and was present over the critical Thanksgiving and Christmas–New Year holidays (Gibson and Evans 2002). The mountain manager at Snowbowl has reported that if the resort is not open over

Table 7.1 *Ski Data for Snowbowl, Sunrise, and State of Arizona Correlated with ENSO Phases, 1981–1982 to 2005–2006*

Season	ENSO phase	Snowbowl			Visitors		
		Snowfall (cm)	Days open	Reliability score	Snowbowl	Sunrise	Arizona
1981–1982	Neutral	673	123		63,000		
1982–1983	Warm	701	135		99,626		
1983–1984	Neutral	193	64		28,913		
1984–1985	Cold	676	118		114,707		
1985–1986	Neutral	533	124		105,252		
1986–1987	Warm	737	112		125,026		
1987–1988	Warm	462	92		119,259		
1988–1989	Cold	432	79		120,132		
1989–1990	Neutral	610	74		99,280		
1990–1991	Neutral	592	112	6	106,000		
1991–1992	Warm	914	134	6	173,000		
1992–1993	Neutral	1,168	130	6	181,000		
1993–1994	Neutral	559	114	7	116,388		
1994–1995	Warm	658	122	7	176,778		
1995–1996	Cold	287	25	6	20,312	74,037	102,575
1996–1997	Neutral	686	109	6	153,176	191,350	365,787
1997–1998	Warm	838	115	7	173,962	189,633	384,665
1998–1999	Cold	381	60	7	35,205	190,562	246,941
1999–2000	Cold	457	45	7	66,152	159,780	243,685
2000–2001	Cold	691	138	7	162,175	174,245	355,780
2001–2002	Neutral	221	4	6	2,875	190,134	214,135
2002–2003	Warm	523	96	5	87,354	171,006	277,361
2003–2004	Neutral	368	120	5	72,000	149,778	238,420
2004–2005	Warm	1,168	133	5	190,000	194,023	405,581
2005–2006	Neutral	338	16	5	~31,000	38,470	73,744
Average		595	96	6	104,903	156,638	264,425

Notes: ENSO phases: Neutral = neutral, Cold = La Niña, Warm = El Niño; 1 centimeter (cm) = 0.39 inches

Source: For Snowbowl through 2004–2005, USDA FS (2005), and for 2005–2006, estimates from local news coverage; estimated Arizona ski visitor data for 1995–1996 through 2005–2006 supplied by Troy Hawks, managing editor, *NSAA [National Ski Areas Association] Journal*.

Christmas, the resort loses 30% of its seasonal business (Murray 2006). This variability in a season's start date and length is a key reason to invest in snowmaking infrastructure. Snowmaking facilitates greater ski season consistency, enabling a resort to open even when natural snowpack would be inadequate.

THE SNOW-BASED ECONOMY

The ski industry is a notable winter season local economic driver in the rural White Mountains, where Sunrise is located, and in Flagstaff, near Snowbowl. Gibson and Evans (2002) estimated resort revenues from skiing at Sunrise at $5.1 million to $5.95 million. According to an anonymous ski industry spokesperson, Sunrise resort revenues for the very good 2004–2005 season were $8.4 million. This figure includes only ski ticket sales and equipment rental. Skiers also spent money on accommodations, food, off-resort ski equipment hire or purchase, and gas. If we use the figure of $100 per skier per day, as estimated in the Snowbowl Environmental Impact Study (EIS; USDA FS 2005), then direct expenditures by visitors at Sunrise and in surrounding areas in the 2004–2005 season were around $20 million. This figure does not include indirect expenditures resulting from increased income in the region. Sunrise attracts a larger proportion of overnight skiers than Snowbowl, because of its rural location, and therefore it is likely that the economic multiplier (the local economywide impact of the resort) for Sunrise is higher than for Snowbowl. This means that every dollar spent by a skier in the White Mountains likely generates more local economic impact than a dollar spent at Snowbowl.

Sunrise is one of the largest nongovernmental employers in the region. Employment peaks in winter months at around 518 employees, but many jobs are seasonal, so the full-time equivalent (FTE) payroll is a much lower 198. Gibson and Evans (2002) used public information on tax revenue and survey data to determine a recreation dependency ratio for the entire White Mountain region, which includes the off-reservation towns of Show Low, Pinetop-Lakeside, Greer, and Springerville-Eager, as well as Sunrise and Whiteriver on the reservation. A recreation dependency ratio is the percentage of total income in the region generated by the recreation sector. The White Mountain area is not as developed as Flagstaff, and thus Sunrise is proportionately more important than Snowbowl to the local community. Gibson and Evans estimated that the recreation dependency ratio in the White Mountains is 44%. However, there is considerable variation among communities. For example, the recreation dependency ratio is 90% for the town of Sunrise and 79% for nearby Greer. The authors also calculated a specific winter dependency ratio, the percentage of annual sales occurring in the winter season, defined in their study as January to March. This definition underestimates winter dependency, as in a good year, a large proportion of ski visits occur during Thanksgiving, over the Christmas–New Year period, and into April. Nevertheless, the data show that although winter recreation is important to some communities in the region, overall this season generates just 27% of annual sales, compared with the summer season's 33%. But Sunrise is highly dependent on winter recreation, with a winter dependency ratio of 60%. The White Mountains' winter recreation dependency ratio is more than three times higher than that at Snowbowl (USDA FS 2005). These data show that although the ski industry in Arizona is small, it is economically important to the White Mountain Apaches, as well as to the surrounding communities in the White Mountains. Importantly, the ski resorts bring winter tourism revenues that balance out peak summertime visitation patterns.

Local economic data for Snowbowl were provided in the Snowbowl EIS (USDA FS 2005), as required by the NEPA. Snowbowl managers estimate that in the 1996–1997 to 2002–2003 seasons, the resort supported an average 172 FTE (this breaks down to 22.1 persons employed full-time, 272.4 employed on a full-time seasonal basis, and 204.3 employed on a part-time seasonal basis). The EIS report quoted average direct expenditures of $9.79 million per season during this same period in Coconino County. This total was calculated by multiplying the average ski visitation of 97,900 by an average expenditure of $100 per skier. The report also estimated total direct and indirect economic impacts in Coconino County of 232 FTE jobs and $12.08 million in economic output (in 2003 US$, adjusted for inflation). From these data, we can determine the employment and output multipliers (1.35 and 1.24, respectively) used in the modeling described in the report. In addition, the expanded resort will contribute higher property and sales tax revenues to the surrounding community, as well as service fees to the USFS.[4]

Without detailed information on the assumptions used in the modeling, it is difficult to directly challenge the data used in the EIS. However, we have reasons to be skeptical about the data. The EIS report estimated that skiers spend $100 per day per person, saying that this figure is based on surveys in Colorado and Utah and at Snowbowl. We believe that it may be somewhat high for Snowbowl, however, for two reasons: the resort estimates that two-thirds of its visitors are day trippers, and according to Snowbowl manager J.R. Murray, 75% of all visitors travel from Phoenix (Benally 2005). We think it is likely that many skiers from Phoenix drive directly to the ski area, ski all day, and then drive directly home. Furthermore, many of these skiers probably rent ski equipment and buy gas in Phoenix. So if a representative Phoenix-based skier bought a $10 lunch on the mountain and a ski pass, he or she might spend $36 to $63 a day, depending on ticket type. The Bed, Board and Booze (BB&B) tax records provide some weight to our argument that the ski resort is not as large an economic driver in Coconino County as was estimated in the EIS. The EIS calculated that winter tourism accounts for 8.6% of the Flagstaff area's overall economy and stated that three linear correlations were tested using BB&B tax data and Snowbowl visitation for the years 1990–2003 (USDA FS 2005, 3-188, 3-125n). The authors reported that they found no statistically significant correlation between skier visitation and BB&B tax collections because Snowbowl is just a small segment of the BB&B tax generated.[5] Another explanation not explored is that no significant relationship was found because the assumptions used in the model were inaccurate. Specifically, that the dominance of Phoenix skiers as well as local Flagstaff skiers, and the consequently low proportion of overnight visitors in Flagstaff, means that indirect expenditure in the Flagstaff area is likely much lower than that calculated in the EIS. Nevertheless, the ski resort is an important local business, and therefore its vulnerability to climate change impacts are of interest to local and state policymakers.

IMPACTS OF CLIMATE CHANGE ON WINTER RECREATION

The predictions of climate change models—less snow, reduced snowpack, and shorter and more erratic snow seasons, with more pronounced effects at lower elevation and lower latitudes—are worrisome for ski industries around the world. The implications are significant for the industry, because ski resorts are often concentrated at low, accessible elevations. Of particular economic concern is the impact climate change will have on small and otherwise marginal communities and regions that are heavily dependent on winter recreation (Breiling and Charamza 1999). Winter recreation research has shown that these small, "community" ski resorts are essential to developing the industry (McBoyle and Wall 1992), as they cater to novices, families, and skiers getting in shape before a longer vacation at a larger ski resort. For example, Scott et al. (2003) reported that 45% of skiers in Toronto, the largest single market of active skiers in Canada, travel less than one hour to ski. Elsasser and Bürki predicted that climate change will be a "catalyst that will reinforce and accelerate the pace of structural change" in the ski industry (2002, 255). For the smaller ski areas in Arizona, a couple of consecutive poor seasons combined with landownership and water resource constraints may lead to more rapid restructuring, which in turn would have a more immediate impact on the surrounding local economies.

Scott et al. noted a lack of current understanding of "how recreational users and tourists respond to climate variability (whether or not to participate or purchase equipment, activity substitution, use patterns, destination choice)" (2003, 180). Bürki et al. (2003) investigated how different types of Swiss skiers might respond to climate change. Their survey results found that during a poor season, 49% of skiers would switch to a more snow-reliable resort, but more worrisome for the industry, 32% of respondents said they would ski less often. Only 4% of the survey respondents said that they would abandon skiing. The broader issue is the differential impact of such demand changes. It is likely that smaller ski resorts would suffer most, as young, day, and novice skiers, their predominant clientele, are the most likely to respond negatively to poor seasons (König 1998). Therefore, a large uncertainty in any modeling of the impact of climate change on winter recreation is the response of local skiers to changes in ski season length, timing, and relative snowpack quality. Scott et al. (2003) argued that one scenario that would leave resorts as well off is if skiers adjusted their recreation behavior by skiing more frequently during the shorter seasons. A sensitivity analysis could address this uncertainty in consumer response and estimate future ski recreation demand.

A systematic regional impact assessment of climate change (Scott et al. 2003) on winter tourism would be necessary to understand the relative impacts on Snowbowl and Sunrise. Scott and colleagues listed several factors that will determine how any single ski resort fares in the future: the relative impact of climate change on the resort and its competitors (a function of elevation, aspect, humidity, snow patterns, and other factors) and any resultant changes in intra- and interregional skiing market share, the costs of additional snowmaking, how adaptation by skiers could alter skiing demand, and the impact of other adaptation strategies, such as business diversification and weather derivatives and insurance. In fact, there are many competitive differences between the two resorts: Snowbowl is better located in respect to large skier markets but has no water rights, whereas Sunrise has sufficient water and slightly better snow conditions, as well as the flexibility afforded by its three peaks. However, it is highly possible that despite their snowmaking investments, both resorts will be somewhat marginal by the end of the century, because the increased winter temperatures predicted as a result of climate change are likely to often be too high to allow for the manufacture of snow.

SNOWMAKING AS AN ADAPTATION TO CLIMATE CHANGE

The decision to invest in snowmaking, the current ski industry climate change adaptation strategy deployed worldwide, is complex. Each resort needs to weigh the costs against the benefits, given the likely responses of consumers and the snow conditions at nearby competing resorts that might be superior as a result of aspect, altitude, or some other factor. Other possible strategies to adapt to climate change include extending slopes to higher elevations, merging with another ski resort to diversify geographic and financial risk (Scott 2003), trading in weather derivatives (financial instruments used by companies to hedge against the risk of weather-related losses), developing summer recreation and conference businesses, and in the Southwest, using enhanced and targeted ENSO-based marketing.

The Intergovernmental Panel on Climate Change (IPCC) uses General Circulation Models (GCMs) to forecast how precipitation and temperature will likely change under various greenhouse gas scenarios. For illustrative purposes, we show predictions for temperatures at Snowbowl in Table 7.2, using the Hadley Center for Climate Prediction and Research/Met Office's UKMO_HADCM3 model and the IPCC's A2 scenario (IPCC n.d.). The first model was chosen because it performs well in representing the climate of the Southwest, with its North American monsoon- and ENSO-dominated winter precipitation; the second is a worst-case scenario. Table 7.2 shows historical minimum mean monthly temperatures at Snowbowl, as well as projected warming and minimum mean monthly temperatures for three future time periods (see also Bark et al. 2009). We chose minimum temperatures because snowmaking is the current adaptation strategy at Snowbowl (as well as Sunrise), and snow is generally manufactured overnight, when temperatures are coldest.

Snowmaking is a temperature-constrained adaptation. The shaded boxes highlight those months when the minimum monthly temperature is forecast to exceed −5 degrees C, the current threshold temperature for snowmaking with newer equipment (Scott et al. 2003). The first impacts may be felt as soon as 2030–2050, when snowmaking in April may become suboptimal. However, because interest in skiing tends to wane in spring (Scott et al. 2003; Murray 2006), the financial

Table 7.2 *Snowbowl Average Minimum Monthly Temperatures, Historical and Projected (°C)*

	Nov	Dec	Jan	Feb	Mar	Apr
Historical mean						
1997–2006	−7.5	−10.0	−9.3	−9.7	−8.1	−6.0
Projected warming						
2030–2050	1.2	2.0	2.6	2.1	1.5	2.2
2051–2080	2.8	3.6	3.2	2.5	1.8	3.1
2081–2098	4.3	3.8	4.3	4.1	4.0	4.9
Projected mean						
2030–2050	−6.3	−8.0	−6.7	−7.6	−6.6	−3.8
2051–2080	−4.7	−6.4	−6.2	−7.2	−6.3	−2.9
2081–2098	−3.2	−6.2	−5.0	−5.6	−4.1	−1.1

repercussions of this warming are likely to be minimal. In 2051–2080, snowmaking in both shoulder seasons may be technically difficult, with possible further contractions in season length by 2081–2098. Snowbowl's mountain manager will have to respond to these changes by optimizing snowmaking during the peak season, because of the resort's water constraints, and later in the century by ensuring that each good snow day is a good revenue day at the resort.

Snowmaking as an adaptation strategy has a number of drawbacks. It can improve season reliability, but it is expensive. Costs, excluding capital costs, range from \$407 to \$3,252 per 1,000 m^3 of snow, depending on system efficiency (Northern Watertek 2007), approximately 15% to 25% of total resort operating costs. Snowmass, in Colorado, provides a higher estimate of snowmaking costs in the southwestern United States, at \$4,065 per 1,000 m^3 inclusive of capital costs (*Snowmass Village Sun* n.d.). However, all these costs are only illustrative. At temperatures warmer than − 2 degrees C (the darker gray shaded cell in Table 7.2), snowmaking is not currently technically viable (Scott et al. 2003), and even at temperatures below this threshold, it may not be financially viable.

Energy costs constitute a large portion of snowmaking costs, second only to labor costs in a typical ski resort's annual operating expenses (Harris 2000). Ski resorts are often located near the edge of electric power grids, and thus their requirement for large amounts of reliable power can present a major challenge for their electricity providers (Troxell et al. 2005). Snowmaking also requires large volumes of water—139,222 gallons of water are needed to make 1 acre-foot of snow (Ratnik Industries 2010)—and water availability is a challenging issue throughout the Southwest. To illustrate, a typical size ski resort produces around 615,000 to 1,230,000 m^3 of snow per season, at a cost between \$0.25 million and \$4 million, and uses 264,000 to 527,000 m^3 of water per season to make this snow.

Table 7.3 *Snowbowl Snowmaking Volumes and Water Contract (10^3 m^3) by Season Type, and Snowmaking Economics*

	El Niño	*La Niña*	*Neutral*
Snowmaking plan			
Snow volume	526	1,053	789
Water contract	230	449	600
Skiable days, average			
Preinvestment	117	68	101
Postinvestment	156	94	125
Additional skiable days	39	26	24
Snowmaking economics			
Snowmaking costs[a]	\$2,138,190	\$4,280,445	\$3,207,285
Additional revenues[b]			
Maximum capacity	\$12,480,000	\$8,320,000	\$7,680,000
Three-fourths capacity	\$9,360,000	\$6,240,000	\$5,760,000
Half capacity	\$6,240,000	\$4,160,000	\$3,840,000

[a]Assuming \$4,065 per 1,000 m^3
[b]Assuming an average lift ticket price of \$40
Source: USDA FS (2005)

In Arizona, Snowbowl's investment plan envisages three distinct season types: wet, dry, and normal, corresponding to El Niño, La Niña, and neutral seasons. The snowmaking and water requirements and expected average skiable days vary by season type (see Table 7.3).

With investment in snowmaking, Snowbowl's management forecasts that the resort will be open longer during every season type. Daily capacity after the terrain and chairlift expansion will be 8,000 skiers. Therefore, each additional skiable day could boost daily revenues by $208,000 to $424,000, depending on the mix of lift tickets sold. For Table 7.3, we assume snowmaking costs are equivalent to those reported at Snowmass, and that on average, lift ticket prices are $40. Using these estimates, we calculate the costs of snowmaking and also the additional seasonal revenues (postinvestment minus preinvestment) under three scenarios: full capacity, three-fourths capacity, and half capacity. In all cases except a La Niña season with half capacity, the snowmaking investment pays for itself. (We assume that a similar exercise was completed for Sunrise's investment plan, and that it too was deemed profitable.) This table highlights how important both the supply and demand sides are: a resort needs adequate runs with sufficient snow, as well as pricing and marketing policies designed to ensure that every good snow day is a good revenue day.

Investment in snowmaking is currently the most popular adaptation strategy, but in the future, it may become more difficult to finance. Bürki et al. (2003) found evidence that Swiss banks are becoming more wary of funding infrastructure investments at ski resorts located below 1,500 meters (4,921 feet). Scott et al. (2003) found that ski area managers in southern Ontario, Canada, have to assess and address climate change impacts in financing negotiations with lenders. There are few other infrastructure-based alternatives. In Switzerland, the ski industry has used climate change as an argument to open new ski runs at high Alpine regions above 3,000 meters (9,843 feet), despite environmental concerns (Bürki et al. 2003). These types of adaptation draw attention away from the demand side; demand for skiing is influenced not only by snowpack, but also by relative snow conditions (Moeltner and Englin 2004), prices at competing resorts, and the availability of other recreation activities. Marketing to skiers in the shoulder seasons is one way resorts can capitalize on their snowmaking investments.

CONCLUSIONS

Ski resorts are a proverbial "canary in a coal mine" for the costs of climate change adaptation and the local economic impacts of climate change (Bustic et al. 2008). Snowmaking adaptation also illustrates the dependence on water and energy resources, sharing this role with the types of water-based recreation discussed in Chapter 6. Climate change models predict shorter snow seasons, which would undermine the longer-term financial viability of ski resorts. However, snowmaking investments enable resorts to adapt to warmer winters in the short and medium term, as well as to more general climate variability (Adger et al. 2007). By the end of the century, this adaptation may no longer be technically viable, forcing a new round of closures and pressures to redirect resources to spring, summer, and fall recreation and tourism. Developing a strategy to attract visitors and generate revenues during these other seasons is an important resilience strategy not only for snow-dependent businesses, but also for local economies that rely on them.

Water and energy requirements render an adaptation strategy based primarily on snowmaking vulnerable to resource cost and availability uncertainties that climate variability will accentuate, as discussed in earlier chapters of this book. In recognition of this vulnerability, the state of New York has been considering legislation specifically to assist ski resorts with the energy cost burden of increased reliance on snowmaking (Assembly Bill A00519 and Senate Bill S836, March 2010).

The increased probability of poor snowfall as a result of climate change is likely to accelerate restructuring in the ski industry worldwide, favoring resorts at higher elevations and latitudes, more geographically diversified companies, and those able to implement and afford snowmaking investments. In the U.S. Southwest, the winners are likely to be resorts in Colorado and Utah, and the losers those in New Mexico and Arizona. Policymakers need to develop a comprehensive regional development plan, such as in the White Mountain area, to ease future structural adjustment resulting from a drier and warmer climate. This might include investment in alternative sectors, such as nonwinter recreation (Scott 2003; EPA 2003), second homes, and retirement (Gibson 1997).[6]

We have focused on Sunrise and Snowbowl in this chapter. However, the overall impact of changing climate on these two resorts should be assessed within a larger framework that takes account of climate impacts elsewhere in the region. One possible outcome is that even if ski seasons are shorter at Sunrise, if snowfall declines are worse at Snowbowl, then Sunrise might benefit, as some skiers who focus on quality switch to Sunrise for powder (Moeltner and Englin 2004). Such a substitution effect would be the best-case scenario, although individual skiers would be displaced from their usual recreation area. However, another likely scenario is that skiing declines overall in the state and region. Future research could model climate change scenarios for Sunrise and Snowbowl and the other southwestern ski areas. Such modeling, when combined with minimum snow requirements for skiing, management goals, snowmaking costs, water requirements, and the relative position of a resort's competitors, would enable better assessment of the most efficient timing of investments in increased snowmaking.

In the meantime, resort management can use other adaptation strategies to ensure that every good snow day at the resort is fully utilized by skiers and profits are maximized (Bark et al. 2009). Modern, engaging websites with links to onsite webcams allow Internet users to see snow and weather conditions before they travel to the resort. This can counter the "backyard hypothesis" that "urban snow conditions significantly affect skier activity" (Hamilton et al. 2007). For desert cities, this means that nearby ski resorts need to convince their urban populations, living in warm, sunny valleys, that the resorts' mountain snow is adequate for good skiing.

The main ski industry adaptation technology, snowmaking, is constrained by climate thresholds, availability of water and energy, costs, and environmental concerns. The demand for adaptation has generated incentives to improve the water and energy efficiency of snowmaking equipment, as well as the technology to manufacture snow at higher temperatures. Investments in adaptation to climate variability and climate change can mitigate or postpone the regional economic dislocation that would be associated with a climate-vulnerable resort's closure and can reduce individual skiers' recreation displacement.

A complementary climate adaptation strategy is to more fully employ climate forecasts that have a high level of predictive skill, such as ENSO forecasts. Vulnerable low-elevation resorts, from the southwestern United States to New Zealand, could use ENSO forecasts to optimize resort management. Using such forecasts, resort operators in the Southwest could plan ahead to increase advertising and hire additional employees to capitalize on strong El Niño conditions, and budget for increased snowmaking in La Niña years. The ski industry and communities that depend on winter recreation can both lead the way in pioneering adaptation strategies and play a prominent role in crafting a more resilient regional economy in mountainous areas.

ACKNOWLEDGEMENTS

This research was funded by the National Oceanic and Atmospheric Administration's Climate Assessment in the Southwest (CLIMAS) program.

NOTES

1. A consortium of tribes and environmental groups opposed using reclaimed water on the sacred San Francisco Peaks and won their case in a Ninth Circuit Court of Appeals ruling (*Navajo Nation v. USFS*) in 2007. Snowbowl operators appealed this decision, arguing that snowmaking was necessary to ensure the future of the resort and the economic benefits it brings to northern Arizona. In August 2008, the Ninth U.S. Circuit Court of Appeals overturned the previous ruling.

2. $F = 41.1$, $p > F = 0.002$. The parameter estimates are CONSTANT $-16{,}022$, $p > |t| = 0.279$; SNOW 120, $p > |t| = 0.002$; and DAYS_OPEN 536, $p > |t| = 1.024$. For example, if the resort receives 10 centimeters (3.9 inches) more snow over the season, visitor numbers are expected to be 1,200 higher, and an additional day open is predicted to raise visits by 536 skiers.

3. This data had to be estimated because the tribe does not make such data publicly available. To estimate Sunrise data, we used lift capacity data and an assumption that Sunrise is open more often than the two smaller resorts in the state. We assumed that Sunrise accounted for 90% of the non-Snowbowl Arizona ski visitor data, slightly more than its 85% share of the non-Snowbowl Arizona lift capacity.

4. The report estimates that the expansion of Snowbowl will create an estimated 232 FTE construction jobs and $21.24 million in economic output in Coconino County. At the end of the 10-year planning period, when Snowbowl anticipates it will attract on average 215,000 visitors a year, local economic output attributable to Snowbowl is forecast to be $23.7 million. Snowbowl also contributes revenues to the USFS through service fees and to its community through property and sales taxes.

5. A separate analysis by city of Flagstaff tax and licensing administrator Steve Shafer (in Benally 2005) compared retail sales tax revenues in Flagstaff with the number of days open at

Snowbowl. In the 2000–2001 season, Snowbowl was open for 138 days, and retail sales tax revenue was the highest recorded in Flagstaff. However, a year later, Snowbowl was open for just 4 days, yet sales tax revenue hit another record. Shafer concluded that the relationship between sales tax revenues and Snowbowl was minimal, and that one reason a bad snow year may be good for retail sales tax revenue is that the roads are open and more shoppers from around the region visit Flagstaff's retail stores.

6. Arizona's ski resorts already have strong nonwinter recreation programs. Snowbowl anticipates that its improved facilities will attract 30,000 summer visitors a year. Sunrise generates around $1 million in nonwinter revenues from Sunrise Lake Marina, a hotel, a general store, and nonskier users of chairlifts (Gibson and Evans 2001). Data from the White Mountains show that summer recreation is the biggest contributor to the local economy (Gibson and Evans 2001). A change in climate could further shift recreation visitation to the summer months, as Phoenix and Tucson valley residents seek respite from even higher temperatures. Scott and McBoyle (2001) cautioned that a change in recreation seasonality could have other implications. For example, current tourism infrastructure might be inadequate to meet higher demand, or additional visitation might increase environmental stress.

REFERENCES

Adger, W.N., S. Agrawala, M.M.Q. Mirza, C. Conde, K. O'Brien, J. Pulhin, R. Pulwarty, B. Smit, and K. Takahashi. 2007. Assessment of Adaptation Practices, Options, Constraints and Capacity. Climate Change 2007: Impacts, Adaptation and Vulnerability. In *Contribution of Working Group II to the Fourth Assessment Report of the Intergovernmental Panel on Climate Change*, edited by M.L. Parry, O.F. Canziani, J.P. Palutikof, P.J. van der Linden and C.E. Hanson. Cambridge, UK: Cambridge University Press, 717–743.

Bark, R.H., B.G. Colby, and F. Dominguez. 2009. Snow Days? Snowmaking Adaptation and the Future of Low Latitude, High Elevation Skiing in Arizona, USA. *Climatic Change* doi:10.1007/s10584-009-9708-x.

Benally, Klee ed. and dir.. 2005. *The Snowbowl Effect: When Recreation and Culture Collide, Who Pays the Price?* Video recording. Flagstaff, AZ: Indigenous Action Media.

Bloomfield, J., and S.P. Hamburg. 1997. *Seasons of Change: Global Warming and New England's White Mountains*. New York: Environmental Defense Fund.

Breiling, M., and P. Charamza. 1999. The Impact of Global Warming on Winter Tourism and Skiing: A Regionalised Model for Austrian Snow Conditions. *Regional Environmental Change* 1 (1):4–14.

Breiling, M., P. Charamza, and O. Skage. 1997. *Klimasensibilitat Osterreichischer Bezirke mit besonderer Berucksichtigung des Wintertourismus [The Sensitivity to Climate of Austrian Districts with Special Consideration of Winter Tourism]*. Report 97:1. Alnarp, Austria: Institute for Landscape Planning.

Brown, D.P., and A.C. Comrie. 2004. Winter Precipitation "Dipole" in the Western United States Associated with Multidecadal ENSO Variability. *Geophysical Research Letters* 31:L09203, doi:10.1029/2003GL018726.

Bürki, R., H. Elsasser, and B. Abegg. 2003. Climate Change: Impacts on the Tourism Industry in Mountain Areas. Paper presented at the 1st International Conference on Climate Change and Tourism April 9–11, 2003, Djerba, Tunisia.

Bustic, V., E. Hanak, and R.G. Valletta. 2008. Climate Change and Asset Prices: Hedonic Estimates for North American Ski Resorts. Federal Reserve Bank of San Francisco Working Paper Series, WP 2008-12, www.frbsf.org/publications/economics/papers/2008/wp08-12bk.pdf (accessed September 21, 2010).

Elsasser, H., and R. Bürki. 2002. Climate Change as a Threat to Tourism in the Alps. *Climate Research* 20:253–257.

EPA (U.S. Environmental Protection Agency). 2003. EPA Research and Development, National Center for Environmental Research (NCER) Research Project Search. http://cfpub.epa.gov/ncer_abstracts/index.cfm (accessed September 21, 2010).

———. 2010. National Environmental Policy Act (NEPA). www.epa.gov/compliance/nepa (accessed September 21, 2010).

Galloway, R.W. 1988. The Potential Impact of Climate Changes on Australian Ski Fields. In *Greenhouse: Planning for Climate Change*, edited by G.I. Pearman. Melbourne, Australia: Commonwealth Scientific and Industrial Research Organisation (CSIRO), 428–437.

Gibson, L.J. 1997. Using Economic Base Analysis to Solve Development Planning Problems: A Best Practice Approach. *Applied Geographic Studies* 1 (3):169–186.

Gibson, L.J., and B. Evans. 2001. *White Mountain Winter Tourism Study: Evaluating the Efficacy of Regional Investment Opportunity.* Tucson: University of Arizona.

———. 2002. Regional Dependence on Tourism: The Significance of Seasonality. *Yearbook of the Association of Pacific Coast Geographers* 64:112–127.

Goodrich, G.B. 2004. Influence of the Pacific Decadal Oscillation on Arizona Winter Precipitation during Years of Neutral ENSO. *Weather Forecasting* 19:950–953.

Gutzler, D.S., D.M. Kann, and C. Thornbrugh. 2002. Modulation of ENSO-Based Long-Lead Outlooks of Southwestern U.S. Winter Precipitation by the Pacific Decadal Oscillation. *Weather Forecasting* 17:1163–1172.

Hamilton, L.C., C. Brown, and B.D. Keim. 2007. Ski Areas, Weather and Climate: Time Series Models for New England Case Studies. *International Journal of Climatology* doi:10.1002/joc.1502.

Harris, T. 2000. How Snow Makers Work. http://adventure.howstuffworks.com/outdoor-activities/snow-sports/snow-maker.htm (accessed September 21, 2010).

Harrison, S.J., S.J. Winterbottom, and C. Sheppard. 1999. The Potential Effects of Climate Change on the Scottish Tourist Industry. *Tourism Management* 20:203–211.

IPCC (Intergovernmental Panel on Climate Change). n.d. Emissions Scenarios: A2 Storyline and Scenario Family, www.ipcc.ch/ipccreports/sres/emission/094.htm (accessed September 21, 2010).

König, U. 1998. Tourism in a Warmer World: Implications of Climate Change due to Enhanced Greenhouse Effect for the Ski Industry in the Australian Alps. *Wirtschaftsgeographie und Raumplanung [Economic Geography and Spatial Planning]* 28. Zurich, Switzerland: University of Zurich.

König, U., and B. Abegg. 1997. Impacts of Climate Change on Tourism in the Swiss Alps. *Journal of Sustainable Tourism* 5 (1):46–58.

Lamothe and Périard Consultants. 1988. Implications of Climate Change for Downhill Skiing in Quebec. In *Climate Change Digest* 88-03. Ottawa, Ontario: Environment Canada.

McBoyle, G.R., and G. Wall. 1992. Great Lakes Skiing and Climate Change. In *Mountain Resort Development*, edited by A. Gill and R. Hartmann. Burnaby, Canada: Simon Fraser University, Center for Tourism Policy and Research, 77–81.

Moeltner, K., and J. Englin. 2004. Choice Behavior under Time-Variant Quality: State Dependence versus "Play-It-by-Ear" in Selecting Ski Resorts. *Journal of Business and Economic Statistics* 22 (2):214–224.

Murray, J.R. 2006. Snowbowl to Open Dec. 8 if We Had Snowmaking. www.arizonasnowbowl.com/news/snowmaking.html (accessed August 21, 2007; site discontinued).

National Assessment Synthesis Team. 2000. *Climate Change Impacts on the United States: The Potential Consequences of Climate Change Variability and Change.* U.S. Global Change Research Program. Cambridge, UK: Cambridge University Press.

NOAA (National Oceanic and Atmospheric Administration). 2010. Cold & Warm Episodes by Season, 1950–2010. NOAA National Weather Service Climate Prediction Center, www.cpc.ncep.noaa.gov/products/analysis_monitoring/ensostuff/ensoyears.shtml (accessed September 21, 2010).

Northern Watertek. 2007. SkiFax #1 Newsletter. www.watertek.org/Skifax1.htm (accessed January 5, 2005; site discontinued).

OECD (Organisation for Economic Co-operation and Development). 2006. Climate Change in the European Alps: Adapting Winter Tourism and Natural Hazards Management. www.oecd.org/document/45/0,2340,en_2649_34361_37819437_1_1_1_1,00.html#Summaries (accessed September 21, 2010).

Patten, J.M., S.R. Smith, and J.J. O'Brien. 2003. Impacts of ENSO on Snowfall Frequencies in the United States. *Weather and Forecasting* 18:965–980.

Public Law No.111–291, title III, White Mountain Apache Tribe Water Rights Quantification Act of 2010.

Ratnik Industries. 2010. Snowmaking 101. www.ratnik.com/snowmaking.html (accessed September 21, 2010).

Scott, D. 2003. Climate Change and Tourism in the Mountain Regions of North America. Paper presented at the 1st International Conference on Climate Change and Tourism. April 9–11, 2003, Djerba, Tunisia.

Scott, D., and G. McBoyle. 2001. Using a "Tourism Climate Index" to Examine the Implications of Climate Change for Climate as a Natural Resource for Tourism. In *Proceedings of the First International Workshop on Climate, Tourism and Recreation*, edited by A. Matzarakis and C. de Frietas. International Society of Biometeorology, Commission on Climate, Tourism and Recreation. October 5–10, 2001, Halkidi, Greece, 69–98.

Scott, D., G. McBoyle, and B. Mills. 2003. Climate Change and the Skiing Industry in Southern Ontario (Canada): Exploring the Importance of Snowmaking as a Technical Adaptation. *Climate Research* 23:171–181.

Snowmass Village Sun. n.d. The Art of Making Snow. www.snowmassvillagesun.com/home.php?content=article&article=1453 (accessed August 23, 2007; site discontinued).

Troxell, W., A. Davis, J. Dean, and S. Jansen. 2005. *Final Report: Energy Management Innovation in the US Ski Industry.* Washington, DC: U.S. Environmental Protection Agency, National Center for Environmental Research.

USDA FS (U.S. Department of Agriculture, Forest Service). 2005. *Final Environmental Impact Statement for Arizona Snowbowl Facilities Improvements,* vol. 1, *Coconino National Forest, Coconino County, Arizona.* Flagstaff, AZ: USDA FS.

Weiher, R., 2000. *Improving El Niño Forecasting: The Potential Economic Benefits.* Washington, DC: National Oceanic and Atmospheric Administration.

PART III

INFORMATION, TECHNOLOGY, AND ADAPTATION

CHAPTER 8

Irrigator Demand for Information, Management Practices, and Water Conservation Program Participation: The Role of Farm Size

George B. Frisvold and Shailaja Deva

Irrigation is the principal water user in the western United States. In the 11 westernmost contiguous states, irrigation accounts for 84% of freshwater withdrawals (Hutson et al. 2005). Small changes in irrigator behavior have important implications for availability of water for other uses. A 10% reduction in irrigation water use would increase water available for all other uses by 50%. With rapid population growth in the West, competition for water for agricultural, riparian habitat protection, residential, and commercial purposes will intensify.

In addition to being the primary user of western water, irrigation is a primary contributor to nonpoint-source water pollution and to salinity problems (Frisvold 2004; CAST 1992; EPA 1998). In the western states, irrigation accounts for 89% of river and 40% of lake pollution from sediment and chemical runoff (NRC 1996). Salinity is a particular problem in the Southern Plains, California, and the Colorado River Basin. Lee and Howitt (1996) estimate that an optimal policy to reduce salinity in the Colorado Basin would reduce salt from the river by more than 1.2 million tons/year, reduce damages to municipal and industrial users in the Lower Basin by nearly $93 million/year, and provide net social benefits of $55 million/year. Agriculture will be at the center of debates over the allocation of western water and control of nonpoint-source water pollution. Important policy questions thus become, how and how well is irrigation water managed?

IRRIGATION EFFICIENCY

Improving irrigation efficiency is seen as an important step toward reducing water pollution and easing competition for scarce water in the West (Caswell and Zilberman 1985; Dinar et al. 1992; Green et al. 1996). Improved irrigation efficiency is also seen as a means to adapt to climate change (e.g., Mendelsohn and Dinar 2003; Parry et al. 1998).

Major policy initiatives in North America have aimed at improving irrigation efficiency. The 2002 Farm Bill set ambitious water conservation goals, relying on farm payments to encourage water conservation on nearly 25 million acres, primarily in the western United States. The U.S. Department of Agriculture's Natural Resources Conservation Service (USDA NRCS) projected that agricultural water applications could be reduced 5.41 inches per acre by switching to dryland production, less water-intensive crops, and more water-efficient practices, and stated that "any water saved would be available for alternative uses such as by municipalities, utility generation, and wildlife habitat restoration" (2003, 56). Assuming 20% loss in storage and transmission, the program would free 8.9 million acre-feet (ac-ft) of water per year over 10 years. This is sizable, considering that total U.S. offstream water withdrawals for domestic and commercial use are 50 million to 55 million ac-ft/year (Frisvold 2004). Similar ambitious goals have been set for irrigation efficiency improvements in western Canada (Bjornlund et al. 2009).

Despite the importance placed on improved irrigation management, federal agencies do not collect comprehensive annual data about irrigator behavior. The U.S. Geological Survey (USGS) has published *Water Use in the United States* at five-year intervals (e.g., 2005, 2000, back to 1950). Data are published by state, and more recently (starting in 2000) by county, with data on acres irrigated and water withdrawn by irrigation method used (flood, sprinkler, or micro) and by source (groundwater versus surface water). Mendelsohn and Dinar (2003) have used this data to examine interrelated effects of climate and irrigation on agricultural land values and explore the influence of climate, economic factors, and agronomic factors on use of irrigation and choice of irrigation technologies. The USGS reports this data with long time lags, however. The 2005 report was not available until October 2009.

As a follow-on survey to the Census of Agriculture, the USDA conducts a Farm and Ranch Irrigation Survey (FRIS) roughly every five years, with the most recent survey conducted in 2008. The FRIS provides the most detailed, comprehensive picture of irrigation practices and water use at the national and state levels. Although FRIS data is often used for descriptive outreach publications, it has been used relatively rarely in econometric analyses of irrigator behavior. The USDA publishes FRIS data as cross-tabulations emphasizing relationships between two variables, such as water use intensity by crop or by method of irrigation. Because the tables focus on 2 × 2 relationships, they do not lend themselves to standard multivariate regression analysis. To protect respondent confidentiality, access to raw survey data is restricted. For this reason, relatively little econometric analysis has been carried out using the FRIS data. This research has been conducted almost exclusively by USDA economists and collaborators, who have easier access to the farm-level survey data (Negri and Brooks 1990; Schaible et al. 1991; Moore et al. 1993, 1994; Schaible 1997; Schaible and Aillery 2003; Negri et al. 2005). These studies are done well and offer important insights into irrigator behavior, but they are few given the size and scope of the FRIS data.

Other empirical analyses of irrigator behavior (Caswell and Zilberman 1985; Dinar et al. 1992; Parker and Zilberman 1996; Green et al. 1996; Knapp and Schwabe 2008; Just et al. 1983; Antle and Hatchett 1986; Skaggs and Samani 2005) have often relied on smaller, localized surveys. Research questions and data collected vary from study to study, which can make comparisons across studies challenging. In addition, a large share of these studies were conducted in California, raising questions over whether relationships found in California's production conditions—a warm, arid climate, emphasizing high-value specialty and tree crops—hold over a wider geographic area.

This chapter demonstrates that, despite limitations, it is possible to use publicly available USDA FRIS data to test economic hypotheses concerning irrigator behavior. We begin by introducing a simple economic model of the returns to information acquisition. Information and information technologies allow irrigators to better match water applications to plant water needs. From the model, we then derive hypotheses about the relationship between farm size and use of irrigation information. Next, we apply nonparametric, univariate measures of association (Gibbons 1993) to analyze FRIS cross-tab data for Arizona and New Mexico. We examine the relationships between farm size and (a) sources and uses of water management information; (b) barriers to improving irrigation systems for water or energy conservation; and (c) participation in government conservation programs.

THE ROLE OF FARM SIZE

We focus on the role of farm size for two reasons. First, the USDA Economic Research Service (ERS) special tabulation stratifies the FRIS data by farm size. That is the data we have. One could accuse us of being like the person who looks under a street lamp for lost keys, not because the keys are there, but because the light is better. Yet both economic theory and previous empirical findings suggest systematic relationships between scale of operation and use of water management (or other agricultural management) information.

Clearfield and Osgood (1986) determined that larger farms are more likely to adopt conservation practices generally. Leib et al. (2002) discovered significant positive relationships between farm size and adoption of scientific irrigation scheduling methods (use of crop evapotranspiration data and soil moisture testing) among Washington State farmers. Skaggs and Samani, in a study of Elephant Butte Irrigation District in New Mexico, reported a "lack of interest in making improvements to current irrigation systems or methods on the smallest farms" (2005, 43). Comparing irrigation districts in Alberta, Canada, Bjornlund et al. (2009) found evidence of greater adoption of information-intensive irrigation management in areas with larger farms.

Smit and Skinner (2002) have emphasized the importance of both information systems and improved water management as adaptations to climate variability. Improved irrigation practices can protect growers against drought and moisture deficiencies brought about by climate change. Weather and climate information systems can serve as early warning systems and allow for adaptations on different temporal scales, from short-term weather forecasts to seasonal forecasts to longer-term climate projections. Smit and Skinner also emphasized that climate change does not simply mean changes in long-run mean temperatures or precipitation levels; it also means changes in climate variability and in probabilities of extreme events. Improved irrigation scheduling can enhance adaptation to climate change and variability. Improved climate and weather information can, in turn, aid irrigation and other agronomic scheduling.

VALUE OF IRRIGATION INFORMATION: A SIMPLE MODEL

In this section, we present a simple model of grower demand for information to make irrigation (and other water management) decisions. Irrigators face a fundamental trade-off: irrigation information can be valuable, but acquiring and applying information involve certain costs. The model is used to examine factors that encourage or discourage irrigation information use. It is then also used to derive economic hypotheses that we later test.

One may express the value of information to an individual irrigator, *V*, as

$$V = \{\pi[x(s),\ z(s)] - \pi_0(x_0, z_0)\}A - c(s, k, A) \tag{8.1}$$

where

A	= acreage;
x_0	= variable inputs chosen when information is not accessed;
z_0	= parameters characterizing irrigation management technologies or practices chosen when information is not accessed;
$\pi_0\ (x_0,\ z_0)$	= per-acre profits when information is not accessed;
s	= the number of information sources accessed or the intensity of use from a given source;
$x(s)$	= variable inputs chosen given new information;
$z(s)$	= technologies or practices chosen given new information;
$\pi\ [x(s),\ z(s)]$	= per-acre profits given information;
c	= cost of processing information;
k	= index of human capital or technical capacity; and
c_i	= first derivative of information cost function with respect to $i = s$, k, or A.

The value of information is the increase in farm profits from using the information minus any costs entailed in processing the information to make it useful for farm-level decisions. Information has positive value only if it alters an operator's choice of inputs x or technology/practices z. In a dynamic extension of the model, information could affect the optimal timing of irrigation and other practices. One could also extend the model to account for crop choice and multioutput production. This model is similar in approach to those of Feder and Slade (1984), who considered active information acquisition and technology adoption generally; Johnson and Holt (1997), who examined the value of weather information; and Parker and Zilberman (1996), who studied use of the California Irrigation Management Information System (CIMIS).

Per-acre profits, π, are the difference between per-acre revenues, R and per-acre costs, C. The value of information is then

$$\begin{aligned} V &= \{R[x(s), z(s)] - C[x(s), z(s)]\}A - \{R[x_0, z_0] - C[x_0, z_0]\}A - c(s, k, A) \\ &= \{R[x(s), z(s)] - R[x_0, z_0]\}A - \{C[x(s), z(s) - C[x_0, z_0]\}A - c(s, k, A) \end{aligned} \tag{8.2}$$

Now, let the gains from information be expressed as percent increases in revenues per acre, $\Delta_R(s)$; percent reductions in costs per acre, $\Delta_C(s)$; or both. The value of information is

$$V = \{R[x_0, z_0][1 + \Delta_R(s)] \;-\; C[x_0, z_0][1 + \Delta_C(s)]\}A - c(s, k, A) \tag{8.3}$$

where

$$\partial V^2/\partial s \partial R > 0;\; \partial V^2/\partial s \partial A > 0;\; \partial V^2/\partial s \partial RA > 0 \tag{8.4}$$

Irrigators with larger baseline sales revenues, $R_0A = R(x_0, z_0)A$, will have greater marginal benefits from processing information. There is an acreage effect and a revenue per-acre (price) effect. Producers growing high-value crops may have limited acreage but still be "large farms" in terms of revenue. The benefits from accessing information increase with farm size where acreage, value of crop, or both contribute to size. If $c_A = 0$, then the cost of processing information is independent of scale, while the benefits of information increase with scale. If processing information is independent of scale, larger irrigators will process more information.

If $c_k < 0$, then irrigators with more human capital or technical capacity can process information at lower cost. If k is positively correlated with sales revenue RA, then we will observe larger irrigators processing more information or using technical sources of information. If $c_s > 0$, the cost of processing information increases with the amount of information or the number of sources accessed. In their model of information acquisition and technology adoption, Feder and Slade (1984) treat cumulative knowledge as a stock variable. Similarly, here irrigators can actively access more information, but at increasing costs. As in Feder and Slade, operations below some critical size may not acquire information.

Consider a case where irrigators can access two types of information. USDA or extension specialists provide lower-cost, but more general, information. Private consultants provide specific information. Consultants tailor this higher-cost information to particular operation conditions and needs. In this case, both large and small operators rely on lower-cost, but more general, public information, whereas large irrigators are more likely than small ones to seek tailored information from private sources. Feder and Slade (1984) characterized information obtained from other farmers as passive and relatively costless. If this were the case, one would expect relatively high use of information from neighbors, and that smaller irrigators would rely on their neighbors (relative to other more costly sources) more than larger irrigators would. Thus, although adoption of low-cost information would be broad, adoption of tailored, private information would be concentrated among larger irrigators, and we may therefore expect that larger irrigators will use more information sources in total than small irrigators will.

Larger farms, in terms of revenues, may be more likely to access more sources of information for another reason. Irrigation information takes time to process. How much time operators have to collect and process information depends on whether farming is a part-time or full-time endeavor. Operators more actively engaged in off-farm work will have less time to devote to farming operations. Studies have found that off-farm work discourages adoption of management-intensive integrated pest management (McNamara et al. 1991; Fernandez-Cornejo et al. 1994; Fernandez-Cornejo 1996, 1998; Fernandez-Cornejo and Jans 1996) and information acquisition for livestock feeding practices (Wozniack 1993). Hoppe (2001) and Fernandez-Cornejo (2007) found evidence that greater off-farm work was negatively associated with adoption of precision farming techniques. Hoppe noted that operators of smaller farms (in terms of sales) worked off-farm more. In Arizona and New Mexico, only 55% of small operators report farming as their principal occupation, whereas more than 87% of operators in the larger farm size classes report farming as their principal occupation (USDA NASS 2002).

Fernandez-Cornejo (2007) has pointed out that some innovations are management time-saving (e.g., herbicide tolerant soybeans), while others are time-intensive (e.g., integrated pest management, precision farming). Management-intensive practices include scheduling irrigation based on soil- or plant-moisture sensing devices, commercial irrigation scheduling services, or computer simulation models. Parker and Zilberman (1996) have argued that basic weather information has aspects of a public good, and that it substitutes (imperfectly) for soil-moisture monitoring devices. If off-farm work declines with farm sales class, larger farms will be more likely to adopt management-intensive irrigation practices. Parker and Zilberman found that the main users of California's CIMIS system were growers of high-value crops.

Our simple model suggests that farm size may affect information acquisition for irrigation management through multiple pathways. It also suggests that farm size may be a proxy for other important factors, such as human capital or off-farm work.

The foregoing suggests some testable hypotheses about irrigator use of information:

H1: The probability that an irrigator accesses any given source of information increases with farm size (with size measured in terms of acreage, sales, or both).

H2: Both small and large irrigators will rely on low-cost sources of information (e.g., government, extension agencies, neighbors). Reliance on private, tailored sources of information (e.g., from private consultants) will increase with farm size.

H3: Following directly from H2, larger irrigators will access a greater total number of information sources.

H4: Use of management-intensive methods to schedule irrigation will increase with farm size, if reliance on off-farm work decreases with farm size.

FARM DATA FOR ARIZONA AND NEW MEXICO

Data used here come from special tabulations of the 1998 Farm and Ranch Irrigation Survey made available by the ERS (USDA ERS 2004). Data are available in cross-tab form and contain several categorical variables related to farmer and rancher irrigation practices. These data include several yes/no responses about use of information, technology, investments to conserve water or energy, and participation in conservation programs. Although the regular FRIS report does not report detailed data by farm sales class, the ERS special tabulation breaks down responses by four farm sales classes:

- small farms, with sales less than $100,000;
- medium farms, with sales from $100,000 to $249,999;
- large farms, with sales from $250,000 to $499,999; and
- very large farms, with sales of $500,000 or greater.

We are using data from 1998 instead of one of the more recent surveys from 2003 or 2008 because data stratified by farm size are not yet available for those survey

Table 8.1 *Distribution of Farms and Irrigation Water Applied in Arizona and New Mexico, by Farm Sales Class*

	Small	*Medium*	*Large*	*Very large*
Arizona				
Percent of farms	64%	8%	11%	18%
Percent of water applied	4%	7%	12%	77%
Cumulative percent of farms	64%	72%	82%	100%
Cumulative percent of water applied	4%	11%	23%	100%
Total number of farms (Arizona total)	2,637			
Total water applied (thousand acre-feet) (all farms)	4,118			
Acre-feet applied/farm (all farms)	1,562			
New Mexico				
Percent of farms	88%	5%	3%	4%
Percent of water applied	26%	12%	15%	48%
Cumulative percent of farms	88%	93%	96%	100%
Cumulative percent of water applied	26%	38%	52%	100%
Total number of farms (New Mexico total)	6,035			
Total water applied (thousand acre-feet) (all farms)	1,730			
Acre-feet applied/farm (all farms)	287			

Note: Reported values may not sum exactly as a result of rounding years. Efforts are currently under way at the ERS to report FRIS data by farm size class using data from the more recent surveys, and it is our hope that analyses such as those carried out here and in Chapter 9 of this book demonstrate the usefulness of stratified special tabulations and encourage their further development.

Data for Arizona and New Mexico reveal that a small share of irrigators account for a large share of water applied (Table 8.1). In Arizona, small farms account for 64% of farms, but only 4% of the water applied. Small, medium, and large farms combined account for 82% of irrigators, but only 23% of water applied. In New Mexico, small farms account for 88% irrigators, but 26% of water used. In contrast, very large farms account for 18% of irrigators and 77% of water used in Arizona, and 4% of irrigators and nearly half of water used in New Mexico. The table also illustrates that farm operations in New Mexico are more numerous and smaller on average than in Arizona. One reason for this difference is New Mexico's tradition of small-scale production and community-based water management organizations called acequias. This system of small-scale, community water management relies on earthen canals for gravity flow irrigation. It dates back to institutions established during Spanish colonization (Rivera 1998). The acequia system took hold more significantly in New Mexico and southern Colorado than elsewhere in the U.S. Southwest and continues today.

MEASURES OF ASSOCIATION FOR FARM SIZE CLASS AND FARMER RESPONSES

The data from the special FRIS tabulations can be arranged as a set of 4 × 2 contingency tables, with four rows for farm size class and two columns representing no and yes responses to a survey question. For example, Table 8.2 shows Arizona farmer responses about whether they received cost-share payments for irrigation or drainage improvements through the USDA's Environmental Quality Incentive Program (EQIP).

There appears to be a definite positive association between farm size and EQIP participation. Only 1% of small farms reported participating in EQIP, but the percentage increases for each larger farm size class. Another question asks Arizona farmers if they rely on extension agents or university specialists for water management information (Table 8.3)

Table 8.2 *Arizona Farms Participating in the EQIP Program in Previous Five Years*

Farm size	*No*	*Yes*	*Percent responding yes*
Small	1,661	18	1%
Medium	203	13	6%
Large	246	33	12%
Very large	343	120	26%

Source: USDA NASS (1998)

Table 8.3 *Arizona Farmers Relying on Extension Agents or University Specialists for Water Management Information*

Farms	*No*	*Yes*	*Percent responding yes*
Small	1,163	515	31%
Medium	118	98	45%
Large	140	139	50%
Very large	278	186	40%

Source: USDA NASS (1998)

Here a relationship between farm size and information use is not obvious. The small farm category has a lower use rate than the other, larger farm size classes, but the very large farm category has a lower use rate than the middle two size classes.

It would be useful to have a method for measuring the strength of the association between farm size and different yes/no responses. The commonly used Pearson correlation coefficient assumes that two variables have a bivariate normal distribution and are measured numerically. The variables in the tables above are categorical (farm size and yes/no). We could also treat the variables as ordinal. This is easy to see for farm size classes, moving from small to larger classes. The yes/no responses could be thought of as binary, with 0 for no and 1 for yes, as in a logit or probit model. We can give an economic interpretation to binary responses, considering the yes category "higher" than the no. For example, we would expect farmers to seek out a particular source of water management information if the expected benefit, in terms of profit or utility, were positive and zero otherwise. The yes response to using an information source would then indicate a positive expected payoff from information use, and a no would indicate zero or a negative expected payoff.

A common test of association for contingency tables presented in mathematical statistics texts is the chi-square test. The test's null hypothesis is that there is independence (no association) between two variables represented in a contingency table. Under the alternative hypothesis, variables are not independent. Gibbons (1993) points out several problems with the chi-squared test, however. First, even for large sample sizes, the test statistic's distribution is only approximately chi-squared; the true distribution is unknown. Second, the chi-square test does not allow for testing a particular type of association against a null. For example, it does not specifically test a positive (or negative) relationship between variables against the null hypothesis. Third, when some cells have few observations relative to the total (i.e., small expected frequencies), the test statistic becomes inflated. Gibbons warns that "the test almost always leads to rejection of the null if the sample size is large" (1993, 61). And fourth, the chi-square test is not appropriate for ordinal data, because the test is independent of the ordering of rows and columns. For example, in Table 8.1, the chi-square test statistic would be the same no matter what order the farm size classes were placed in. For a test of association, we would want the order of variables to matter.

An alternative to the chi-square test is the Goodman-Kruskal gamma (γ) coefficient. The γ coefficient is a nonparametric measure of association between two ordinal variables. As with the correlation coefficient, values range from $+1$

(for positive association) to -1 (for negative association) (Goodman and Kruskal 1980). For complete independence, $\gamma = 0$, but γ can also equal 0 for U-shaped or inverted U-shaped relationships between variables. Appendix 8.1 describes how values of γ are computed.

The γ coefficient measuring the association between farm size and reliance on extension or university specialists (from Table 8.3) is 0.22. For the values in Table 8.2 relating farm size to EQIP participation, γ is 0.82. So the γ coefficient makes intuitive sense. There is a stronger positive association between farm size and EQIP participation ($\gamma = 0.82$) than between farm size and getting information from extension agents or university specialists ($\gamma = 0.22$). But can we say anything more about the differences besides noting that 0.82 is greater than 0.22?

One can interpret γ in terms of proportionate reduction in error (PRE) (Garson 2006). Consider two randomly chosen farm size–EQIP participation pairs. Suppose we know that the farm size class of the second observation is greater than the first observation (in a lower row). Suppose now we want to predict the rank of the column of the second observation (where rank increases moving to the right), and we know the column rank of the first observation. What if we were to make a prediction based on the assumption that because the row of the second observation is below that of the first observation, the column of the second observation will be to the right of that of the first observation? If $\gamma = 0.82$, following this strategy (predicting based on information about the rows) will reduce our errors in predicting the rank of the columns by 82% compared with assuming no association.

Table 8.4 shows different values for γ for six different 3×3 contingency tables, labeled (a) through (f).

For Table 8.4(a) (where $\gamma = 1$), we can see that the decision rule above would reduce prediction error to 0. The same would apply for Table 8.4(d) (where $\gamma = -1$), except that we would predict based on negative association. In Table 8.4(b), observations are uniformly distributed and $\gamma = 0$. There is no gain in prediction accuracy by assuming a positive or negative association. Note that γ also equals 0 in Table 8.4(e). It appears that there is a quadratic relationship between variables. The middle row has the highest column rank. Thus γ is a measure of monotonic association (positive or negative) and not one of every kind of

Table 8.4 *Values for the Goodman-Kruskal Gamma (γ) Coefficient for Different Contingency Tables*

(a)	$\gamma = 1$		(b)	$\gamma = 0$		(c)	$\gamma = 0.4$	
3	0	0	1	1	1	2	0	1
0	3	0	1	1	1	1	1	1
0	0	3	1	1	1	1	0	2
(d)	**$\gamma = -1$**		**(e)**	**$\gamma = 0$**		**(f)**	**$\gamma = -0.4$**	
0	0	3	3	0	0	1	2	2
0	3	0	0	3	0	0	1	0
3	0	0	3	0	0	2	1	1

association. Tables 8.4(c) and (f) show cases where knowing the relative ranks of rows of observations has less of an effect on reducing prediction error.

Exact tests of the significance level of γ require special tables (Goodman and Kruskal 1980). Gibbons (1993), however, presents an approximate, normally distributed test statistic, Z, where

$$Z = 3\gamma[n(r-1)(c-1)]^{1/2}/2[(r+1)(c+1)]^{1/2} \tag{8.5}$$

where n = number of observations, r = number of rows, and c = number of columns.

Another method for testing the significance of association is the Cochran-Armitage trend test. It is analogous to testing the significance of the slope coefficient β in a linear probability regression model (Agresti 2002). Let π_i be the probability the ith farmer responds yes to a question. Then a linear probability model would be

$$\pi_i = \beta(\text{category score})_i + \varepsilon_i \tag{8.6}$$

In our case, the categories are small farms, medium-size farms, large farms, and very large farms, which would be assigned numeric scores. The null hypothesis is $\beta = 0$; there is no linear trend between the category score (farm size) and the proportion of yes responses. The Cochran-Armitage trend test has a normally distributed z-test statistic. This test is for both significance and sign of the trend. This test statistic can be calculated using statistical analysis software (SAS) (see Tables 8.5–8.7 and 8.9–8.10, below). Results of the Cochran-Armitage trend test are analogous to a test of a zero slope coefficient in a logistic regression. The SAS output z score is the *negative* of the trend when the binary response is in columns. We report $-z$ scores so that a positive score represents a positive relationship. In general, the significance test for γ and the Cochran-Armitage z-scores provide consistent results for the test conducted.

TESTING THE FARM SIZE ASSOCIATIONS

We now examine the farm size associations by applying the Goodman-Kruskal gamma (γ) coefficient and the Cochran-Armitage trend test. Specifically, we examine relationships between farm size and irrigation information sources used, information sources used to schedule irrigation, scheduling decisions, barriers to improving irrigation systems and practices, and participation in water conservation cost-share programs.

Farm Size and Sources of Irrigation Information

Farmers were asked, "What are the sources of information you rely on for guidance in reducing irrigation costs or to conserve water used to for irrigation?

(mark all that apply)" (USDA NASS 1998). Options for responses were as follows:

1. Extension agents or university specialists
2. Government specialists
3. Irrigation equipment dealers
4. Irrigation district or water supplier
5. Media reports/press
6. Neighboring farmers
7. Private irrigation specialists or consultants

Choices were not mutually exclusive; respondents could list more than one information source.

One striking result is that no single information source was used by more than half of irrigators. This suggests there is no "one-stop-shopping" source of irrigation management information. For federal and cooperative extension agencies, this means that their directly provided information is not being directly accessed by most irrigators. Public entities may have to consider delivering information both directly and indirectly. For example, cooperative extension agencies may need to target news media, trade journals, irrigation district managers, and input suppliers as clientele. These sources can serve as "retail outlets" for extension messages that irrigators fail to access directly.

Results are largely consistent with our hypothesis H1, that farms in larger sales classes are more likely to access any given source of information. Based on tests of γ, there is a positive and significant association between sales class and use of most information sources. Exceptions are irrigation districts, where there is a negative association, and neighboring farmers, where there is a negative association in New Mexico and no association in Arizona. Smaller farms are more likely to have their irrigation scheduling determined by irrigation districts, taking some water management decisions out of their hands. Smaller operations may thus be relatively more reliant on irrigation districts for information.

Our hypothesis H2 predicted both large and small operators rely on lower-cost, but more general, public information, while large irrigators are more likely to seek tailored information from private sources. Reliance on university/extension specialists and neighbors (sources of lower-cost, general information) tends to be higher than for other sources. There is no significant relationship between sales class and reliance on extension/university specialists in New Mexico and a relatively weak positive association in Arizona. In both states, there was a relatively strong positive association between farm size and reliance on private irrigation specialists and consultants, with $\gamma = 0.81$ for Arizona and $\gamma = 0.93$ for New Mexico. While use was still relatively low for even very large farms, it was virtually nonexistent for small farms. These results conform generally to hypothesis H2, suggesting that smaller irrigators may face barriers to accessing private, tailored information.

Table 8.5 also reports the row sums of the percentages of irrigators accessing different sources of information. One may think of the row sum as an index of the

Table 8.5 *Sources of Irrigation Information Accessed by Farms*

Farm size	*Extension agents or university specialists*	*Government specialists*	*Irrigation equipment dealers*	*Irrigation district/ water supplier*	*Media reports/press*	*Neighboring farmers*	*Private irrigation specialists or consultants*	*Row sum*
				Arizona				
Small	31	5	7	42	3	36	0	124
Medium	45	16	3	16	14	40	1	135
Large	50	30	6	9	10	32	10	147
Very large	40	41	13	21	11	38	17	170
Gamma	0.22	0.72	0.21	−0.49	0.45	0.02	0.81	
Gamma z value	7.58	24.69	7.30	−16.72	15.37	0.61	25.61	
Cochran-Armitage −z	5.59	20.55	3.99	−11.20	7.06	0.35	14.98	
				New Mexico				
Small	32	19	8	22	12	49	0.2	142
Medium	23	22	29	7	20	44	16	161
Large	36	29	22	10	16	35	11	159
Very Large	36	40	44	8	12	42	28	210
Gamma	−0.02	0.30	0.67	−0.52	0.15	−0.14	0.93	
Gamma z value	−1.67	22.47	49.10	−38.22	10.83	−10.46	68.68	
Cochran-Armitage −z	−0.62	8.65	19.75	−7.65	1.54	−3.53	30.85	

Note: Values are percentage yes responses as a share of irrigators investigating improvements. Significance level = 5% for $|z| = 1.96$; 1% for $|z| = 2.576$; 0.1% for $|z| = 3.291$

number of information sources used per farm for each size class. The row sums suggest that at least some irrigators rely on multiples sources of information, and that larger farm size classes tend to rely on more information sources. This is consistent with hypothesis H3, that larger farms will access more information sources.

Farm Size and Irrigation Scheduling

In the 1998 FRIS survey, growers were asked, "How did you decide when to apply water in 1998?" Response options were as follows:

1. Condition of crop by observation
2. Feel of the soil
3. Soil-moisture sensing devices
4. Commercial scheduling services
5. Media reports on crop water needs
6. Water delivered in turn
7. Calendar schedule
8. Computer simulation models
9. Other practices

ERS also tabulated an additional aggregate category, "Most water-management-intensive and water-conserving means to decide when to apply water." Farms were assigned to this category if they used any one of the following methods: soil-moisture sensing devices, commercial irrigation scheduling services, or computer simulation models.

Consistent with Leib et al.'s 2002 study of Washington farmers, the dominant methods to decide irrigation timing are observation of the condition of the crop and using the feel of the soil. Next in importance was calendar scheduling. Our hypothesis H4 suggested that larger irrigators would use relatively management-intensive techniques to schedule irrigation. There is a significant positive association between farm size and use of the most water-management-intensive and water-conserving means to decide when to apply water (Table 8.6). The association is stronger in New Mexico ($\gamma = 0.60$) than in Arizona ($\gamma = 0.36$). Adoption rates of soil-moisture testing or commercial services were relatively low for each farm size class. Both Arizona and New Mexico had positive and significant γ coefficient and Cochrane-Armitage test statistics for soil-moisture testing. In New Mexico, the γ for commercial scheduling was high (0.96), but adoption among even the very large farms was low (8%). In Arizona, the γ for commercial scheduling was weakly negative (-0.08). Although significant at the 5% level for the γ test, it was insignificant under the Cochrane-Armitage test. However, overall use of management-intensive methods is positively associated with farm size. Again, this is consistent with Leib's finding of a positive association between farm size and use of scientific scheduling methods. In their comparison of two Canadian irrigation districts, Bjornlund et al. (2009) found that adoption of management-intensive methods (use of soil-moisture monitoring devices, private

Table 8.6 *Methods Used to Determine When to Apply Irrigation Water*

Farm sizes	*Condition of crop by observation*	*Feel of the soil*	*Soil-moisture sensing devices*	*Commercial scheduling services*	*Media reports on crop water needs*	*Water delivered in turn*	*Calendar schedule*	*Computer simulation models*	*Other practices*	*Water-management-intensive and water-conserving*
					Arizona					
Small	63	5	7	7	0	7	28	0	0	7
Medium	88	30	0	1	0	3	25	0	3	0
Large	66	27	12	—	0	—	51	0	5	12
Very large	91	50	14	8	6	5	30	1	5	20
Gamma	0.45	0.76	0.23	−0.08	1	−0.28	0.13	1	0.78	0.36
Gamma *z*	15.55	26.14	7.85	−2.36	34.45	−8.90	4.37	38.51	26.87	12.49
Cochran-Armitage −*z*	10.76	23.25	4.70	−0.06	9.57	−2.94	3.40	4.49	8.86	7.86
					New Mexico					
Small	61	23	5	0.07	0.20	25	31	*	2	5
Medium	91	19	8	—	2	2	10	*	9	9
Large	93	39	8	3	—	—	13	*	6	15
Very large	84	58	21	8	10	9	15	*	5	28
Gamma	0.66	0.34	0.48	0.96	0.90	−0.75	−0.51	*	0.55	0.60
Gamma *z* value	48.89	24.96	35.58	64.81	60.70	−50.27	−37.49	*	40.77	44.06
Cochran-Armitage −*z*	12.65	11.73	10.36	18.51	16.12	−10.81	−8.87	*	6.04	15.18

Note: Values are percentage yes responses. Significance level = 5% for $|z| = 1.96$; 1% for $|z| = 2.576$; 0.1% for $|z| = 3.291$

*Insufficient data for publication by USDA

consultants, etc.) was greater in the district with more large operations. Smaller irrigators are less likely to be engaged primarily in agriculture. Less than 55% of small farms reported farming as their primary occupation in Arizona and New Mexico. In contrast, more than 87% of medium, large, and very large farms reported farming as their primary occupation in both states. Thus results are consistent with previous conjectures that reliance on off-farm work can discourage adoption of management-intensive farming techniques.

The association between having water delivered in turn and farm size was negative. In New Mexico, 25% of small farms responded that water was delivered in turn, suggesting little discretion over when to apply water. Use of media reports is slight outside the very large farm size category and still low (6% for Arizona and 10% for New Mexico). Smaller farms were more likely to receive their water delivered in turn by irrigation districts and to rely on irrigation district staff for information. This suggests that government and extension specialists could fruitfully target irrigation district staff for water management information. Public institutions can act as wholesalers of information to irrigation district staff, while irrigation districts may transfer public information as retailers of water management information.

Farm Size and Barriers to Improving Irrigation Systems and Practices

Kislev and Shchori-Bachrach (1973) and Feder and Slade (1984) developed models where information acquisition increases productivity. Kislev and Shchori-Bachrach (1973) considered how passively acquired information enhances factor-neutral productivity. Feder and Slade (1984) extended their approach by treating information acquisition as a decision variable, where information is actively acquired at a cost. They also allowed information to have different productivity effects on different inputs. In their model, actively acquired information increases returns to adopting modern technologies. In the context of our model, this formulation would imply that not only the level of inputs, x, but also their productivity could be enhanced by information. It would also imply that acquiring information about new irrigation technologies or systems would be an important precursor to investments in irrigation improvements.

The 1998 FRIS survey asked producers, "What were barriers to implementing improvements that might reduce energy and/or conserve water in your irrigation system?" In an earlier question, the survey instructed respondents to consider changes in equipment or management practices as improvements, considering any improvements since 1994. The survey gave respondents a choice of barriers and instructed them to mark all that applied:

1. Have not investigated improvements (since 1994)
2. Risk of reduced yield or poorer quality crop yield from not meeting water needs
3. Physical field/crop conditions limit system improvements
4. Improvements will reduce costs, but not enough to cover the installation costs
5. Cannot finance improvements, even if they reduce costs

6. Landlords will not share in the cost of improvements
7. Uncertainty about future availability of water
8. Will not be farming this place long enough to justify new improvements

Table 8.7 shows the results for the first response, "Have not investigated improvements." Both the Feder and Slade model and our model imply that smaller farms would be less likely to investigate, or acquire information about, improvements. Results are consistent with this prediction. Of small farms in Arizona, 59% responded that they had not investigated improvements, while the response was 23% among small farms in New Mexico. In both states, there was a significant negative association between farm size and a "no investigation" response. The relationship was particularly strong in Arizona ($\gamma = 0.814$).

Initial γ and Cochrane-Armitage tests also suggested a positive association between nearly all of the barriers and farm size. This result runs counter to the vast bulk of literature that maintains that larger farms are more likely to adopt irrigation improvements. What accounts for the counterintuitive result? One possibility is that farmers who responded that they had not investigated improvements did not mark any other barriers to adoption. This assumes that only farmers seeking to implement improvements would encounter the other barriers. We have no way of knowing whether this is a reasonable assumption without looking at the original raw data files. Table 8.8, however, shows the share of each farm size class of the total "Have not investigated improvements" responses. Small farms account for more than 90% of farms not investigating improvements in each state.

We carried out hypothesis tests, removing irrigators not investigating improvements. Table 8.9 shows farms experiencing adoption barriers as a share of farms investigating improvements. The positive associations were weakened or reversed in some cases. However, in most cases, there is still a positive and significant association between farm sales class and adoption barriers. This is a puzzle. Even excluding farms not investigating improvements, larger farms still reported facing more barriers to adopting irrigation system improvements. A question for future research (unfortunately, the data do not allow us to explore this more fully) is whether larger farmers had more difficulty in *overcoming* the barriers they faced. Some evidence exists that larger irrigators were more likely to adopt more efficient sprinkler irrigation systems (see Chapter 9).

Table 8.7 *Percent of Farms Not Investigating Irrigation Improvements*

Farm size	*Arizona*	*New Mexico*
Small	59%	23%
Medium	14%	4%
Large	10%	23%
Very large	9%	9%
Gamma	−0.814	−0.396
Gamma z	−28.00	−29.19
Cochran-Armitage −z	−22.80	−5.85

Table 8.8 *Farm Size Class Shares of Total Responses "Have Not Investigated Improvements" in Irrigation*

Farm category	Arizona		New Mexico	
	Number of farms	Percentage	Number of farms	Percentage
Small	989	90.9%	1,216	94.1%
Medium	31	2.8%	11	0.9%
Large	27	2.5%	43	3.3%
Very large	41	3.8%	22	1.7%
Total	1,088	100%	1,292	100%

Economic constraints (installation costs outweighing benefits and lack of financing) are the dominant ones. Interestingly, lack of landlord financing appears to be more of a constraint for larger than for smaller farms. In addition, more than 20% of Arizona's small farms and New Mexico's medium-size farms responded that they did not expect to be farming long enough to make investments pay off. This figure was 7% in each state for the very large farm class. This may be more significant, considering that these larger farms account for a relatively larger percentage of total water use.

Farm Size and Participation in Water Conservation Cost-Share Programs

Federal, state, and local agriculture conservation programs provide cost-share payments to farmers to encourage adoption of capital and technology to improve irrigation efficiency. The 1998 FRIS collected data on farm-level participation in cost-share programs, asking farmers whether in the previous five years (1994–1998) they had received irrigation-related cost-share payments for irrigation improvements from one or more of the following sources:

1. USDA conservation cost-share programs [including the Environmental Quality Incentive Program (EQIP) and other earlier USDA cost-share programs]
2. Non-USDA federal cost-share programs [including those from the U.S. Environmental Protection Agency (EPA), the Bureau of Reclamation, or other programs]
3. State programs and local water management or supply district programs
4. Other cost-share programs
5. Any federal program
6. Any program of any source

Among very large farms, 30% of those in New Mexico and 34% in Arizona participated in some form of cost-share program (Table 8.10, last column). Participation rates for small farms were lower, 8% in both states. There is a significant positive association between cost-share program participation and farm size. In Arizona, the positive association is greater than in New Mexico, except for state and local programs. Here there is a weak negative association, $\gamma = -0.06$,

Table 8.9 *Barriers to Adopting Irrigation Improvement*

Farm size	*Risk of reduced yield or quality*	*Physical field or crop conditions*	*Installation costs greater than benefits*	*Lack of financial ability*	*Lack of landlord cost-sharing*	*Uncertainty about future water availability*	*Will not be farming in the future*	*Other reasons*
				Arizona				
Small	8	11	28	22	—	3	—	8
Medium	10	8	37	57	4	7	22	32
Large	15	17	50	52	13	19	0	12
Very large	15	14	28	26	23	12	7	22
Gamma	0.25	0.14	0.06	0.14	0.63	0.42	0.38	0.31
Gamma z	8.51	4.72	2.20	4.84	17.66	14.54	13.28	10.80
Cochran-Armitage −z	4.11	2.45	1.46	2.88	10.54	6.57	3.78	5.53
				New Mexico				
Small	6	12	19	27	3	15	21	6
Medium	5	6	8	20	5	18	—	5
Large	26	16	28	34	16	17	4	7
Very large	19	14	37	23	16	13	7	6
Gamma	0.46	−0.04	0.14	−0.06	0.61	0.01	−0.64	−0.01
Gamma z	33.72	−3.00	9.95	−4.76	45.22	1.16	−42.82	−0.68
Cochran-Armitage −z	9.75	−0.39	5.44	−0.93	11.84	0.23	−8.34	−0.04

Note: Values are percentage responses as a share of irrigators investigating improvements. Significance level = 5% for $|z| = 1.96$; 1% for $|z| = 2.576$; 0.1% for $|z| = 3.291$

Table 8.10 *Participation by Farms in Water Conservation Cost-Share Programs*

Farm size	*USDA cost-share payments*	*Non-USDA federal programs*	*State/local programs*	*Other programs*	*Any federal program*	*From any program source*
			Arizona			
Small	1	1	7	1	1	8
Medium	6	0	0	4	6	11
Large	12	4	5	6	12	18
Very large	26	5	8	5	27	34
Gamma	0.82	0.68	−0.06	0.57	0.83	0.55
Gamma *z* value	28.26	23.51	−2.10	19.54	28.53	18.88
Cochran-Armitage −*z*	18.47	6.94	−0.02	6.54	19.03	13.96
			New Mexico			
Small	6	5	4	3	7	8
Medium	15	5	5	5	15	15
Large	14	6	6	3	16	16
Very large	22	5	9	3	24	30
Gamma	0.49	0.08	0.28	0.04	0.49	0.48
Gamma *z* value	36.17	6.16	20.58	3.25	36.09	35.15
Cochran-Armitage −*z*	10.38	0.81	4.01	0.26	11.03	11.74

Note: Values are percentage responses. Significance level = 5% for $|z| = 1.96$; 1% for $|z| = 2.576$; 0.1% for $|z| = 3.291$

while the trend coefficient is insignificant using the Cochrane-Armitage test. The positive association is relatively strong for USDA program participation in Arizona, $\gamma = 0.82$. Only 1% of Arizona small farms reported having USDA contracts, while this number was 6% for New Mexico. In New Mexico, there was a significant but weaker association between farm size and USDA program participation. Table 8.10 provides information about the targeting of EQIP (USDA) and state/local cost-share programs. In Arizona, EQIP payments are targeted more toward larger farms and less than 10% of small farms reported receiving EQIP payments. In contrast, in New Mexico, nearly three-quarters of small farms reported receiving EQIP payments.

States have latitude in administration of EQIP programs, and Arizona appears to be targeting larger irrigators. Here the very large farm class accounts for 65% of farms receiving EQIP payments and 77% of irrigation water applied. In contrast, the small farm class in New Mexico accounts for 74% of farms receiving EQIP payments but less than 26% of irrigation water applied. Although participation rates increase with farm size in New Mexico, small farms account for a large share of total farms, over 88%, so they account for a large share of total contracts.

In both Arizona and New Mexico, small farms account for the great majority of farms receiving state or local cost-share payments (Table 8.11). In both states, state and local programs target small farms, although these programs reach a small share of irrigators. In Arizona, only 7% of small irrigators received state or local payments; in New Mexico, this figure was 4% (Table 8.10). However, in Arizona, small farms accounted for nearly 70% of farms receiving this assistance, even though they account for only 4.4% of water use. In New Mexico, small farms accounted for 81% of farms receiving state or local cost-share payments (Table 8.11) but less than 26% of water use.

These results have implications for program targeting. Schaible (2004) has argued that the efficiency of water conservation programs could be increased by targeting those farms that account for relatively more water use. It appears that the administration of EQIP in Arizona does just this. In New Mexico, a large number of small farms still account for a large share of EQIP contracts. In both states, state and local programs, though very limited, appear to be targeting irrigators that account for a small share of overall irrigation water use.

DATA FROM THE 2008 FRIS

The USDA released data from the 2008 Farm and Ranch Irrigation Survey on November 30, 2009. Although published tables do not report differences in irrigator behavior by farm sales class, we may still examine certain patterns regarding irrigator scale of operation and water management. As in 1998, no single information source was used by more than half of irrigators or by irrigators that accounted for more than half of irrigated acres (Table 8.12). Table 8.12 lists information sources by the percent of irrigators that use them and the percent of acres these producers irrigate. If the percent of farms is greater than the percent of irrigated acres for an information source, this suggests that smaller farms rely more

Table 8.11 *Farms Receiving Cost-Share Payments for Irrigation or Drainage Improvements*

	Small		*Medium*		*Large*		*Very large*		*All farm size classes*	
State	*Farms*	*Percentage*	*Farms*	*Percentage*	*Farms*	*Percentage*	*Farms*	*Percentage*	*Farms*	*Percentage*
Farms receiving EQIP payments										
Arizona	17	9.3%	14	7.7%	33	18%	119	65%	183	100%
New Mexico	333	73.7%	42	9.3%	25	5.5%	52	11.5%	452	100%
Farms receiving payments from state and local programs										
Arizona	120	69.8%	0	0%	15	8.7%	37	21.5%	172	100%
New Mexico	206	81.4%	15	5.9%	11	4.3%	21	8.3%	253	100%

Source: USDA NASS (2007)

Table 8.12 *Source of Irrigation Information Relied on to Reduce Irrigation Costs or Conserve Water (% of Farms Using Each Source)*

	Arizona		*New Mexico*	
Information source	*Percent of farms*	*Percent of irrigated acres*	*Percent of farms*	*Percent of irrigated acres*
Extension/university specialists	24%	47%	30%	28%
NRCS, local conservation district, other federal or state agencies	25%	36%	25%	29%
Neighboring farmers	38%	29%	36%	25%
Private irrigation specialists or consultants	15%	30%	6%	37%
Irrigation equipment dealers	10%	16%	6%	36%
Irrigation district or water supplier	27%	16%	6%	9%
Media reports	7%	12%	6%	8%
Electronic information services (Internet links)	11%	13%	3%	12%

Note: Farms may rely on more than one information source
Source: USDA NASS (2007)

on that source. Here scale is measured in terms of irrigated acres. Conversely, if the percent of farms is smaller than the percent of acres, this suggests that larger farms (in terms of irrigated acreage) rely more on the information source. Following this approach, it appears that larger farms (in terms of irrigated acreage) in Arizona rely more on extension specialists, government agencies, private specialists, and equipment dealers. Smaller farms rely more on neighboring farms or their irrigation district or other water suppliers. In New Mexico, both large and small farms appear to rely similarly on extension specialists, but otherwise patterns are similar to Arizona.

Larger farms appear to rely more on media reports and electronic media (the Internet) in both states. Just et al.'s (2002) model of agricultural demand for information posits that farmers with more human capital will be better able to interpret data, which comes as raw statements of fact or numbers. Just et al. distinguish between *data* and *information*, which is manipulated, summarized, or interpreted in some way by intermediaries. Producers with sufficient human capital can process raw data more easily and may process it themselves. Thus media or electronic media may have some interpretation but less information provided through intermediaries (e.g., extension specialists, consultants, USDA specialists). If human capital is positively correlated with scale of operation, the results regarding use of media reports and Internet information appear consistent with Just et al.'s hypothesis.

The dominant methods to decide irrigation timing are still observing the crop's condition, feel of the soil, and calendar scheduling. A significant share of producers still has water delivered in turn by irrigation districts. Adoption rates of scientific irrigation scheduling methods, such as soil- or plant-moisture sensing devices, computer simulations, or reports of crop evapotranspiration (ET), remain low (Table 8.13). Less than 5% of irrigators in Arizona or New Mexico use these methods. Roughly 5% of irrigators rely on government or private scheduling services, however, and it is quite possible that these intermediaries use scientific scheduling methods.

Table 8.13 *Methods Used in Deciding When to Irrigate (% of Farms Using Each Method)*

Method	*Arizona*	*New Mexico*
Condition of crop	68%	74%
Feel of soil	30%	36%
Personal calendar schedule	43%	35%
Scheduled by delivery organization	17%	25%
When neighbors begin to irrigate	4%	8%
Commercial or govt. scheduling service	4%	5%
Soil-moisture sensing device	2%	2%
Reports on daily crop water ET	1.8%	0.3%
Computer simulation models	0.1%	0.2%
Plant-moisture sensing device	0.2%	0.1%

Note: Farms may use more than one method
Source: USDA NASS (2007)

We now turn to barriers to making improvements to reduce energy use or conserve water. In Arizona, growers who did not report any barriers accounted for 48% of farms and 58% of water applied (Table 8.14). In New Mexico, these figures were 22% of farms and 58% of water applied. This suggests that large farms (where scale is measured in terms of water use) are less likely to report barriers. In contrast, improvements to conserve water or energy were not a priority for 28% of Arizona farms, accounting for 13% of applied water, or for 33% of New Mexico farms, accounting for 17% of applied water (Table 8.14). These results are consistent with the analysis of the 1998 data above, suggesting that smaller-scale operators are less likely to seek out water- or energy-conserving improvements. For relatively large operations (in terms of water use), relatively important barriers appear to be economic (e.g., expected cost savings from energy or water conservation are less than installation costs; lack of landlord cost-sharing). For relatively smaller operations, lack of financing and uncertainty about future water supplies appear to be relatively important barriers. A relatively higher proportion of smaller-scale operations also do not expect to continue farming long enough to make improvements worthwhile. These operators account for 4% of Arizona's irrigation water and 3% of New Mexico's. This suggests that significant amounts of water may be reallocated to other uses. It also suggests that one adaptation to climate change, especially for smaller-scale irrigators, may be to leave agriculture altogether.

Results from the 2008 FRIS suggest changes in government funding of irrigation and drainage improvements (Table 8.15). First, the share of irrigators receiving funding from sources other than the USDA (primarily EQIP) has declined since 1998. Second, compared with 1998, in 2008 there appears to be less difference between Arizona and New Mexico in targeting of irrigators under USDA programs. In both states, growers accounting for more than 40% of state irrigated acreage received USDA payments. These were less than 20% of growers in each state. This suggests that USDA programs in New Mexico, as in Arizona, are targeting irrigators that account for relatively larger shares of irrigated acreage.

Table 8.14 *Barriers to Making Improvements to Reduce Energy Use or Conserve Water (% of Farms Facing Each Barrier)*

Respondents	Arizona Percent of farms	Arizona Percent of acre-feet applied	New Mexico Percent of farms	New Mexico Percent of acre-feet applied
Not reporting any barriers	48%	58%	22%	58%
Facing some barrier(s)	62%	42%	78%	42%
Improvements not a priority	28%	13%	33%	17%
Risk to crop yield or quality	10%	4%	4%	7%
Physical constraints	6%	6%	7%	5%
Cost reductions < installation costs	4%	9%	18%	6%
Lack of financing	28%	12%	28%	10%
Landlord will not share costs	2%	13%	1%	3%
Uncertainty about future water availability	12%	5%	20%	9%
Will not be farming long enough	12%	4%	9%	3%

Note: Farms may face more than one barrier
Source: USDA NASS (2007)

Table 8.15 *Irrigators Receiving Payments for Irrigation or Drainage Improvements*

Funding source	Arizona Percent of farms	Arizona Percent of irrigated acres	New Mexico Percent of farms	New Mexico Percent of irrigated acres
USDA	19%	45%	14%	43%
Other federal agency	0%	2%	0%	0%
State or local	1%	3%	0%	0%
Other	1%	3%	1%	1%

Source: USDA NASS (2007)

CONCLUSIONS

This chapter introduced a simple model of irrigator demand for information to develop and test hypotheses concerning scale of farm operations and use of water management information. Smit and Skinner (2002) noted that climate and weather information systems and improvements in water management practices, such as irrigation scheduling, are key components of climate change adaptation in agriculture. Smit and Skinner did not view information provision itself as an adaptation, but something that encourages it. Here we have emphasized the distinction between the supply of information for water management and the demand for information. Adaptation occurs when information is used, thus our emphasis on farm-level information demand.

Univariate, nonparametric tests of data from Arizona and New Mexico generally support the hypotheses derived from the model. Larger operations (in terms of sales) were more likely to use irrigation management information from any given source. Reliance on lower-cost sources of general information was more common, while larger operations relied more heavily than did smaller farms on private, tailored sources of information. Thus larger operations relied on more sources of information in total. Smaller farms were less likely to investigate irrigation improvements or use management-intensive methods for irrigation scheduling. Larger operations were also more likely to use data provided more directly (e.g., from media and Internet reports) than were smaller operators, which relied more on information provided by intermediaries. Adoption of scientific irrigation scheduling methods (such as soil-moisture testing devices) was low across all producer groups, but especially low for smaller-scale irrigators. These results suggest that smaller operations face significant barriers to information acquisition and use of management-intensive irrigation practices.

The data from the USDA's Farm and Ranch Irrigation Survey (FRIS) revealed that irrigators rely on diverse sources of information about water management. No single institutional source of irrigation information was relied on by more than half of any farm size group. A significant share of irrigators did not report relying on public sources of irrigation information (university/extension specialists or government specialists). The data also revealed the important role of irrigation districts. In many cases, districts decided the timing of irrigation and were a source of irrigation information, particularly for smaller farms. This has important implications for outreach and technology transfer. Many irrigators are not directly seeking out public information providers, while irrigation districts play a central role in scheduling and information provision. Public agencies might fruitfully target irrigation district staff for their information services, acting as wholesalers of information to the districts, which in turn interact with irrigators directly.

Finally, results indicated a significant positive association between farm size and participation in federal cost-share programs to encourage adoption of improved irrigation practices. In 1998, this positive association was more pronounced in Arizona than in New Mexico, perhaps because New Mexico has many more small-scale operators in total. Relatively larger operations in both states account for the vast majority of water use. Schaible (2004) has argued that one could increase the efficiency of cost-share programs by targeting farms that account for relatively more water use. FRIS data from 1998 suggested that USDA's EQIP program was doing this type of targeting to a greater extent in Arizona than in New Mexico. More recent 2008 FRIS data suggest that EQIP payments in both states are now being directed to larger irrigators that account for larger shares of total water use.

APPENDIX 8.1: COMPUTING THE GOODMAN-KRUSKAL γ COEFFICIENT

The γ coefficient is the number of concordant pairs, C, minus the number of discordant pairs, D, divided by the sum of concordant and discordant pairs (excluding tied pairs):

$$\gamma = (C - D)/(C + D) \tag{8.7}$$

To illustrate, let's look at the numbers in Table 8.3 again. To calculate concordant pairs for an $r \times 2$ contingency table, where $r =$ the number of rows and $2 =$ the number of columns, we will use the cell in the left column and uppermost row. Concordant pairs with these 1,163 observations are the sum of observations that are in a (lower row) and a rightward column, $1{,}163 \times (98 + 139 + 186) = 491{,}949$. Repeat for the next row down in the left column. Concordant pairs are $118 \times (139 + 186) = 38{,}350$. Repeat again for the left column, one row down. Concordant pairs are $140 \times 186 = 26{,}040$. The sum of all concordant pairs is $491{,}949 + 38{,}350 + 26{,}040 = 556{,}339$. Concordant pairs are pairs between cells where one cell is to the right and down from the other.

Discordant pairs are pairs between cells where one cell is to the left and down from the other. So the discordant pairs are $515 \times (118 + 140 + 278) = 276{,}040$; $98 \times (140 + 278) = 40{,}964$; and $139 \times 278 = 38{,}642$. The sum of all discordant pairs is $276{,}040 + 40{,}964 + 38{,}642 = 355{,}646$. Pairs are considered tied on columns if they share the same column and tied on rows if they share the same row (Gibbons 1993). These pairs are excluded from the γ calculation: $\gamma = (C - D)/(C + D)$. The γ measuring the association between farm size and reliance on extension or university specialists is $\gamma = (556{,}339 - 355{,}646)/(556{,}339 + 355{,}646) = 0.22$. For the values in Table 8.2 relating farm size to EQIP participation, $\gamma = (336{,}305 - 33{,}232)/(336{,}305 + 33{,}232) = 0.82$.

REFERENCES

Agresti, A. 2002. *Categorical Data Analysis*. Wiley Series in Probability and Statistics. Hoboken, NJ: Wiley.

Antle, J.M., and S. Hatchett. 1986. Dynamic Input Decisions in Econometric Production Models. *American Journal of Agricultural Economics* 68:939–949.

Bjornlund, H., L. Nicol, and K. Klein. 2009. The Adoption of Improved Irrigation Technology and Management Practices: A Study of Two Irrigation Districts in Alberta, Canada. *Agricultural Water Management* 96:121–131.

CAST (Council for Agricultural Science and Technology). 1992. Water Quality: Agriculture's Role. Task Force Report 120. Ames, IA: CAST.

Caswell, M., and D. Zilberman. 1985. The Choices of Irrigation Technologies in California. *American Journal of Agricultural Economics* 67:224–324.

Clearfield, F., and B.T. Osgood. 1986. Sociological Aspects of the Adoption of Conservation Practices. Washington, DC: Soil Conservation Service.

Dinar, A., M. Campbell, and D. Zilberman. 1992. Adoption of Improved Irrigation and Drainage Reduction Technologies under Limiting Environmental Conditions. *Environmental and Resource Economics* 2:373–398.

EPA (U.S. Environmental Protection Agency). 1998. National Water Quality Inventory: 1996 Report to Congress. EPA 841-R-97-008. Washington, DC: EPA Office of Water.

Feder, Gershon, and Roger Slade. 1984. The Acquisition of Information and the Adoption of New Technology. *American Journal of Agricultural Economics* 66:312–320.

Fernandez-Cornejo, J. 1996. The Microeconomic Impact of IPM Adoption. *Agricultural and Resource Economics Review* 25:149–160.

———. 1998. Environmental and Economic Consequences of Technology Adoption: IPM in Viticulture. *Agricultural Economics* 18:145–155.

———. 2007. *Off-Farm Income, Technology Adoption, and Farm Economic Performance.* Economic Research Report No. ERR-36. Washington, DC: United States Department of Agriculture, Economic Research Service.

Fernandez-Cornejo, J., E.D. Beach, and Wen-Yuan Huang. 1994. The Adoption of IPM Techniques by Vegetable Growers in Florida, Michigan, and Texas. *Journal of Agricultural and Applied Economics* 26:158–172.

Fernandez-Cornejo, J., and S. Jans. 1996. The Economic Impact of IPM Adoption for Orange Producers in California and Florida. *Acta Horticulturae* 429:325–334.

Frisvold, G.B. 2004. How Federal Farm Programs Affect Water Use, Quality, and Allocation Among Sectors. *Water Resources Research* 40:W12S05, doi:10.1029/2003WR002753.

Garson, D.G. 2006. Statnotes: Topics in Multivariate Analysis, North Carolina State University, Department of Public Administration. www2.chass.ncsu.edu/garson/pa765/statnote.htm (accessed August 15, 2009)

Gibbons, J.D. 1993. *Nonparametric Measures of Association.* Newbury Park, CA: SAGE Publications.

Goodman, L.A., and W.H. Kruskal. 1980. *Measures of Association for Cross Classification.* New York: Springer-Verlag.

Green, G., D. Sunding, D. Zilberman, and D. Parker. 1996. Explaining Irrigation Technology Choices: A Microparameter Approach. *American Journal of Agricultural Economics* 78:1064–1072.

Hoppe, R.A. 2001. *Structural and Financial Characteristics of U.S. Farms: 2001 Family Farm Report.* AIB-768. Washington, DC: U.S. Department of Agriculture, Economic Research Service.

Hutson, S., N. Barber, J. Kenny, K. Linsey, D. Lumia, and M. Maupin. 2005. *Estimated Use of Water in the United States in 2000.* USGS Circular 1268. Reston, VA: U.S. Geological Survey.

Johnson, S.R., and M.T. Holt. 1997. The Value of Weather Information. In *Economic Value of Weather and Climate Forecasts*, edited by R.W. Katz and A.H. Murphey. Cambridge, UK: Cambridge University Press.

Just, D.R., S.A. Wolf, S. Wu, and D. Zilberman. 2002. Consumption of Economic Information in Agriculture. *American Journal of Agricultural Economics* 84:39–52.

Just, R.E., D. Zilberman, and E. Hochman. 1983. Estimation of Multicrop Production Functions. *American Journal of Agricultural Economics* 65:770–780.

Kislev, Y., and N. Shchori-Bachrach. 1973. The Process of an Innovation Cycle. *American Journal of Agricultural Economic* 55:28–37.

Knapp, K., and K. Schwabe. 2008. Spatial Dynamics of Water and Nitrogen Management in Irrigated Agriculture. *American Journal of Agricultural Economics* 90:524–539.

Lee, D.J., and R.E. Howitt. 1996. Modeling Regional Agricultural Production and Salinity Control Alternatives for Water Quality Policy Analysis. *American Journal of Agricultural Economics* 78:41–53.

Leib, B.G., M. Hattendorf, T. Elliott, and G. Matthews. 2002. Adoption and Adaptation of Scientific Irrigation Scheduling: Trends from Washington, USA as of 1998. *Agricultural Water Management* 55:105–120.

McNamara, K.T., M.E. Wetzstein, and G.K. Douce. 1991. Factors Affecting Peanut Producer Adoption of Integrated Pest Management. *Review of Agricultural Economics* 13:129–139.

Mendelsohn, R., and A. Dinar. 2003. Climate, Water, and Agriculture. *Land Economics* 79:328–341.

Moore, M.R., N.R. Gollehon, and M.B. Carey. 1994. Multicrop Production Decisions in Western Irrigated Agriculture: The Role of Water Price. *American Journal of Agricultural Economics* 76:859–874.

Moore, M.R., N.R. Gollehon, and D.H. Negri. 1993. Alternative Forms for Production Functions of Irrigated Crops. *Journal of Agricultural Economics Research* 44:16–32.

Negri, D.H., and D.H. Brooks. 1990. Determinants of Irrigation Technology Choice. *Western Journal of Agricultural Economics* 15:213–223.

Negri, D.H., N.R. Gollehon, and M.P. Aillery. 2005. The Effects of Climatic Variability on US Irrigation Adoption. *Climatic Change* 69:299–323.

NRC (National Research Council). 1996. *A New Era for Irrigation*. Washington, DC: National Academy Press.

Parker, D., and D. Zilberman. 1996. The Use of Information Services: The Case of CIMIS. *Agribusiness* 12:209–218.

Parry, M., N. Arnell, M. Hulme, R. Nicholls, and M. Livermore. 1998. Adapting to the Inevitable. *Nature* 395:741.

Rivera, J.A. 1998. *Acequia Culture: Water, Land, and Community in the Southwest*. Albuquerque: University of New Mexico Press.

Schaible, G.D. 1997. Water Conservation Policy Analysis: An Interregional, Multi-output, Primal-Dual Optimization Approach. *American Journal of Agricultural Economics* 79:163–177.

———. 2004. Irrigation, Water Conservation, and Farm Size in the Western United States. *Amber Waves* 8 (June):8.

Schaible, G.D., and M.P. Aillery. 2003. Irrigation Technology Transitions in the Mid-plains States: Implications for Water Conservation/Water Quality Goals and Institutional Changes. *International Journal of Water Resources Development* 19:67–88.

Schaible, G.D., C.S. Kim, and N.K. Whittlesey. 1991. Water Conservation Potential from Irrigation Technology Transitions in the Pacific Northwest. *Western Journal of Agricultural Economics* 16:194–206.

Skaggs, R., and Z. Samani. 2005. Farm Size, Irrigation Practices, and On-Farm Irrigation Efficiency. *Irrigation and Drainage* 54:43–57.

Smit, B., and M.W. Skinner. 2002. Adaptation Options in Agriculture to Climate Change: A Typology, Climatic Change. *Mitigation and Adaptation Strategies for Global Change* 7:85–114.

USDA ERS (U.S. Department of Agriculture, Economic Research Service). 2004. Data Sets: Western Irrigated Agriculture: Tables. http://ers.usda.gov/Data/WesternIrrigation/ShowTables.asp (accessed September 27, 2010).

USDA NASS (U.S. Department of Agriculture, National Agricultural Statistical Service). 1998. 1997 Census of Agriculture, AC97-SP-1, vol. 3, Special Studies, part 1, 1998 Farm and Ranch Irrigation Survey. Washington, DC: USDA NASS. www.agcensus.usda.gov/Publications/1997/Farm_and_Ranch_Irrigation_Survey/index.asp (accessed September 27, 2010).

———. 2002. 2002 Census of Agriculture. www.agcensus.usda.gov/Publications/2002/index.asp (accessed September 27, 2010).

———. 2007. Census of Agriculture: 2008 Farm and Ranch Irrigation Survey. www.agcensus.usda.gov/Publications/2007/Online_Highlights/Farm_and_Ranch_Irrigation_Survey/index.asp (accessed September 27, 2010).

USDA NRCS (U.S. Department of Agriculture, Natural Resources Conservation Service). 2003. 2002 Farm Bill: Environmental Quality Incentives Program. www.nrcs.usda.gov/programs/eqip/index.html#2003 (accessed September 22, 2010).

Wozniack, G.D. 1993. Joint Information Acquisition and New Technology Adoption: Late versus Early Adoption. *Review of Economics and Statistics* 75:438–445.

CHAPTER 9

Irrigation Technology Choice: The Role of Climate, Farm Size, Energy Costs, and Soils

George B. Frisvold and Shailaja Deva

Many studies (including Chapter 8 of this volume) view improvements in irrigation efficiency as a fundamental adaptation to climate change (e.g., Burton 2000; Cavagnaro et al. 2006; Jackson et al. 2009; Joyce et al. 2009; Kurukulasuriya and Rosenthal 2003; Smit and Skinner 2002). Many in turn see improved irrigation technology as a primary way to improve irrigation efficiency. Compared with flood (or gravity) irrigation, improved technologies such as sprinkler or drip irrigation allow for better control over the quantity and timing of water applications to crops. Thus water applications may be better matched to plant needs. Improved timing and levels of irrigation protect against climate extremes, helping prevent frost damage as well as drought stress. Modern irrigation technology, however, can require large capital investments. The costs of such investments can take years to recoup, so growers have an incentive to select new irrigation methods that are best suited to dominant climate patterns they face. As Chapter 8 illustrated, financial constraints are often critical barriers in making investments in improved irrigation efficiency.

IRRIGATION TECHNOLOGY CHOICE AS CLIMATE ADAPTATION

Although researchers have thoroughly studied factors affecting adoption of improved irrigation technology, and the importance of climate in irrigation choice is often acknowledged, relatively few studies have formally focused on the role of climate or included climate variables. Most empirical studies of irrigation technology adoption have concentrated on small geographic areas, such as a single irrigation district or relatively small production region. The geographic scope of such analyses is too limited to effectively measure effects of climate over the long term. By their very nature, long-term climate averages change little over time, and these limited studies have had insufficient variation in climate to allow for econometric analysis. Studies must have enough spatial, cross-sectional variation to capture the effects of changes in climate, and to measure the effects of long-term climate variables, studies must have large enough geographic scope to have measurable differences in these variables across observations.

This chapter adds to the relatively small list of studies that have formally focused on the role of climate in adoption of irrigation technology. These include Negri et al. (2005), who examined the effect of climate on the decision whether or not to irrigate; Negri and Brooks (1990), who considered effects of climate variables on choice of irrigation technology among irrigators; Mendelsohn and Dinar (2003), who examined climate effects on the share of irrigated acreage and share of irrigated acreage under different technologies across U.S. counties; and Caswell et al. (2001), who also examined climate effects on the choice of whether to irrigate along with the choice of irrigation method for individual producers. These studies found that climate variables had important impacts on irrigation adoption decisions.

In our study, we consider how climate, farm size, energy costs, and soil characteristics affect irrigator choice of either gravity flow or the more efficient sprinkler irrigation in 17 western states. The geographic scope of this chapter is broader than that of the other chapters in this book to provide sufficient spatial variation in climate variables so that we are able to measure their effects. As in Chapter 8, we make use of data from special cross-tabulations of the Farm and Ranch Irrigation Survey (FRIS) (USDA NASS 2008) made available by the U.S. Department of Agriculture's Economic Research Service (USDA ERS 2004). These cross-tabs allow us to stratify observations into 48 state and farm size pairs in order to conduct a western regionwide analysis of irrigator behavior, examining irrigator choice of sprinkler versus gravity flow irrigation.

The chapter begins with a discussion of the importance of irrigation to western agriculture. Next, we compare the salient features of sprinkler and gravity flow irrigation technologies and discuss trends in the diffusion of sprinkler technology. We then review the empirical literature on sprinkler irrigation adoption. This literature has examined the role of field and soil characteristics (especially those affecting water-holding capacity of soils); costs of water, energy, and labor; farm size; and climate as important factors influencing adoption of modern irrigation technologies. The next section introduces a regression specification and data sources to examine how variation in sprinkler adoption across state and farm sales classes is affected by soil characteristics, climate, reliance on surface water (versus groundwater), energy pumping costs, and labor costs. We then analyze the results of the multivariate regression and conclude by discussing the implications of our main findings.

To preview our main findings, indicators of soil attributes explain sprinkler adoption, even at state levels of aggregation. State average sheet and rill erosion (erosion caused by water) may act as an index capturing the effects of multiple variables found important in microlevel studies, including rainfall, field slope, and soil water-holding capacity. Sprinkler adoption increased with energy pumping costs and with relative reliance on higher-cost groundwater. Adoption also increased with farm sales class. The effects of temperature were complex. A longer growing season decreased sprinkler adoption. Adoption first increased and then decreased with cropland-weighted mean temperature. Results suggest lower sprinkler adoption in warmer states with long growing seasons.

IRRIGATION IN WESTERN AGRICULTURE

Irrigation is enormously important to western agriculture. In the 17 westernmost contiguous states, about three-quarters of the value of crops grown come from the 25% of the cropland that is irrigated (Gollehon and Quinby 2000). Improved irrigation technology has been seen as a necessary, if not sufficient, part of relieving pressure on water supplies and quality in the West. Improved technologies such as sprinkler or drip irrigation allow producers better control over water applications and increase the share of applied water that is taken up by plants. Such improved technologies have been seen as means to improve farm yields and incomes (Caswell and Zilberman 1986; Aillery and Gollehon 2003), allow producers to better adapt to drought or climate change (Schuck et al. 2005; Mendelsohn and Dinar 2003), and reduce water pollution from soil erosion and chemical leaching (Caswell et al. 1990, 2001).

The role of improved irrigation technology in water conservation has proven more controversial. Although improved application efficiency implies that less water is needed to generate a given level of crop yield, improved efficiency does not necessarily reduce demand for water (Caswell and Zilberman 1986; Peterson and Ding 2005). Even when increased farm-level efficiency does reduce demand for water diversions, it may still increase consumptive use of water. Increasing the percentage of diverted water consumed by crops can reduce return flows and aquifer recharge. Thus, as discussed in more detail in Chapter 3, improved farm-level efficiency may not necessarily conserve water at the basin level (Skaggs 2001; Skaggs and Samani 2005; Ward and Pulido-Velazquez 2008; Huffaker and Whittlesey 2000, 2003).

IRRIGATION TECHNOLOGY

Traditional irrigation systems have used gravity to distribute water. Water is conveyed to the field using open ditches or pipes, then released along the upper end of the field. Furrows control water movement and channel the flow down or across the field. Gravity systems are best suited to soils with higher moisture-holding capacities and relatively flat fields to prevent excessive water runoff.

With sprinkler systems, water is sprayed over the field, usually using aboveground pipes. Sprinkler irrigation systems use pressure to distribute water, which requires energy for pumping. As Negri and Brooks note, "Sprinkler irrigation technologies save water relative to gravity-flow systems by distributing

Table 9.1 *Shares of Irrigated Acreage in the United States by Method of Irrigation*

Year	*Drip and subsurface*	*Sprinkler*	*Gravity*
1979	1%	36%	63%
1998	5%	45%	50%
2003	6%	51%	43%

Source: Aillery and Gollehon (2003); USDA NASS (2002)

water evenly on the field, reducing percolation below the root zone, and eliminating field runoff" (1990, 214). Sprinkler systems can be used on steeper slopes unsuited to gravity systems and on soils with lower water-holding capacity. Sprinkler irrigation also can have much higher application efficiencies than traditional gravity irrigation (Sloggett 1985). Whereas field application efficiencies usually range from 40% to 65% for gravity systems, they are more typically 75% to 85% for sprinkler systems (Aillery and Gollehon 2003). Sprinkler systems have higher capital costs than gravity systems, however, which may act as a barrier to adoption for farms below some critical size. Sprinkler systems tend to be energy-using and labor-saving relative to gravity systems. Thus their relative profitability will depend on labor and energy costs.

The amount of U.S. acreage irrigated with sprinkler systems increased from 36% to 51% between 1979 and 2003, while acreage irrigated by gravity systems declined from 63% to 43% over the same period (Table 9.1). Low-pressure drip and subsurface irrigation has increased from 1% to 6% of irrigated acreage. Drip irrigation is most commonly used for vegetables, orchards, vineyards, nuts, and other perennial crops. About 82% of acreage under drip irrigation is in California and Florida, primarily on citrus and specialty crops. Because of limited adoption of drip systems outside these two states, we focus here on the choice between sprinkler and gravity systems.

REVIEW OF THE LITERATURE ON IRRIGATION

Caswell and Zilberman (1986) introduced a theoretical model of irrigation technology choice that characterized modern irrigation systems (sprinkler and drip systems) as land quality augmenting. Modern systems, they argued, enhanced the water-holding capacity of soils. Thus improved irrigation technologies would be relatively more profitable to adopt on land with poorer water-holding capacity. Subsequent empirical literature supports Caswell and Zilberman's theoretical specification, suggesting that soil and field characteristics are important determinants of irrigation technology choice, and that modern irrigation technologies are more likely to be adopted on soils with lower water-holding capacity (Dinar and Yaron 1990; Lichtenberg 1989; Negri and Brooks 1990; Mendelsohn and Dinar 2003; Dinar et al. 1992; Schuck and Green 2001; Green and Sunding 1997). Sprinkler adoption rates are generally higher on fields with steeper slopes, which also reduce water-holding capacity.

Sprinkler systems require fixed capital investments, so average fixed costs fall with the scale of operation. One might expect, then, that adoption rates would increase with scale. Chapter 8 illustrated that small farms were less likely even to investigate improvements in irrigation efficiency. Evidence of the impact of farm size on sprinkler adoption is mixed, however. Skaggs (2001) and Schuck et al. (2005) found evidence of a positive relationship between farm size and adoption among New Mexico and Colorado irrigators, but Negri and Brooks (1990) reported the opposite result in a large sample of western groundwater-using farms from the 1984 FRIS. Green et al. (1996), Green and Sunding (1997), and Dinar et al. (1992) observed positive relationships between sprinkler adoption and farm field size.

Several studies have examined the role of water cost on sprinkler adoption. Negri and Brooks (1990) and Dinar et al. (1992) found a positive relationship between water costs and sprinkler adoption. Caswell and Zilberman (1985) did as well, but only at a 12% level of significance. Green et al. (1996) reported that water price had a negative but statistically insignificant effect, whereas Green and Sunding (1997) maintained that water price had a positive effect on sprinkler adoption in citrus production, but not in vineyards.

Empirical findings for the relationship between water source and technology source were also mixed. Caswell and Zilberman (1985) found some positive association between farmer use of groundwater and sprinkler adoption. However, this positive relationship was significant at only the 24% level. According to Negri and Brooks (1990), greater reliance on surface water decreased the probability of sprinkler adoption. Dinar et al. (1992) noted that a greater surface water quota reduced the share of farmland irrigated with sprinklers, but the effect was insignificant. In contrast, Green and Sunding (1997) reported that access to surface water increased sprinkler adoption. Analyzing U.S. county-level data, Mendelsohn and Dinar (2003) found a positive association between a county's surface water withdrawals per acre and its share of acres irrigated with sprinkler technology.

The number of studies on the effects of climate variables is limited, because smaller-scale studies at, for example, the irrigation district level do not exhibit sufficient cross-sectional variation in climate. Studies covering wider geographic areas, however, have found climate to have important effects on irrigation technology choice. Both Negri and Brooks (1990) and Mendelsohn and Dinar (2003) found higher rates of sprinkler adoption in areas with greater rainfall. Negri and Brooks argued that sprinkler adoption would be greater in areas with greater rainfall. This is because growers in high-rainfall areas face a greater risk of crop damage from unexpected rainfall following flood irrigation; here sprinklers provide growers with greater control over applications. In contrast, in hot, arid regions, evaporation losses are large with sprinkler technology. Evaporation losses can reach levels close to 50% under the hot, arid conditions found in Arizona or Southern California (McLean et al. 2000). Mendelsohn and Dinar also pointed out the problem of large evaporation losses and noted that "sprinkler systems are more frequently adopted in cooler locations with a lot more rainfall" (2003, 338). Dinar et al. (1992) did not find a significant rainfall effect, but this analysis was confined to the San Joaquin Valley of California, which has less variation in rainfall than the other two studies.

Negri and Brooks (1990) reported that sprinkler adoption was lower in areas with more frost-free days and a longer growing season (measured in growing degree days) and greater in areas with more frost. They argued that sprinklers are better suited for irrigation for frost protection, whereas longer growing seasons are associated with warmer climates, where sprinkler evaporation losses are greater. Mendelsohn and Dinar (2003) also pointed out that sprinkler adoption rates were inversely related to temperature. Dinar et al. (1992) again found no significant effect, but as before, this may be because of lower variation in their sample.

Finally, because sprinkler systems tend to be labor-saving, some studies have considered impacts of labor costs on sprinkler adoption. Negri and Brooks (1990) found that higher farm labor wages were associated with greater adoption of sprinkler irrigation systems. Mendelsohn and Dinar (2003) reported greater use of sprinkler systems in counties with higher farm wage rates in January, but lower wage rates in April.

EMPIRICAL SPECIFICATION AND IRRIGATION DATA

As in Chapter 8, irrigation data come from special cross-tabulations of the 1998 Farm and Ranch Irrigation Survey (FRIS) made available by USDA's Economic Research Service (USDA ERS 2004). Although the regular FRIS report does not report detailed data by farm sales class, the ERS special cross-tabs report data for each of the 17 westernmost contiguous states by four farm sales classes:

- small farms, with sales less than $100,000;
- medium farms, with sales from $100,000 to $249,999;
- large farms, with sales from $250,000 to $499,999; and
- very large farms, with sales of $500,000 or greater.

The 17 states in the database are North Dakota, South Dakota, Nebraska, Kansas, Oklahoma, Texas, Montana, Wyoming, Colorado, New Mexico, Idaho, Utah, Arizona, Nevada, Washington, Oregon, and California.

The proportion of acres irrigated with sprinkler systems (as a share of acreage under sprinkler plus gravity irrigation) by irrigators in sales class i and state j is P_{ij}. Drip and subsurface irrigation acreage were not included, as data were not reported for many states and farm sales classes because of insufficient observations. More than 85% of drip and subsurface irrigation acreage is in California, and more than 91% is in California, Oregon, and Washington.

Following previous work on irrigation adoption using proportions data (Caswell and Zilberman 1985; Schaible et al. 1991; Mendelsohn and Dinar 2003), the regression equation explaining adoption is specified as a logistic function:

$$\ln(P_{ij}/(1-P_{ij})) = \alpha_0 + \alpha_1\textbf{Small} + \alpha_2\textbf{Medium} + \alpha_3\textbf{Large} + \beta'\mathbf{X}_{ij} + u_{ij} \quad (9.1)$$

where

Small = 1 for small farms, = 0 otherwise;

Medium $= 1$ for medium farms, $= 0$ otherwise;
Large $= 1$ for large farms, $= 0$ otherwise;
X_{ij} $=$ a vector of other explanatory variables (discussed below); and
u_{ij} $=$ a stochastic error term.

A regression intercept α_0 is included, so one farm class dummy (for very large farms) is omitted. Coefficients for the other sales class dummy variables, then, represent differences from the very large farm class. Operations may be in the larger sales classes if they have more acreage, grow higher-value crops, or both. The sales class variables may therefore capture some combination of acreage and crop value effects.

The variable *Energy Pumping Costs* is the average energy costs (in $ per acre) to pump water by farm size and state, for 1998 FRIS irrigated farms. Irrigators use pumps to bring well water to the surface; relift or boost water within irrigation systems; discharge water from ponds, lakes, reservoirs, and rivers; or discharge water from tailwater pits. Pumping costs increase with well depth and energy prices. Energy pumping costs are costs averaged over all energy sources: electricity, natural gas, diesel, and so on.

The variable *Surface Water* is the share of surface water to total irrigation water applied by farm size and state, for 1998 FRIS irrigated farms. Irrigators' relative reliance may affect sprinkler adoption in at least two ways. First, as discussed above, marginal costs of surface water are usually lower than costs for groundwater. Previous research suggests that sprinkler adoption increases with water costs. Second, surface water diversions naturally complement gravity flow systems.

We also include three climate variables: *TEMP*, *TEMPSQ*, and *BELOW32*. *TEMP* is the annual 40-year mean temperature in degrees Fahrenheit of cropland in a state, and *TEMPSQ* is the square of *TEMP*. *BELOW32* is the total number of months the 40-year mean monthly temperature is less than 32 degrees F. The variables come from Teigen and Singer (1988), who discussed in detail how average weather station measurements are weighted by harvested cropland. The climate variables thus give more weight to temperature readings where crops are grown. So, for example, readings from Death Valley or sites high in the Rocky Mountains with little or no agricultural production would receive little or no weight. In contrast, readings from major agricultural production areas would receive great weight.

The variable *SREROSION* measures average annual sheet and rill erosion on cultivated cropland in each state, measured in tons of soil per acre per year. Data come from the 1997 Natural Resources Inventory of the USDA's Natural Resources Conservation Service (USDA NRCS 2000). Sheet and rill erosion is caused by water. *SREROSION* is derived from the universal soil loss equation (USLE) (Wischmeier and Smith 1978; IWR MSU 2002). It is included because it comprises several variables that past studies have found affect irrigation technology choice.

$SREROSION = RKLSCP$, where R is a rainfall factor, K is a soil erodibility factor, L is a slope length factor, S is a slope steepness factor, C is a cover and management factor, and P is a conservation practice factor. R is a factor measuring erosivity of soil from rainfall runoff. It increases with the total amount and peak intensity of rainfall.

The factor *K* assigns values to different types of soil based on susceptibility to erosion and rate of runoff. *K* is thus related to the water-holding capacity of soils. Sprinkler irrigation technology enhances the water-holding capacity, and previous studies have found higher adoption rates on soils with low water-holding capacity. For example, clay soils have *K* values ranging from about 0.05 to 0.15, because they resist detachment. Loam soils tend to have moderate *K* values, and sandy soils have higher *K* values. Both Negri and Brooks (1990) and Mendelsohn and Dinar (2003) reported that farming on sandy soils encouraged sprinkler adoption, while farming on clay soils discouraged sprinkler adoption, relative to farming on loam soils. These results also suggest that sprinkler adoption would increase with an increase in *SREROSION*, via the relationship with *K*.

Land with steeper slopes has also been associated with greater adoption of sprinkler irrigation relative to gravity irrigation. Mendelsohn and Dinar (2003), however, argued that slope length should be positively associated with gravity irrigation, because it implies flatter fields. They found empirical evidence to support this argument. Slope length appears to be the only variable that increases *SREROSION* and has been associated with less sprinkler irrigation use. All the other components of the USLE appear to both increase *SREROSION* and contribute to greater sprinkler adoption. Thus slope length may have a confounding effect. However, computed values for *SREROSION* are not very sensitive to slope length, particularly on flat landscapes (IWR MSU 2002). In sum, all the individual factor components that increase *SREROSION* appear to be positively associated with greater sprinkler adoption except for slope length, *L*. Slope length, however, exerts relatively minor influence on *SREROSION*. Overall, then, we might expect a positive association between *SREROSION* and sprinkler adoption.

Finally, following Mendelsohn and Dinar (2003), we include separate variables *JANWage* and *APRWage*, representing January and April farm labor wages in dollars per hour for hired agricultural field laborers. Wage rates are reported by

Table 9.2 *Descriptive Statistics for Variables Used in Irrigation Choice Regression*

Variable name	*Variable description*	*Mean*	*Min.*	*Max.*
P_{ij}	Acreage irrigated with sprinkler systems as a proportion of the sum of area to irrigate with sprinkler and gravity systems	0.51	0.02	0.88
Energy Pumping Costs	Average energy pumping costs ($ per acre)	$34.34	$11.00	$147.12
Surface Water	Proportion of surface water to total irrigation water from all sources	0.59	0.007	0.99
TEMP	Mean annual temperature	51.23	40.28	68.42
SREROSION	Average annual sheet and rill erosion on nonfederal land	1.99	0.20	4.70
BELOW32	Number of months the mean monthly temperature was less than 32 degrees F	2.12	0.00	5.00
JANWage	January farm wages for hired labor	7.23	6.53	8.34
APRWage	April farm wages for hired labor	6.85	6.2	7.55

multistate region rather than state in the USDA's annual *Agricultural Statistics* (USDA NASS 1999).

Descriptive statistics for the variables are presented in Table 9.2. Adoption rates for sprinkler irrigation range across state–farm size pairs from 2% to 88%. There are also wide ranges between minimum and maximum values for pumping costs, soil erosion, relative reliance on surface water, and number of months with mean temperature below 32 degrees F. At a regional level of aggregation, far less variation occurs in agricultural wage rates.

RESULTS OF THE MULTIVARIATE REGRESSION ON SPRINKLER SYSTEM ADOPTION

The regression equation was estimated using ordinary least squares (Table 9.3). With the log proportions transformation of the dependent variable, we failed to reject the null hypothesis of homoskedasticity. The coefficients for small, medium, and large farms were negative and statistically significant, indicating that these operations had a lower percentage of acreage under sprinkler systems than did very large farms. The coefficient for small farms was the most negative.

The proportion of acres under sprinkler systems increased with energy pumping costs and decreased with greater reliance on surface water. This result is inconsistent with earlier findings suggesting that sprinkler irrigation increases with water costs (Caswell and Zilberman 1985; Negri and Brooks 1990; Dinar et al. 1992). Negri and Brooks have argued that pumping costs may be a better measure of water costs than reported purchase prices of water, because surface water is often quantity-rationed rather than price-rationed. In many western surface water projects, average costs can deviate substantially from marginal costs. Indeed, in many cases,

Table 9.3 *Factors Affecting the Adoption of Sprinkler Irrigation, Ordinary Least Squares Results*

Variable	*Parameter coefficient*	*Standard error*	*P value*
Intercept	−23.74	3.422	< 0.0001
Small	−0.72	0.188	0.0003
Medium	−0.45	0.184	0.0183
Large	−0.49	0.183	0.01
Energy Pumping Costs	0.01	0.004	0.0241
Surface Water	−2.37	0.272	< 0.0001
TEMP	0.91	0.128	< 0.0001
TEMPSQ	−0.008	0.001	< 0.0001
BELOW32	0.27	0.058	< 0.0001
SREROSION	0.62	0.068	< 0.0001
JANWage	0.36	0.228	0.1183
APRWage	−0.44	0.313	0.1598

Note: Dependent variable: ln $[(P_{ij}/1 - P_{ij})]$ where P_{ij} is acreage irrigated with sprinkler systems as a proportion of the sum of area irrigated with sprinkler and gravity systems. Adjusted R-squared: 0.84. Number of observations: 68

irrigators are charged per acre of irrigated land, rather than per acre-foot of water applied. In these cases, the marginal cost of water (per acre-foot) is zero. The fact that sprinkler irrigation adoption rates increase with reliance on groundwater in the regression may also reflect the lower marginal cost of surface water.

The relationship between temperature and sprinkler adoption is complex. The coefficient for mean annual temperature was positive, while the temperature squared term was negative. Both were statistically significant. In the sample, temperature ranged between 40 and 68 degrees F. Sprinkler adoption increased in mean annual temperature for temperatures up to 57 degrees F. Above that temperature, sprinkler adoption declined in mean annual temperature. Sprinkler adoption was positively associated, however, with the number of months of mean annual temperature below 32 degrees F. Thus, consistent with Negri and Brooks (1990), we found that having fewer days of frost or freezing temperatures or a longer growing season appears to discourage sprinkler adoption. For states with the coolest climates, then, a rise in temperatures would have countervailing effects on sprinkler adoption. In low-temperature states, mean temperature increases from global warming would increase sprinkler adoption. However, warming could also reduce the number of months with average temperatures below freezing, in turn decreasing sprinkler adoption. Warming in warmer areas would tend to decrease sprinkler use. This suggests that moderate warming of a few degrees would discourage the use of sprinkler irrigation relative to gravity irrigation in states with warmer average temperatures (Arizona, California, Oklahoma, and Texas), but increase adoption in states with medium-range temperatures and milder winters (Oregon and New Mexico). It would have ambiguous effects in states with more months of freezing temperatures (Washington, other mountain states, Northern Plains states) because of the countervailing effects of mean temperature and months with freezing temperatures.

These results regarding temperature apply only to the choice of sprinkler irrigation relative to gravity irrigation (i.e., changes at the intensive margin). They say nothing about whether sprinkler adoption may encourage more land to be brought under irrigated cultivation. Caswell and Zilberman (1986) discuss conditions where introduction of sprinklers would increase total acreage under irrigation. Caswell et al. (2001) found evidence of warmer temperatures increasing sprinkler irrigation adoption, relative to dryland production.

The coefficient on average annual sheet and rill erosion is positive and significant, indicating that in areas with high erosion, sprinkler irrigation systems are more likely to be adopted. Because sheet and rill erosion is more likely on areas with steeper slopes, greater rainfall, and poorer water-holding capacity, this variable may be picking up the effects of combinations of these factors.

Similar to Mendelsohn and Dinar's (2003) findings, the coefficient for January wages is positive and that for April wages is negative. Mendelsohn and Dinar provided no explanation for the negative relationship between April wages and sprinkler adoption. The wage coefficients are not significant (at the 10% level), however. Because sprinkler systems tend to be labor-saving, we would expect that higher wages would encourage sprinkler adoption. Negri and Brooks (1990), using annual wage rates, find a positive relationship.

CONCLUSIONS

Many commentators have suggested that adoption of improved irrigation technology, such as sprinkler irrigation, will be an important part of agricultural adaptation to climate change. Between 1994 and 2008, the share of U.S. acres irrigated using only gravity-flow flood methods fell from 36% to 24%. This chapter illustrates how special tabulations of irrigation practices by farm size class developed by the USDA's Economic Research Service can by used for multivariate regression analysis. Results suggest that a relatively small number of economic and environmental variables explain a considerable portion of producer choice of the relatively efficient sprinkler technology over gravity irrigation systems. A simple model that divides irrigators into state–farm size pairs explains 84% of the variation in the extent of adoption of sprinkler irrigation relative to gravity irrigation across 17 western states. The proportion of acres irrigated with sprinkler irrigation was greater for very large farms and lowest among farms in the smallest sales class. Adoption of sprinkler irrigation was also positively influenced by the extent of sheet and rill erosion, which may proxy for effects of steeper slopes, soils with less water-holding capacity, and greater rainfall. Sprinkler adoption also decreased with reliance on surface water and increased with energy costs.

Few studies have had enough geographic variation to consider how climate variables affect irrigation technology choice. This analysis has enough of such variation to reach some tentative conclusions regarding climate and sprinkler adoption. The effects of temperature were complex. A longer growing season decreased sprinkler adoption, whereas adoption increased and then decreased with cropland-weighted mean temperature. Results suggest lower sprinkler adoption in warmer states with long growing seasons. Overall, then, the results indicate there may be limited scope for farmers to adapt to climate warming by switching from gravity to sprinkler technology. This may be especially true in areas with relatively high average temperatures, such as Arizona or Southern California.

REFERENCES

Aillery, M., and N. Gollehon. 2003. Irrigation Water Management. In *Agricultural Resources and Environmental Indicators, 2003*, edited by Ralph Heimlich. Agriculture Handbook No. AH722. Washington, DC: Department of Agriculture, Economic Research Service, Chapter 2.2.

Burton, I. 2000. Adaptation to Climate Change and Variability in the Context of Sustainable Development. In *Climate Change and Development*, edited by L. Gómez-Echeverri. New Haven, CT, and New York: Yale School of Forestry and Environmental Studies and United Nations Development Programme (UNDP), 153–173.

Caswell, M., K. Fuglie, C. Ingram, S. Jans, and C. Kascak. 2001. *Adoption of Agricultural Production Practices: Lessons Learned from the U.S. Department of Agriculture Area Studies Project.* Agricultural Economic Report No. AER792.

Caswell, M., E. Lichtenberg, and D. Zilberman. 1990. The Effects of Pricing Policies on Water Conservation and Drainage. *American Journal of Agricultural Economics* 72:883–890.

Caswell, M., and D. Zilberman. 1985. The Choices of Irrigation Technologies in California. *American Journal of Agricultural Economics* 67:224–324.

———. 1986. The Effects of Well Depth and Land Quality on the Choice of Irrigation Technology. *American Journal of Agricultural Economics* 68:798–811.

Cavagnaro, T., L. Jackson, and K. Scow. 2006. Climate Change: Challenges and Solutions for California Agricultural Landscapes. White Paper CEC-500-2005-189-SF. California Climate Change Center, www.energy.ca.gov/2005publications/CEC-500-2005-189/CEC-500-2005-189-SF.PDF (accessed September 22, 2010).

Dinar, A., M. Campbell, and D. Zilberman. 1992. Adoption of Improved Irrigation and Drainage Reduction Technologies under Limiting Environmental Conditions. *Environmental and Resource Economics* 2:373–398.

Dinar, A., and D. Yaron. 1990. Influence of Quality and Scarcity of Inputs on the Adoption of Modern Irrigation Technologies. *Western Journal of Agricultural Economics* 15:224–233.

Gollehon, N., and W. Quinby. 2000. Irrigation in the American West: Area, Water and Economic Activity. *International Journal of Water Resources Development* 16:187–195.

Green, G., and D. Sunding. 1997. Land Allocation, Soil Quality and Irrigation Technology Choices. *Journal of Agricultural and Resource Economics* 27:267–275.

Green, G., D. Sunding, D. Zilberman, and D. Parker. 1996. Explaining Irrigation Technology Choices: A Microparameter Approach. *American Journal of Agricultural Economics* 78:1064–1072.

Huffaker, R., and N. Whittlesey. 2000. The Allocative Efficiency and Conservation Potential of Water Laws Encouraging Investments in On-Farm Irrigation Technology. *Agricultural Economics* 24:47–60.

———. 2003. A Theoretical Analysis of Economic Incentive Policies Encouraging Agricultural Water Conservation. *Water Resource Development* 19:37–55.

IWR MSU (Institute of Water Research, Michigan State University). 2002. RUSLE Online Soil Erosion Assessment Tool. www.iwr.msu.edu/rusle/ (accessed September 22, 2010).

Jackson, L.E., F. Santos-Martin, A.D. Hollander, W.R. Horwath, R.E. Howitt, J.B. Kramer, A.T. O'Geen, B.S. Orlove, J.W. Six, S.K. Sokolow, D.A. Sumner, T.P. Tomich, and S.M. Wheeler. 2009. Potential for Adaptation to Climate Change in an Agricultural Landscape in the Central Valley of California. Final Paper CEC-500-2009-044-F. California Climate Change Center, www.energy.ca.gov/2009publications/CEC-500-2009-044/CEC-500-2009-044-F.PDF (accessed September 22, 2010).

Joyce, B., V. Mehta, D. Purkey, L. Dale, and M. Hanemann. 2009. Climate Change Impacts on Water Supply and Agricultural Water Management in California's Western San Joaquin Valley, and Potential Adaptation Strategies, Final Paper CEC-500-2009-051-F. California Climate Change Center, www.energy.ca.gov/2009publications/CEC-500-2009-051/CEC-500-2009-051-F.PDF (accessed September 22, 2010).

Kurukulasuriya, P., and S. Rosenthal. 2003. *Climate Change and Agriculture: A Review of Impacts and Adaptations*. Climate Change Series No. 91. Washington, DC: World Bank.

Lichtenberg, E. 1989. Land Quality, Irrigation Development, and Cropping Patterns in the Northern High Plains. *American Journal of Agricultural Economics* 71:187–194.

McLean, R.K., R. Sri Ranjan, and G. Klassen. 2000. Spray Evaporation Losses from Sprinkler Irrigation Systems. *Canadian Agricultural Engineering* 42:1–8.

Mendelsohn, R., and A. Dinar. 2003. Climate, Water, and Agriculture. *Land Economics* 79:328–341.

Negri, D.H., and D.H. Brooks. 1990. Determinants of Irrigation Technology Choice. *Western Journal of Agricultural Economics* 15:213–223.

Negri, D.H., N.R. Gollehon, and M.P. Aillery. 2005. The Effects of Climatic Variability on US Irrigation Adoption. *Climatic Change* 69:299–323.

Peterson, J.M., and Y. Ding. 2005. Economic Adjustments to Groundwater Depletion in the High Plains: Do Water-Saving Irrigation Systems Save Water? *American Journal of Agricultural Economics* 87:147–159.

Schaible, G., C. Kim, and N. Whittlesey. 1991. Water Conservation Potential from Irrigation Technology Transitions in the Pacific North-West. *Western Journal of Agricultural Economics* 16:194–206.

Schuck, E.C., W.M. Frasier, R.S. Webb, L.J. Ellingson, and W.J. Umberger. 2005. Adoption of More Technically Efficient Irrigation Systems as a Drought Response. *International Journal of Water Resources Development* 21:651–662.

Schuck, E.C., and G.P. Green. 2001. Field Attributes, Water Pricing, and Irrigation Technology Adoption. *Journal of Soil and Water Conservation* 56 (4):293–298.

Skaggs, R.K. 2001. Predicting Drip Irrigation Use and Adoption in a Desert Region. *Agricultural Water Management* 51:125–142.

Skaggs, R.K., and Z. Samani. 2005. Farm Size, Irrigation Practices, and On-Farm Irrigation Efficiency. *Irrigation and Drainage* 54:43–57.

Sloggett, G. 1985. *Energy and U.S. Agricultural Irrigation Pumping, 1974–1983.* Agricultural Economics Report No. 545. Washington, DC: U.S. Department of Agriculture, Economic Research Service.

Smit, B., and M.W. Skinner. 2002. Adaptation Options in Agriculture to Climate Change: A Typology, Climatic Change. *Mitigation and Adaptation Strategies for Global Change* 7:85–114.

Teigen, L., and F. Singer. 1988. *Weather in U.S. Agriculture: Monthly Temperature and Precipitation by State and Farm Production Region, 1950–86.* Statistical Bulletin No. 765. Washington, DC: U.S. Department of Agriculture, Economic Research Service.

USDA ERS (U.S. Department of Agriculture, Economic Research Service). 2004. Data Sets: Western Irrigated Agriculture: Tables. http://ers.usda.gov/Data/WesternIrrigation/ShowTables.asp (accessed September 27, 2010).

USDA NASS (U.S. Department of Agriculture, National Agricultural Statistical Service). 1998. 1997 Census of Agriculture, AC97-SP-1, vol. 3, Special Studies, part 1, 1998 Farm and Ranch Irrigation Survey. Washington, DC: USDA NASS. www.agcensus.usda.gov/Publications/1997/Farm_and_Ranch_Irrigation_Survey/index.asp (accessed September 27, 2010).

———. 1999. *Agricultural Statistics 1999.* Washington DC: U.S. Government Printing Office.

———. 2002. Census of Agriculture: 2003 Farm and Ranch Irrigation Survey. www.agcensus.usda.gov/Publications/2002/FRIS/index.asp (accessed September 30, 2010).

USDA NRCS (U.S. Department of Agriculture, Natural Resources Conservation Service). 2000. *Summary Report: 1997 National Resources Inventory.* Rev. December 2000. Washington, DC, and Ames, IA: Natural Resources Conservation Service and Iowa State University Statistical Laboratory.

Ward, F.A., and M. Pulido-Velazquez. 2008. Water Conservation in Irrigation Can Increase Water Use. *Proceedings of the National Academy of Sciences* 105:18215–18220.

Wischmeier, W.H., and D.D. Smith. 1978. *Predicting Rainfall Erosion Losses: A Guide to Conservation Planning.* Agricultural Handbook No. 537. Washington, DC: U.S. Department of Agriculture, Soil Conservation Service.

CHAPTER 10

Using Climate Information to Improve Electric Utility Load Forecasting

Bonnie G. Colby and Paulo Tanimoto

The interaction of climate and electricity generation in the Southwest is a matter for concern. The southwestern United States is characterized by high population growth, with households and businesses relying more heavily over time on electricity-demanding technology. Climate change is expected to amplify this issue even further. Electricity providers are particularly challenged because demand for energy is connected with many factors, including population, weather, regulations, and water. One promising climate change adaptation is improved forecasting of electricity demand so that power generation can occur more cost-effectively. This chapter examines the benefits of using climate information to forecast loads and illustrates the application of advanced statistical methods to address climate change challenges. Electric utilities can benefit from using relatively simple climate information to improve the accuracy of their load forecasts and reduce inefficiencies associated with generating either more power than is demanded or too little power.

This chapter demonstrates how statistical models developed for forecasting short- and medium-term usage of electricity can be applied to reduce waste in electricity generation and facilitate effective adaptation to climate variability and change. Load forecast predictions can be improved significantly when climate variables are included, especially temperature. For the medium term, we find that publicly available climatic forecasts perform well in a seasonal planning model. This chapter complements the findings presented in Chapters 8, 9, and 11 on the use of information and new technology as adaptation strategies to climate change.

Climatic factors have a strong influence on demand for electricity. Localities facing extreme summer or winter temperatures employ more energy for cooling or heating. Use of electricity is also influenced by the occurrence of severe droughts or storms. Changes in climate will undoubtedly affect energy use. For example, in the Southwest, warmer temperatures can result in more energy used for cooling during the summer but less for heating in the winter.

Water and energy use are closely interrelated, as described in Chapter 1. Vast quantities of water are consumed in the process of producing electricity, including in the extraction of fossil fuels, vaporization from hydroelectric power plants, and cooling for nuclear plants. The electric sector is the second-largest user of water in the United States, surpassed only by agriculture (DOE 2006). Conversely, much electricity is also involved in the treatment of water for consumption and the delivery of water for agricultural, municipal, and industrial uses. A study by Powicki (2002) determined that about 75% of the cost of water treatment in municipalities comes from electricity. The California Energy Commission (CEC 2005) has found that electricity usage associated with water consumption (primarily residential consumption) is even greater than energy usage for water supply and treatment.

While electricity loads are linked to many factors, the water-energy nexus is particularly relevant to our study, as demand and supply of both resources are sensitive to climatic and weather conditions. For the Southwest, warmer temperatures are associated not only with increased use of electricity and water at peak periods, but also with greater incidence of long-term drought. These interdependencies amplify the impacts of climate change. Increasing consumption of water will increase usage of electricity and vice versa, creating a problematic cycle.

Electricity is vulnerable to climate change in terms of supply as well as demand. Because electricity cannot be efficiently stored, and transmission is limited by existing infrastructure, the amount to be generated and delivered must be determined in advance. However, the quantity consumed may differ from what was anticipated. If an electric utility overproduces, the surplus is wasted unless it is sold in the spot market. When a utility underproduces, the deficit has to be compensated by purchases from other companies, or costly power outages may occur.

Generally, electricity generators make use of forecasting techniques in which climate has typically been assumed to be constant. This chapter develops statistical models that account for weather and climatic information in forecasting future demand for electricity. These models are tested against models that do not incorporate climate information, using load data provided by an electric utility company in the southwestern region. Evaluating the forecasts resulting from both kinds of models, we find that our more comprehensive models outperform those that do not employ weather or climatic data. Society as a whole gains from improved load forecasts, because better use of resources can result in more reliable and cost-effective provision of electricity, as well as reduce pollution.

IMPROVED CLIMATE AND LOAD FORECASTS AS AN ADAPTATION TOOL

One of the most effective ways of dealing with changing conditions is to develop adaptive responses in advance. Electric utilities face the challenges not only of meeting average daily demands, or loads, but also of forecasting and meeting electricity demand at peak levels. In regions where a large portion of generation is dependent on hydroelectric plants, low storage levels in reservoirs force a trade-off between supplying electricity at peak levels and having capacity to generate electricity for the rest of the year. Tensions also arise over using water for generating electricity versus reserving water in reservoirs for agriculture.

Forecasts can play a central role in the electric sector to estimate demand for energy sufficiently in advance so that preemptive actions can be taken. This has been well summarized by Feinberg and Genethliou: "With supply and demand fluctuating and the changes of weather conditions and energy prices increasing by a factor of ten or more during peak situations, load forecasting is vitally important for utilities" (2005, 270).

As totally accurate foresight is impossible, decisionmakers use available information to attempt to best predict what will happen and improve the plan the utility will implement to achieve or prevent certain outcomes. In a workshop organized by EPRI, in 2005, load specialists agreed that demographic and economic variables, such as fuel prices, population, and economic growth, play a larger role in medium- and long-term load forecasts than do climate variables, and that climate forecasts can become increasingly important as climate change takes place. The timescale of forecasts and utility planning horizons can range from short- to long-term (decadal in scope).

Weather is a short-term manifestation that is highly volatile, and thus weather forecasting is difficult even with the best of current models and technology, and accuracy decreases considerably after one or two weeks. Climate, on the other hand, is more predictable because it is a composite made up of many weather components averaged over the longer term, and it varies slowly in comparison with individual weather conditions and events (Palmer et al. 2005). El Niño and La Niña are examples of recurrent climatic phenomena that make seasonal outlooks more predictable. Despite advances in more predictable seasonal outlooks, researchers have done little work to connect demand for electricity to climate forecasts.

PREVIOUS STUDIES

Only a few previous studies have attempted to rigorously assess the impacts of climate change on electric utilities' operations. Experts note that severe conditions (such as heat waves and extreme droughts) that coincide with high-demand seasons can have major impacts on power generation and cause disruptions of service. There is a consensus that more applied research on the effects of climate change is necessary for improved load forecasting and for long-term planning for generation, transmission and distribution (EPRI 2005).

Electricity load prediction formerly was based on simple estimates accounting for the number of households, typical mix of appliances, and proportion of commercial and industrial sectors in the area served. This method ignored the varying behavior of households and businesses, and electric utilities shifted to a more stochastic approach by employing simple regression models in which system load is a function not only of consumer characteristics, but also of the time of day and weather conditions (Papalexopoulos and Hesterberg 1990). Gross and Galiana noted that "time-of-day models do not accurately represent the stochastically correlated nature of the load process, or its relation to weather variables. As a result, when weather patterns are changing rapidly, the coefficients are not appropriate, except for a short time interval into the future" (1987, 1563).

During the 1980s, time-series models became common to take advantage of recurrent patterns in the data. By specifying load as a function of past load, other factors become implicitly accounted for in the model. Typically, autoregressive (AR) and moving-average (MA) terms are specified, and differencing is used for removing trends in the data (Hagan and Behr 1987). One strong advantage, besides minimal requirements on data for external variables, is the ability of dynamic models to capture nonlinearity.

In the 1990s, artificial intelligence techniques were applied to the problem of load forecasting (Peng et al. 1992; Papalexopoulos et al. 1994). However, coefficients from these models do not have a straightforward interpretation. Our study takes the middle route of creating a dynamic model extended with explanatory variables. From the time-series toolbox, we model current demand for electricity partly based on the evolution of load itself. This approach has the appealing feature that unobservable variables affecting demand are implicitly captured in the previous observations. Usage of energy is highly sensitive to changing conditions, and including additional explanatory variables accounts for this. We have refrained from using artificial intelligence techniques in order to explicitly model the relationship between climatic variables and electricity use. More useful forecasts can be obtained by specifying the structural form of the model, and this increases understanding of the influence of weather and climatic factors on energy consumption.

ECONOMIC VALUE OF FORECASTS

Although consensus exists that forecasts are economically important to electricity providers, not much attention has been put into rigorously quantifying their value. Hobbs (1997) addressed this issue by using survey data from 28 utilities that provided their estimates of savings from switching to better forecasting techniques. The average value reported was $800,000 per year. One utility, with a peak load of 35 gigawatts (GW), reported that a 1.5% reduction in the mean absolute percentage error (MAPE) of forecasts would correspond to an annual savings of $7.6 million.

Hobbs et al. (1999) used data provided by two of those utilities to more rigorously assess those savings. The authors combined forecast simulation, optimization for unit commitment at each hour, and modeling of economic dispatch. Forecast errors incur costs for the generator, as the difference between what is produced and the quantity that is actually demanded may be wasted or sold in the spot market for a low price. When an overforecast occurs—that is, more electricity is produced than demanded—Hobbs et al. enumerated the following costs for the utility:

- Units may have been unnecessarily committed, raising fuel costs
- Expensive power may have been purchased which was not needed, or a profitable opportunity to sell bulk power might have been spurned

- Hydropower may have been produced which would have been more valuable if generated at a later time
- Overly high real-time prices might have been quoted, depressing sales
- Unnecessary interruptions or load controls might be invoked, annoying consumers and lowering revenue (1999, 1342)

On the other hand, when utilities underforecast and generate less electricity than what turns out to be demanded, these consequences can follow:

- Insufficient resources may be available for meeting security constraints, such as spinning reserve margins
- To meet the unanticipated load increase, uneconomic generation or purchases of spot power might be necessary
- Commitments to sell power may have been made at a price less than the value of that power to the utility
- Too low real-time prices might have been quoted, resulting in revenue falling short of the utility's cost (1999, 1342)

Although the potential savings depend on the characteristics of the utilities and the behavior of consumers, Hobbs and colleagues estimated that a 1% decrease in the MAPE can reduce variable costs of generation by 0.1% to 0.3%. Thus, for a utility with 10 GW peak load, the value can reach $1.6 million annually.

Teisberg et al. (2005) extended the Hobbs et al. (1999) models to estimate the economic value of temperature forecasts for electricity generation. The authors employed load data from six electric utilities in diverse locations and climatic conditions across the United States. The weather information used consisted of daily temperature forecasts from the National Weather Service (NWS), as well as their actual values. These were compared with the naïve approach of persistence forecasts—that is, assuming that tomorrow's temperature will be the same as today's. From the calculations, the authors estimated that the nationwide value of the NWS forecasts is $166 million annually, relative to persistence forecasts.

In our study, we estimate the net benefits from improved forecasts by comparing the performance of simple forecast models with new models that include weather variables. We observe the improvement in the load forecasts from including the additional variables.

DATA, METHODOLOGY, AND ESTIMATION USED IN OUR MODELS

In this chapter, we develop both a model to forecast hourly loads and a simple medium-term model to assess the feasibility of incorporating climate information in planning. We compare the medium-term model's performance with that of an alternative model employing actual values for the climate variables used in forecasting.

Electricity providers are protective of their data on system load, as such information may be of strategic value and can also pose privacy issues. This research was made possible by cooperation with the Arizona Electric Power Cooperative (AEPCO), which provided data for one of its member electricity providers. For purposes of confidentiality, the name of the area served is not revealed. AEPCO is a not-for-profit cooperative that provides electricity to seven member cooperatives in Arizona and California. The cooperative functions in three distinct modes: provision of electricity services, transmission, and power generation (AEPCO 2008). Most of the energy is generated by combustion of coal and, to a lesser extent, natural gas. The load data consist of hourly measurements of total electricity provided in a nonidentified member service area. The data provided covers the period from January 1993 through December 2006.

Load data are combined with weather data, because individuals tend to respond quickly to changes in temperature, humidity, and luminosity conditions, especially when those changes are atypical (Willis 2002). Weather data used are from AZMET, a University of Arizona meteorological network that measures a large number of weather-related variables at hourly intervals. The variables used in this study, shown with summary statistics in Table 10.1, include air temperature (temp), relative humidity (humi), vapor pressure deficit (vapor), solar radiation (solar), precipitation (prcp), and wind speed (wind).

For visual inspection, we plot electricity load in different aggregation levels. First, as a broad picture, we present demand aggregated at the monthly interval. In Figure 10.1, we can observe a clear seasonal pattern: the annual peak of electricity usage occurs in the summer, and a second, much smaller peak is seen in the winter. Demand for energy for cooling and, to a lesser extent, for heating is the driving force behind this recurrent phenomenon. In addition to the annual seasonality, it is evident that the time series is nonstationary across all periods. In other words, there is an upward trend in electricity usage for this area that does not revert to the long-term mean.

Table 10.1 *Load Data Summary Statistics, 1995–2006*

Variable	*Unit*	*Mean*	*Standard deviation*	*Minimum*	*Maximum*
load	MW	40.10	19.11	10.40	144.28
temp	°C	20.25	10.00	−7.80	45.80
humi	%	42.62	24.93	2.50	100.00
vapor	kPa	1.81	1.55	0.00	9.50
solar	MJ/m^2	0.86	1.15	0.00	4.29
prcp	mm	0.03	0.41	0.00	42.00
wind	m/s	2.42	1.46	0.20	12.90
holiday	dummy	0.001	0.033	0	1

Note: *T* (total number of periods) = 122,639; MW = megawatts; kPa = kilopascals; MJ = megajoules; m^2 = square meters; mm = millimeters; m/s = meters per second

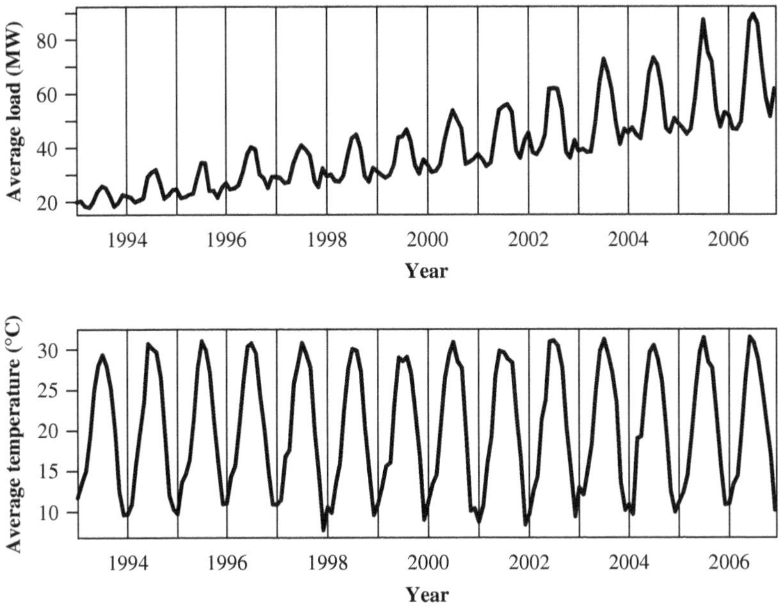

Figure 10.1 *Monthly Electricity Load for 2003 to 2006*

Source: Tanimoto (2008)

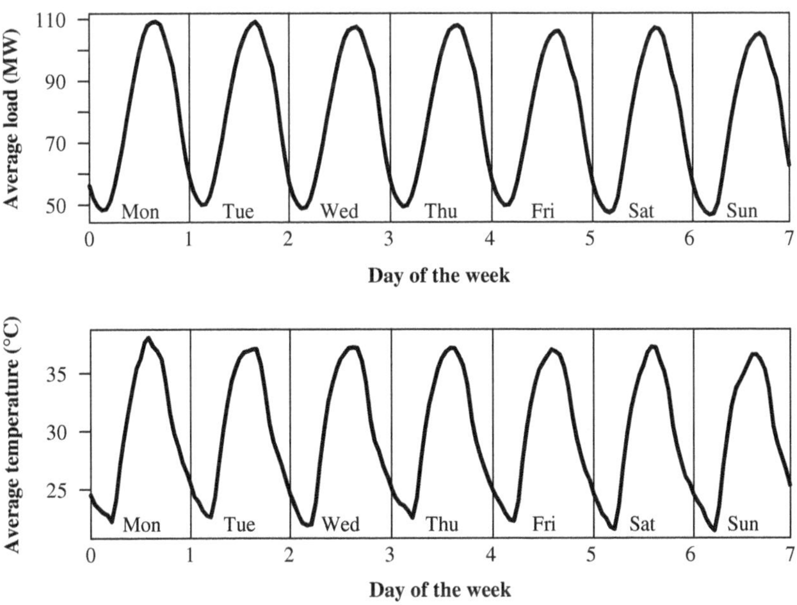

Figure 10.2 *Average of Weekly Electricity Load for Summer 2005*

Source: Tanimoto (2008)

We can gain more insight when we visualize the evolution of hourly load during a week. In Figure 10.2, average load and temperature are plotted for each hour of the day, each day of the week during the summer of 2005 as a way of displaying the recurring fluctuations that occur during a week. In the summer, a strong relationship between load and temperature can be seen. The shape of the load curve is unimodal, with a peak occurring in the early evening. This spike is linked to individuals returning to their houses after work and turning on electric devices, including air conditioners. Consumption of electricity tends to decrease during the weekends and on Fridays, but even between those days the reduction is different.

In the winter, the load curve is bimodal, with one peak in the morning and another, slightly more accentuated, in the evening. Those correspond to hours when the temperature is low but individuals are home and active, demanding energy for heating. The distinction between weekdays and weekends is now even more evident. Note that a base level of electricity continues to be used independent of the daily fluctuation. This base level has been increasing across the years for the region of study.

General inspection of the data reveals that demand for electricity is sensitive to weather conditions, especially temperature. Also, seasonality is present in the series on at least three levels: (1) an annual pattern, including a peak in the summer and another in the winter; (2) a daily periodicity that changes from unimodal in the summer to bimodal in the winter; and (3) a weekly "seasonality," in which weekdays are different from weekends. Finally, the process is nonstationary, with an upward trend.

FORECAST EVALUATION

One of the main goals of this research is to develop models that can make good forecasts. These are desirable because they can help in planning decisions, reducing costs, and increasing economic efficiency. In considering how forecasts can be evaluated, we present measures of how close forecasts are to actual values, and we discuss the costs of making and improving the forecasts.

Forecasts must be evaluated by their statistical properties. Defining the error as the difference between the predicted and actual values, we are interested in a measure that summarizes the error of various predictions, not a single prediction. More formally, the expected value of the error, $E(\varepsilon)$, measures how wrong an estimator is on average. This term is also known as bias, and its reciprocal is the accuracy of an estimator. In the second moment, the variance of the error, $V(\varepsilon)$, can be thought of as a measure of the dispersion of the error, with its reciprocal giving the precision.

Aside from these summary statistics, an important consideration is the weights associated with positive and negative discrepancies. While it is customary to assume that over- and underpredictions have equally severe consequences, this need not be the case. With electricity generation, the decision of underproducing can cause blackouts, while overproducing results in excess generating costs and reduced net revenue.

A commonly used measure of forecast error is the root mean squared error (RMSE). It is given by the square root of the mean squared error (MSE), a closely related measure of error. The formula is

$$RMSE = \sqrt{\frac{1}{T}\sum_{t=1}^{T}\left(Y_t - \hat{Y}_t\right)^2} \tag{10.1}$$

where Y_t is the actual observation for period t, $\hat{Y}_t$ is the corresponding predicted value, and T is the total number of periods.

The RMSE has a few features worth discussing. Values are given in the same units as the variable of interest, thus making it easier to visualize the dispersion of errors. Because of squaring, larger values are given more weight than smaller values. Finally, and also because of squaring, negative and positive errors are treated identically. That is, an underprediction and overprediction of the same magnitude are considered to be the same error in this formula.

In the electricity industry, the most widely employed measure of error is the mean absolute percentage error (MAPE). The formula for MAPE is

$$MAPE = \frac{1}{T}\sum_{t=1}^{T}\frac{\left|Y_t - \hat{Y}_t\right|}{Y_t} \tag{10.2}$$

MAPE exhibits a number of properties. First, its values are given in percentage (unitless), therefore making it comparable across different situations. As in RMSE, no distinction is made between positive and negative errors; all are considered to have equal weights in absolute terms. Unlike RMSE, the impact of small or large deviations is always relative to the actual value, capturing the idea that a forecast error that is severe for a small utility may be less important for a large company. In evaluating our models, we report both RMSE and MAPE, but both measures are largely equivalent: in most cases, an increase or decrease in the forecast error will lead to a change in both scores in the same direction, albeit at a different rate.

An additional measure of the performance of predictions comes from the meteorological literature. Forecast skill, as it is called, is "a statistical evaluation of the accuracy of forecasts or effectiveness of detection techniques" (AMS 2008). The formula for the skill score (SS) is

$$SS = 1 - \left(RMSE_f / RMSE_r\right) \tag{10.3}$$

where $RMSE_f$ is the root mean squared error of the forecasts being evaluated, and $RMSE_r$ is the corresponding measure of error for values that are taken as a reference. A positive value for SS indicates that the model offers some skill over the reference technique. A common practice is to use persistence forecasts (known as "naïve forecasts" in the economics literature) as the reference, in which observations of similar periods are used as the predictions for the period of interest. For example, one can assume that tomorrow's temperature will be the same that was recorded today, or that next summer will have the same climate characteristics as the current one.

In addition to errors in forecasts, costs also result from time and resources allocated to estimating statistical models. In economic terms, the marginal gains from spending one extra hour in modeling should be greater than or equal to the marginal costs of doing so. Even at a more fundamental level, ultimately there is a time constraint within which the decision of how much electricity to produce must be made, or otherwise blackouts can occur. To address this issue, we will include an extra benchmark measure: time of estimation, in minutes. This is merely the time an estimation takes to converge to a solution. Although this does not take into account the time spent in setting up the parametric form of the model, it can be important in determining the feasibility of generating a large number of alternative specifications of models.

MODELING CONSIDERATIONS

Next, we derive the models that will be used for forecasting system load. The first step in time-series modeling is to ensure that the data conform to the assumptions required by the methodology. One of the most important assumptions is that of stationarity. Seasonality, which is technically a violation of this assumption, deserves special attention in our model. Traditional time-series techniques require that the input series be stationary. Intuitively, stationarity implies that in the long term, the series does not wander away indefinitely from its mean. The load series was detrended by taking the first difference of the logged hourly load.

Seasonality

Systematic patterns that appear to be associated with daily, weekly, and annual cycles are commonly referred to as seasonality or periodicity (Chao 1983). One interesting characteristic of our data, and demand for electricity in general, is that the daily periodicity changes over time in a systematic manner, with a unimodal curve during the summer and a bimodal shape in the winter.

Figure 10.3 reveals more details about the different shapes of the load curve, month by month. We can loosely cluster the curves into five groups: (1) a clearly bimodal interval, coinciding with the winter months (December to March); (2) a transitional phase, in which the curve is flat but the two peaks can still be observed (November and April); (3) a first, low unimodal period (October and May); (4) a second, intermediate unimodal period (June, August, and September); and (5) a third unimodal period, which is the highest across the entire year (July). Note that not only does the shape change over the months, but the range of the curves differs significantly as well. The minimum point of July, for example, is larger than the maximum of most winter months.

This peculiarity poses a challenge to modeling for a number of reasons. First, it is not completely captured by weather variables, because the behaviors of individuals exhibit other fluctuations, such as seasonal business cycles and

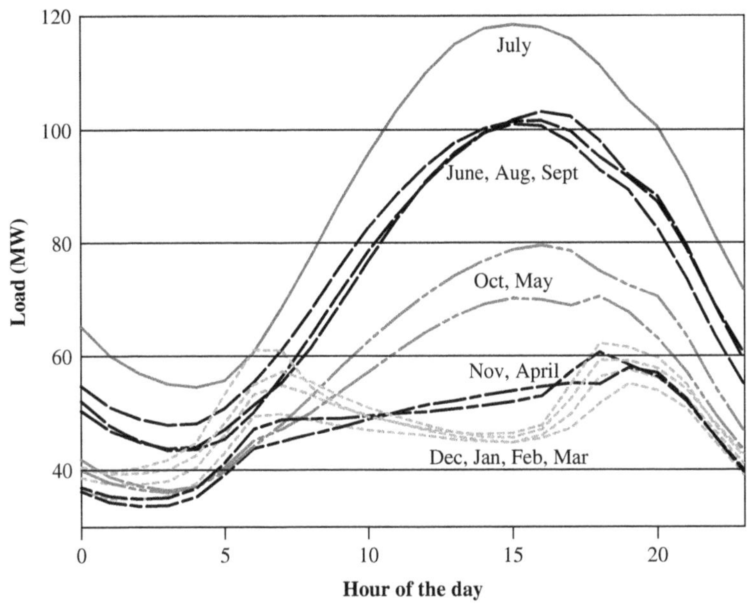

Figure 10.3 *Monthly Averaged Hourly Load for 2005*

Source: Tanimoto (2008)

migrations, that are exogenous to our model. Thus, some form of seasonal modeling is necessary if our goal is forecasting. Second, to the best of our knowledge, a model with changing periodicity has not received attention from the literature, and there is no established method of dealing with it.

One innovation of this study is a way of accounting for a multidimensional seasonality in a time-series dataset that can be estimated in one single pass. We build on the work of Aradhyula and Ergün (2004) and Aradhyula and Tronstad (2006), who developed a general approach for dealing with seasonality. As in those studies, we estimate a time-series model together with high-order polynomials. Different from prior studies, we generalize the framework to handle periodicity in more than one dimension (Roberts 1978; Sims 1992; Hylleberg 1992; Weron 2006). During the winter (beginning and end of the year), demand for electricity follows a bimodal form, with peaks in the morning and evening. In the summer (middle of the year), the load changes into a unimodal form, with much higher minimum and maximum values. The innovation of our modeling technique is to capture the transition between these two seasons in a continuous and smooth manner.

Dependent Variable

The dependent variable of our model is electricity load measured in an hourly interval for a specific region in Arizona. Because of the observed exponential growth in demand, we transform the variable into its natural log form. To ensure stationarity, we include a linear trend variable in our model. Once the model is estimated and forecasts are made, we transform the values back via antilog. This transformation is not without its problems. Nevertheless, the introduced bias is negligible and is not enough to justify a correction, as noted in other studies that have taken a similar approach (e.g., Brocklebank and Dickey 2003).

Explanatory Variables

We include a term for trend due to the increasing usage of electricity for the region of study. Because this growth is steady, we expect to find a robust positive coefficient. Variability across days of the week follows a pattern, and we create a dummy for each day, leaving Monday as the baseline. Demand for electricity is affected by holidays, with a decrease in the usage of electricity when government offices and many commercial businesses are closed. Therefore, we include a dummy for holiday. Although some holidays may be associated with more demand by residential customers, we still expect the coefficient for holiday to be negative.

For this particular region, the relationship between load and temperature follows a quadratic pattern. Temperature and its squared value are both included in the model, with the expectation that the quadratic coefficient will be negative. In other words, we believe that temperature levels that are far from a certain neutral point lead to more consumption of electricity for cooling or heating. As previously discussed, the insensitive level found in exploratory analysis was around 15 degrees C.

Humidity plays a role in the influence of weather on electricity load, because the sensation caused by humidity can compensate for high or low temperature, therefore affecting use of cooling or heating equipment. We used different specifications in the model: a simple linear coefficient, a quadratic relationship as useful for temperature, and the interaction term between temperature and humidity.

The region under study has many days without rain, and the fact that the unit of analysis is measured at hourly intervals also contributes to this sparseness. Hence we recode the variable as a dummy: precipitation takes the value of 1 if any rain fell during that period, and 0 otherwise. We expect less demand for electricity during rainy intervals.

Our final variable, wind speed, is assumed to work analogously to the previous two: rapid air movement generates a chilling sensation that can decrease the need for electric cooling or, conversely, can augment the use of heating during the winter. The coefficient of this variable should appear as negative in our model.

Estimation Process

We use data from 1993 to 2005 to forecast next-day electricity load for the entire year of 2006, in a rolling-sample scheme. More specifically, on each new day, we exclude the oldest 24 hours of our sample and include the actual values of the 24 hours that have passed since the last estimation. We then reestimate the model and produce new forecasts for the next 24 hours. This process is repeated until the hours of the reserved sample have passed.

This procedure mimics how the forecasts are generated in a utility in a day-to-day routine. However, we already have the entire year of 2006 available, so the procedure can be fully automated, whereas in practice, an analyst at the utility would have to feed in the new values. Updating the model as time passes is of key importance, because usage of electricity is very sensitive to a number of factors, including current weather conditions and recent consumption, as a result of inertia. Older load values are discarded in the rolling-sample setup, based on the perspective that recent observations have greater influence on current values, while distant points have an ever-decreasing impact. From a practical perspective, the oldest observations are taken out so that the sample size remains constant across all estimations, thus facilitating comparisons of statistics that are sensitive to the number of observations, such as the Akaike information criterion or the Schwartz Bayesian criterion and the standard errors of coefficients.

Once the entire process is finished, the forecasts are merged with their corresponding actual values, and residuals are calculated. From the residuals, the measures of prediction error are created, in this case MAPE and RMSE. It is also valuable to generate a scatter plot of the residuals, which can be inspected for systematic patterns.

ANALYSIS OF RESULTS

In the following sections, we present and discuss results of our analysis for two types of models: a short-term load forecast model, with three different specifications; and medium-term seasonal load forecast models, presented with five variations.

Short-Term Forecasts

To compare performance, we ran three different specifications of the model: a full version, with all variables; one with only the temperature variables; and the "naïve model," with no weather variables. In all cases, we modeled seasonality as discussed, by including interaction terms for orthogonal polynomials. Estimation results are presented in Table 10.2. (Because of space constraints, the seasonality coefficients have been omitted from the table.) The dependent variable was the log of hourly load, and therefore the magnitude of the coefficients and standard errors cannot be easily interpreted.

The model captured the positive trend in the load for our region. Modeling the trend in this explicit form gave better results than the usual approach of taking the first difference. As expected, Friday, Saturday, and Sunday tend to have lower load, while the effect for the other days is not so clear. The variable for holiday also had a negative coefficient throughout. The quadratic relationship between temperature and load was adequately captured by the model. Also as expected, the second coefficient was positive, indicating higher demand being associated with large deviations from a certain temperature value. Both temperature coefficients were statistically significant at the 0.001 level, suggesting a robust relationship throughout the sample.

Table 10.2 *Short-Term Forecast Model Estimation Results*

Variable	*Full model*	*Temperature only*	*No weather*
constant	2.8686‡	2.8739‡	2.7756‡
	(0.0194)	(0.0194)	(0.0486)
lag1	−0.7822‡	−0.7878‡	−0.8803‡
	(0.0018)	(0.0018)	(0.0013)
lag24	−0.1810‡	−0.1767‡	−0.1070‡
	(0.0018)	(0.0018)	(0.0013)
trend	0.00001‡	0.00001‡	0.00001‡
	(0.000001)	(0.000001)	(0.000001)
tuesday	0.0023*	0.0022*	0.0030†
	(0.0009)	(0.0009)	(0.0011)
wednesday	0.0031†	0.0031*	0.0036*
	(0.0012)	(0.0012)	(0.0014)
thursday	0.0022	0.0020	0.0028
	(0.0013)	(0.0013)	(0.0016)
friday	−0.0030*	−0.0031*	−0.0019
	(0.0013)	(0.0013)	(0.0016)
saturday	−0.0123‡	−0.0124‡	−0.0100‡
	(0.0012)	(0.0012)	(0.0014)
sunday	−0.0123‡	−0.0123‡	−0.0104‡
	(0.0009)	(0.0009)	(0.0011)
holiday	−0.0038*	−0.0052†	−0.0030
	(0.0016)	(0.0017)	(0.0020)
temp	−0.0230‡	−0.0229‡	
	(0.0002)	(0.0002)	
temp^2	0.0006‡	0.0006‡	
	(0.00001)	(0.00001)	
humi	0.0012‡		
	(0.0001)		
humi^2	0.00001‡		
	(0.00001)		
prcp	0.0019		
	(0.0011)		
wind	−0.0009‡		
	(0.0001)		
Forecasts for 2006			
MAPE	4.177	4.276	5.666
RMSE	3.897	4.100	5.989
SS	0.05	0.32	0.22
Time	8′23″	8′04″	8′01″

Note: MAPE = mean absolute percentage error; RMSE = root mean squared error; SS = skill score

* significant at 0.05 level

† significant at 0.01 level

‡ significant at 0.001 level

Humidity presented a similar pattern to that of temperature. To our knowledge, this relationship has not been reported in other studies of load forecasting. In this case, the pattern is the inverse of temperature: extremely high or low levels of humidity are associated with less demand for electricity. As with temperature, this relationship seems to be robust and stable in all runs of the models. We also tested an interaction term between temperature and humidity, in the expectation that the combination of both might have a specific effect. However, we did not find a detectable effect of the interaction terms and do not report the estimation results.

Precipitation, as a dummy variable, was expected to have a negative coefficient sign, but as seen in the table, the variable had the opposite sign and is not significant. Although the results appear to contradict our expectation, a few things should be noted. First, as previously noted, the variable is highly skewed, with more than 98% of cases equal to 0. Second, it is not evident how consumers will respond to rain; although it may serve as a means of cooling, it might also drive individuals indoors and lead to more usage of electricity. We unfortunately cannot identify each case separately, but because the variable does seem to reduce the forecast errors, we decided to keep it in the model. Humidity and precipitation do not appear to suffer from a problem of multicollinearity. In Table 10.3, the correlation with precipitation is only about 0.09 for humidity and −0.15 for humidity squared.

Finally, wind speed met our prior expectations of having a negative relationship with load. Even on a day with a high temperature, wind can cause the sensation of chillness, thereby reducing the demand for electricity for cooling. It is clear in comparing the alternative specifications that weather holds information that is important for modeling and forecasting demand for electricity.

Although our baseline model did not perform so well (MAPE of 5.67%), it is not far from what is reported in the literature and by AEPCO. Merely including temperature in the model is sufficient to reduce the error rate by almost 1.4%. The gains from additional variables, humidity, precipitation, and wind speed, are not as accentuated (only 0.1%), but for a large utility, they may be enough to bring significant savings.

In terms of skill score (SS), we find that each of the specifications brings improvements over the previous one. In the case of the model with no weather, the reference is the RMSE of simple persistence forecasts. In other words, our simplest model brings an improvement in skill of 0.22 over merely assuming that the load for the next 24 hours will be the same as the previous 24 observations. The model with temperature alone possesses a skill score of 0.32 over the model with no weather. Finally, the model with all weather variables offers more skill over

Table 10.3 *Correlation Matrix for Explanatory Variables in Short-Term Forecast Models*

	temp	*temp²*	*humi*	*humi²*	*prcp*	*wind*
temp	1.00					
temp²	−0.87	1.00				
humi	−0.05	0.24	1.00			
humi²	0.09	−0.22	−0.96	1.00		
prcp	−0.03	0.07	0.09	−0.15	1.00	
wind	−0.15	0.11	−0.01	0.04	−0.09	1.00

Note: $T = 122{,}639$

using only temperature, with an improvement of 0.05. Note that performance measured by skill level is in line with the results in terms of MAPE, as discussed just above.

The time of estimation did not vary significantly for each of the models. All of them finished generating the next-day hourly forecasts for 2006 in about 8 minutes. As expected, the models including more variables were slightly slower to generate forecasts, but the difference from best case to worst case is only 22 seconds. These measures include the time that is required to create the polynomial terms; conduct the variable transformations, model estimation, and forecasts; and calculate MAPE and RMSE. Note that the main advantage advocated here is that a single model can be used for the entire year, instead of having separate models for different seasons.

To further illustrate the performance of the forecasts, a plot of actual peak versus predicted values is presented in Figure 10.4. For all days of the out-of-sample period, we selected the daily maximum load and the equivalent value from the forecasts. This aggregation was done partly because the fine scale of one-hour intervals would make the graph overly crowded, to the detriment of understanding. The peak was chosen because it constitutes the most interesting value for AEPCO (Thompson and Cathers 2005). The figure indicates that the predicted values appear to be consistently close to the actual load throughout the year. The model performs relatively well even during the periods of transition between seasons, although initial underforecasts can be seen.

Finally, we plot actual and predicted values averaged by month in Figure 10.5. Each box in that figure presents hourly load averaged over every month. These

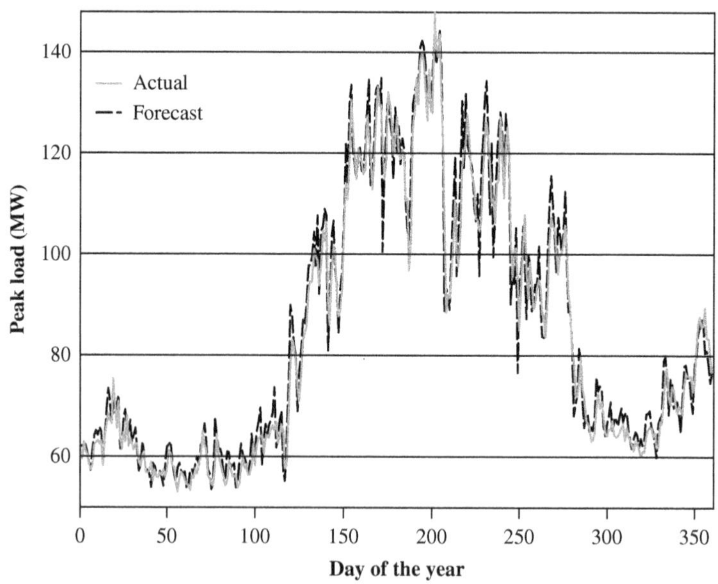

Figure 10.4 *Actual versus Predicted Daily Electricity Peak*

Source: Tanimoto (2008)

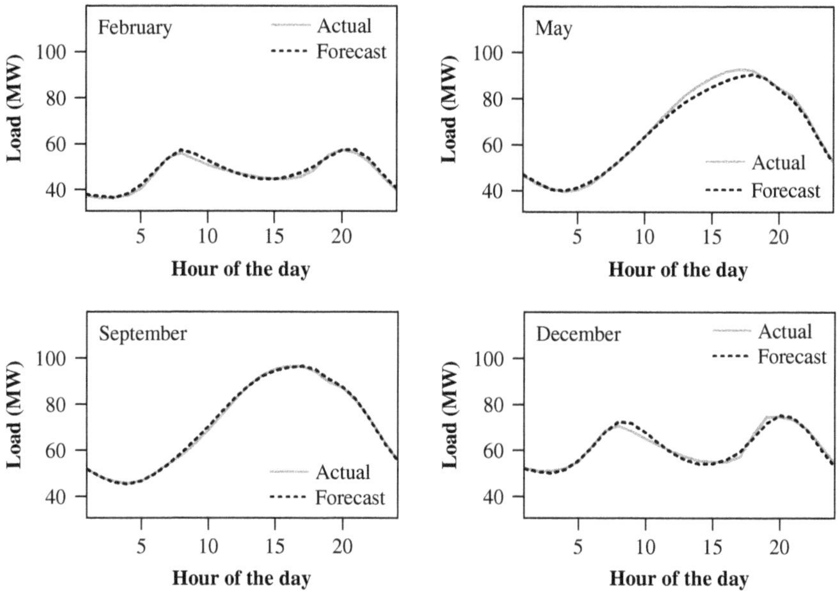

Figure 10.5 *Actual versus Predicted Monthly Averaged Hourly Load*

Source: Tanimoto (2008)

plots provide evidence for the success of polynomials in capturing the change in seasonality that occurs across the year. The technique for dealing with seasonality advocated in this study is able to adjust itself to unimodal, bimodal, and intermediate shapes of the load curve in a smooth manner.

Medium-Term Seasonal Forecasts

To evaluate the value of seasonal forecasts, we model and estimate five specifications of monthly average load, using the same dataset as before. In addition, one-year-ahead seasonal forecasts developed by the Climate Prediction Center (CPC) at the National Oceanic and Atmospheric Administration are included for both temperature and precipitation (NOAA 2010). The descriptive

Table 10.4 *Medium-Term Seasonal Forecasts Models: Variable Summary Statistics*

Variable	*Mean*	*Standard deviation*	*Minimum*	*Maximum*
load-avg	44.39	14.69	24.61	89.39
actual-temp	20.24	7.59	7.76	31.51
actual-prcp	0.02	0.04	0.00	0.25
pred-temp-low	16.80	4.48	7.40	29.55
pred-temp-up	29.34	7.68	16.01	51.92
pred-prcp-low	0.26	0.28	0.00	1.27
pred-prcp-up	0.26	0.28	0.00	1.27

Note: $T = 133$

statistics of the variables employed in the models are shown in Table 10.4. Load, the dependent variable, is averaged at the monthly interval, yielding 133 periods. Actual temperature and precipitation are aggregated in a similar fashion, and at this level, exploratory analysis did not find a nonlinear relationship, in contrast to the short-term models. The CPC forecasts are given in probability terciles—that is, the likelihood that a particular season will observe conditions that are below, near, or above the historical averages. Because our interest lies in measuring how much a unit increase in temperature or precipitation will affect electricity consumption, we transform these predictions by multiplying the terciles by the historical averages.

Again, to achieve stationarity, the dependent variable is logged, and an explicit term for trend is included in the model to control for the growth in demand that the region has been facing. Because we have only 13 years of data and lost some when matching our own data with the CPC forecasts, we refrain from reserving part of the dataset for out-of-sample prediction. Instead, we will evaluate the models by their in-sample properties.

The estimation results for all models are presented in Table 10.5. Model 1 includes only autoregressive terms and the trend variable. Each additional model expands on the previous one with more variables. Models 2 and 4 are estimated with actual values for average temperature and precipitation, whereas Models 3 and 5 use the CPC forecasts.

Table 10.5 *Medium-Term Seasonal Models: Estimation Results*

Variables	*Model 1 trend only*	*Model 2 (1) + temp*	*Model 3 (1) + forecast temp*	*Model 4 (2) + prcp*	*Model 5 (3) + forecast prcp*
constant	3.3069‡	2.9269‡	2.6218‡	2.9167†	2.5983‡
	(0.0425)	(0.0292)	(0.0546)	(0.0285)	(0.0568)
trend	0.0065‡	0.0063‡	0.0060‡	0.0062†	0.0060‡
	(0.0006)	(0.0002)	(0.0003)	(0.0002)	(0.0003)
actual-temp		0.0196‡		0.0199†	
		(0.0012)		(0.0012)	
actual-prcp				0.2406	
				(0.1883)	
pred-temp-low			0.0124‡		0.0131‡
			(0.0024)		(0.0024)
pred-temp-up			0.0174†		0.0174‡
			(0.0014)		(0.0014)
pred-prcp-low					−0.9368
					(1.9529)
pred-prcp-up					0.9721
					(1.9520)
DF	129	128	127	127	125
R^2	0.8829	0.9237	0.9130	0.9229	0.9144
SBC	−196.90	−249.83	−228.02	−243.69	−220.41

Note: DF = degrees of freedom; SBC = Schwartz Bayesian criterion
⋆ significant at 0.05 level
† significant at 0.01 level
‡ significant at 0.001 level

By comparing Model 1 with Models 2 and 3, we can observe a large improvement in the explanatory power due to temperature alone. The improvement is supported by both R^2 and the SBC. For both models, the temperature variables are statistically significant and have the expected sign. Months with higher average temperatures tend to be followed by larger demand for electricity, again in accordance with the literature and with our own previous findings. Interestingly, it appears that the CPC forecasts, despite the long lead of one year, are able to predict the seasonal average with good accuracy. Not only are the coefficients fairly similar, but also the R^2 and SBC values are not distant from each other.

Actual and predicted values for precipitation are introduced in Models 4 and 5. This time, however, the variable did not appear to bring any improvement. Even in Model 4, in which the real monthly average for precipitation was used, the coefficient was not statistically significant at any conventional level. In Model 5, the forecast values for lower and upper terciles had diverging signs. Accordingly, by judging from the SBC scores, we would favor the previous models for their better suitability and parsimony.

However, as before, we make the caveat that the precipitation variables, as available for this study, may not be adequate for capturing the relationship with demand for electricity. To be sure, we have hypothesized that dry periods would pose higher demand for electricity for different reasons, including evaporative, energy-intensive treatment of water, and artificial irrigation (the water-energy nexus). Conversely, periods with more precipitation would lead to less use of electricity by naturally satisfying those needs.

Nevertheless, precipitation as a measure of condensation may not necessarily capture the effect of water scarcity in the region. If this is the case, a variable for drought intensity might prove to be more adequate. The relationship can also be countered by preemptive actions taken by the public. If the seasonal forecasts serve to prepare the government and individuals in general, we would not be able to observe the effect in the model. Because of these limitations, we cannot conclude from our estimations whether precipitation would play an important role in long-term planning.

In general, we have found support for the idea that climate forecasts are reliable enough to be used in medium-term modeling and planning. Predictions made by the CPC can accurately capture the movement in seasonal averages of temperature and, to a lesser extent, precipitation. With good assessments of how the next year's season will be, decisionmakers can have better chances of properly allocating necessary resources and take preemptive measures to handle energy crises. Without these techniques, managers have higher risk of being caught off-guard by unexpected variations in climate. With tighter regulations controlling the expansion of power plants, the room for error becomes narrow, calling for more sophisticated methodology. The simple models presented here can be put to use with little effort, as they simply require the utility to retrieve information that is publicly available on the CPC website. Because the service also provides historical information, going back many decades, more thorough validation can be done if more data on electricity load are available.

CONCLUSIONS

The Southwest faces rapid growth in its population and economy, while access to many natural resources is limited by physical and regulatory constraints. Electricity embodies the essence of the challenge this presents, because of the intimate connection power generation has with water. An additional source of pressure is created by climate change. Electric utilities have long known that usage of energy is strongly influenced by the climate of the locality served. However, with warmer temperatures and more variable precipitation, seasonal energy use patterns can change drastically. Of key significance is the possibility of increased occurrence of extreme conditions, such as heat waves and droughts, that coincide with periods when utilities are operating at their maximum capacity. Service disruptions due to energy crises, such as the ones that occurred in California in 2001, can follow from a lack of preparation for adverse situations.

To help electricity suppliers adapt to climate change, our study investigated the feasibility of developing statistical models that use weather and climatic information to provide planners with better input for decisions. Models for forecasting short- and medium-term electricity load have been created, leading to promising results. For both time horizons, we found that temperature is one of the most influential factors in explaining patterns of demand for electricity. Compared with a naïve univariate specification, next-day forecasts were improved by about 1.4%, measured by MAPE, when temperature was included in the model. Augmenting the model with humidity, precipitation, and wind speed reduced the rate of error by 0.1%. Despite being small percentage changes, such improvements can save millions of dollars annually. These results also add to research examining the economic value of weather and climate forecasts being freely provided by governmental agencies.

For the medium-term seasonal forecast model, availability of only a few years of load data limited the range of possible models that could be developed. However, even a simple autoregressive model was able to generate interesting findings for the study. Perhaps the most outstanding finding is the observation that a model using seasonal forecasts provided by the Climate Prediction Center (CPC) at the National Oceanic and Atmospheric Administration performed similarly to an alternative model using actual values for the average variables. This result should bring comfort to utility planners, who, according to a survey by Changnon et al. (1995), were uncertain about the reliability of seasonal forecasts. The medium-term model was not without its shortcomings, the most severe being the omission of demographic and economic variables, which are not collected at monthly intervals even though they have been found to play major roles in demand for energy in any region. This issue was handled by explicitly modeling trend in electricity consumption in the data.

Usage of electricity has been increasing steadily over the past years, and even more alarmingly, summer peak loads are increasing. Peak load reliability issues can be aggravated by the connection between water and energy, in which demand for one is closely associated with demand for the other. As water shortages occur, decisionmakers will face trade-offs between demand for water for consumption, irrigation, environmental needs, and power generation. The energy sector is responsible for a large share of carbon dioxide emissions in the United States, mostly as a result of combustion of fossil fuels used for electricity generation. The sector can be doubly affected by emissions, with climate change increasing demand for cooling and through carbon emission policies, such as a carbon tax. Preemptive planning can be improved by improving methods of assessing future scenarios such as those demonstrated in this chapter. We have proposed a method of handling multiple dimensions of seasonality that has not been investigated in the literature.

The research presented here can assist with climate change adaptation. Improved load forecasting will be useful in negotiating agreements with major electricity-using industries to decrease their economic activity and demand for electricity under extreme conditions. Water and power exchange agreements can be mutually advantageous, especially when locations and users have peak demand at different times of the year. Electric utility management decisions that can be improved using the techniques presented in this chapter include planning for new generating units, short-term energy purchases and sales, and peak load control. Many of these management decisions are affected by various aspects of climate change. This chapter also demonstrates, along with Chapters 9 and 11, the potential for more sophisticated use of information as a climate change adaptation strategy.

REFERENCES

AEPCO (Arizona Electric Power Cooperative). 2008. Electric Power Cooperative: About Us. www.azgt.coop/about-us/arizona-electric-power-cooperative (accessed September 22, 2010).

AMS (American Meteorological Society). 2008. Glossary of Meteorology. http://amsglossary.allenpress.com/glossary/2008 (accessed January 26, 2008).

Aradhyula, Satheesh V., and Tolga Ergün. 2004. Trading Collar, Intraday Periodicity and Stock Market Volatility. *Applied Financial Economics* 14:909–913.

Aradhyula, Satheesh, and Russell Tronstad. 2006. Estimating Seasonality in the Mean and Variance with ARCH Models. In *Essays in Honor of Stanley R. Johnson*, edited by Matt Holt and Jean-Paul Chavas. Berkeley, CA: Berkeley Electronic Press. Article 20. www.bepress.com/sjohnson/art20/ (accessed September 22, 2010).

Brocklebank, John C., and David A. Dickey. 2003. *SAS for Forecasting Time Series*. 2nd ed. Cary, NC: SAS Institute.

CEC (California Energy Commission). 2005. Integrated Energy Policy Report. November 2005. www.energy.ca.gov/2005publications/CEC-100-2005-007/CEC-100-2005-007-CMF.PDF (accessed October 11, 2010).

Changnon, Stanley A., Joyce M. Changnon, and David Changnon. 1995. Uses and Applications of Climate Forecasts for Power Utilities. *Bulletin of the American Meteorological Society* 76 (5):711–720.

Chao, Hung-po. 1983. Peak Load Pricing and Capacity Planning with Demand and Supply Uncertainty. *Bell Journal of Economics* 14 (1):179–190.

DOE (U.S. Department of Energy). 2006. Energy Demands on Water Resources, Report to Congress on the Interdependency of Energy and Water. www.sandia.gov/energy-water/docs/121-RptToCongress-EWwEIAcomments-FINAL.pdf (accessed October 10, 2010).

EPRI (Electric Power Research Institute). 2005. *Climate Change Impacts and the Electric Sector.* Palo Alto, CA: EPRI.

Feinberg, Eugene A., and Dora Genethliou. 2005. Load Forecasting. In *Applied Mathematics for Restructured Electric Power Systems*, edited by J. Chow, F. Wu, and J. Momoh. Berlin: Springer.

Gross, George, and Francisco D. Galiana. 1987. Short-Term Load Forecasting. *Proceedings of the IEEE* 75 (12):1558–1573.

Hagan, Martin T., and Suzanne M. Behr. 1987. The Time Series Approach to Short Term Load Forecasting. *IEEE Transactions on Power Systems* 2 (3):785–791.

Hobbs, B.F. 1997. *Identification and Quantification of Benefits of Using ANNSTLF.* Palo Alto, CA: Electric Power Research Institute.

Hobbs, Benjamin F., Suradet Jiprapaikulsarn, and Dominic J. Maratukulam. 1999. Analysis of the Value for Unit Commitment of Improved Load Forecasts. *IEEE Transactions on Power Systems* 14 (4):1342–1347.

Hylleberg, S. 1992. General Introduction. In *Modelling Seasonality*, edited by S. Hylleberg. Oxford: Oxford University Press, 3–13.

NOAA (National Oceanic and Atmospheric Administration). 2010. NOAA National Weather Service Climate Prediction Center. www.cpc.ncep.noaa.gov/ (accessed September 22, 2010).

Palmer, T.N., F.J. Doblas-Reyes, R. Hagedorn, and A. Weisheimer. 2005. Probabilistic Prediction of Climate Using Multi-model Ensembles: From Basics to Applications. *Philosophical Transactions of the Royal Society B* 360:1991–1998.

Papalexopoulos, Alex D., S. Hao, and T.M. Peng. 1994. An Implementation of a Neural Network Based Load Forecasting Model for the EMS. *IEEE Transactions on Power Systems* 9:1956–1962.

Papalexopoulos, Alex D., and Timothy C. Hesterberg. 1990. A Regression-Based Approach to Short-Term System Load Forecasting. *IEEE Transactions on Power Systems* 5 (4):1535–1547.

Peng, M., N.F. Hubele, and G.G. Karady. 1992. Advancement in the Application of Neural Networks for Short-Term Load Forecasting. *IEEE Transactions on Power Systems* 7:250–257.

Powicki, C. 2002. *The Water Imperative.* Technical Report Palo Alto, CA: Electric Power Research Institute.

Roberts, H.V. 1978. Comments on "The Analysis of Single and Related Time Series into Components: Proposals for Improving X-11." In *Seasonal Analysis of Economic Time Series*, edited by A. Zellner. Washington, DC: U.S. Department of Commerce, Bureau of the Census, 161–170.

Sims, Christopher A. 1992. Seasonality in Regression. In *Modelling Seasonality*, edited by Svend Hylleberg. Oxford: Oxford University Press, 61–81.

Tanimoto, Paulo Minoru. 2008. Forecasting the Impact of Climate Change for Electric Power Management in the Southwest. MS thesis, University of Arizona.

Teisberg, Thomas J., Rodney F. Weiher, and Alireza Khotanzad. 2005. The Economic Value of Temperature Forecasts in Electricity Generation. *Bulletin of the American Meteorological Society* 86 (12):1765–1771.

Thompson, Gary D., and Clifford A. Cathers. 2005. *Forecasting Short-Term Electricity Load Profiles.* Technical Report. Tucson: University of Arizona.

Weron, Rafal. 2006. *Modeling and Forecasting Electricity Loads and Prices: A Statistical Approach.* West Sussex, UK: Wiley.

Willis, H. Lee. 2002. *Spatial Electric Load Forecast.* 2nd ed. New York: Marcel Dekker.

CHAPTER 11

Use of Climate and Weather Information in Agricultural Decisionmaking

George B. Frisvold and Anand Murugesan

This chapter follows the approach of Dolan et al. (2001) and Meinke and Stone (2005) by considering use of weather data in agricultural decisionmaking as a critical part of climate change adaptation. Seasonal variability and extremes are as much a part of climate change as changes in long-run temperature or precipitation means (Parry and Carter 1985; Smit et al. 2000; Wandel and Smit 2000; Dolan et al. 2001). A large source of anticipated vulnerability to climate change in agriculture is difficulty in coping with weather extremes (Smit et al. 2001). Using weather information to anticipate and respond to shorter-run climate variability will be a crucial way for agricultural producers to adapt to long-run climate change.

The value of weather information is a discrete–continuous choice problem. First, the discrete choice is made whether to use any weather information at all. Once the choice to use data is made, a continuous aspect of the problem involves how much data to use and for how many decisions. The value of information depends on how much and how well information is used. Much of the literature on the value of weather information is concerned with estimating the continuous choice problem of how information is used and what the returns are to that use (Stewart 1997; Wilks 1997). However, as Freebairn and Zillman noted, studies have found that "a significant proportion of potential users do not use meteorological services in decision-making" (2002, 40).

A number of agricultural decisions could potentially benefit from timely weather information (Table 11.1). However, analysis of factors driving use of weather data is complicated by the fact that different data are useful for different management decisions. For example, frost forecasts may be critical to prevent crop damage, whereas data on wind speed and direction would be more important to control pesticide drift.

Table 11.1 *Applications of Weather Data to Agricultural Decisions*

	Agricultural decisions
Temperature	Planting, harvesting, defoliation, crop modeling, disease risk, animal sheltering, pest control, sheep shearing
Precipitation	Planting, harvesting, fertilizer applications, cultivation, spraying, irrigation, disease risk, livestock and poultry protection
Soil moisture	Planting, harvesting, fertilizing, transplants, spraying, irrigation, monitoring of growing conditions, measuring plant stress
Soil temperature	Planting, pest overwintering conditions, transplanting, fertilizing
Frost	Pest overwintering conditions, protecting crops from damage, animal sheltering, irrigation (to avert crop damage)
Degree days	Planting, irrigation, pest control
Relative humidity	Harvesting, pollination, spraying, drying conditions, crop stress potential
Wind speed	Defoliation, harvesting, freeze potential/protection, animal sheltering, shelter, pest control, pruning, spraying or dusting, pollination, dust drift, pesticide drift
Wind direction	Freeze potential/protection, cold or warm air advection over crop areas, pesticide drift, dust drift

Source: Adapted from Pasteris et al. (2004)

This chapter uses data from a survey of Arizona farmers and ranchers to focus on the initial discrete choice of weather data use. In it, we introduce the discrete choice framework, illustrating the relationship between information valuation and use; discuss the survey instrument and weather use data; present the econometric specifications and explanatory variables used in regression analyses; present and discuss our results; and summarize our main findings and suggest areas of future research.

EMPIRICAL SPECIFICATIONS

Here we introduce a discrete choice framework that treats acquisition and use of weather data as part of a farm household's overall utility function. Information use is a choice variable (Stigler 1961), and its acquisition involves cost (Feder and Slade 1984). Using weather data is costly in terms of household time (to acquire and process information) and potentially monetary expense (e.g., for a subscription to information services). For simplicity, we consider one type of weather information. The concepts readily extend to multiple types of information.

Utility when a producer, i, does not use weather information, is

$$U_{i0} = \gamma(\boldsymbol{z}_{i0}, \boldsymbol{x}_i) + \varepsilon(\boldsymbol{z}_{i0}, \boldsymbol{x}, \boldsymbol{e}_{i0}) \tag{11.1}$$

where γ and ε are real valued functions, $\boldsymbol{z}_{i0}$ is a vector of attributes associated with not using information, and $\boldsymbol{x}_i$ is a vector of socioeconomic characteristics of the producer and the operation. The vector $\boldsymbol{e}_{i0}$ represents unmeasured attributes of the producer or the producer's operation. Utility when a producer uses weather information is

$$U_{i1} = \gamma(\boldsymbol{z}_{i1}, \boldsymbol{x}_i) + \varepsilon(\boldsymbol{z}_{i1}, \boldsymbol{x}, \boldsymbol{e}_{i1}) \tag{11.2}$$

where $\boldsymbol{z}_{i1}$ is a vector of attributes associated with using weather information and $\boldsymbol{e}_{i1}$ is a vector of unmeasured attributes of the producer, operation, and weather data. If the population is drawn from a random sample with common socioeconomic characteristics, the vectors $\boldsymbol{e}_{i0}$ and $\boldsymbol{e}_{i1}$ will be random, and the utility function value will be stochastic (Domencich and McFadden 1975). We therefore assume that the random components can be expressed as $\varepsilon(\boldsymbol{e}_{i0})$ and $\varepsilon(\boldsymbol{e}_{i1})$.

The value of weather information, $V^\star$, can be expressed as

$$V^\star = y(\boldsymbol{z}_{i1}, \boldsymbol{x}_i) + \varepsilon(\boldsymbol{e}_{i1}) - y(\boldsymbol{z}_{i0}, \boldsymbol{x}_i) + \varepsilon(\boldsymbol{e}_{i0}) \tag{11.3}$$

A utility-maximizing producer will use weather information if there is a net gain in utility of doing so. In other words, the producer will use information if the value of doing so is positive:

$$V^\star = [y(\boldsymbol{z}_{i1}, \boldsymbol{x}_i) - y(\boldsymbol{z}_{i0}, \boldsymbol{x}_i)] - [\varepsilon(\boldsymbol{e}_{i1}) - \varepsilon(\boldsymbol{e}_{i0})] > 0 \tag{11.4}$$

The $\boldsymbol{x}_i$ vector represents observed operator and operation characteristics. It could also represent relative reliance on farm and nonfarm income. If farm income were a small share of farm household income, then weather-related farm risk would be less important to the household. In contrast, as farm income makes up a greater share of total household income, weather-related agricultural risk grows in importance. The $\boldsymbol{z}_{i0}$ and $\boldsymbol{z}_{i1}$ represent factors specific to the decisions of using or not using weather information. These factors would include expected profits under each scenario, variance of profits, the monetary cost of acquiring and processing weather information, and the cost in producer time to acquire and process information.

The value of information, $V^\star$, is not observed. However, we observe the discrete choice, V (weather information used or not used). If $\varepsilon(\boldsymbol{e}_{i1})$ and $\varepsilon(\boldsymbol{e}_{i1})$ are each normally distributed, then their difference, $u = \varepsilon(\boldsymbol{e}_{i1}) - \varepsilon(\boldsymbol{e}_{i0})$, is also normally distributed. If we assume that $[y(\boldsymbol{z}_{i1}, \boldsymbol{x}_i) - y(\boldsymbol{z}_{i0}, \boldsymbol{x}_i)]$ can be written as a linear function of variables, then

$$V^\star = \alpha' \boldsymbol{x}_i + \beta' \boldsymbol{z}_{i0} + \gamma' \boldsymbol{z}_{i1} + u \tag{11.5}$$

where α, β, and γ are regression coefficients, and $V = 1$ if $V^\star > 0$ and $V = 0$ otherwise.

Assuming u is normally distributed, this is the standard form of a probit regression. Much of the literature on the value of weather information focuses on estimating value of information to users. Our binary choice model produces negative economic information of sorts. By distinguishing between users and nonusers of data, it distinguishes between agricultural producers that expect a net benefit from using weather information and those that do not. While we cannot obtain estimates of the value of $V^\star$ for producers where $V = 1$, we can infer that $V^\star \leq 0$ for nonusers. In other words, we can identify characteristics of producers who do not derive expected benefits from weather data.

A growing number of practitioners are using contingent valuation (CV) surveys to directly elicit user willingness to pay (WTP) for meteorological information and services (Anaman and Lellyett 1996, 2007; Chapman 1992; Rollins and Shaykewich 2003; Stratus Consulting 2002; Teske and Robinson 1994). Freebairn and Zillman (2002) noted that many respondents in these CV studies report a zero WTP and suggested that this is because of the high rate of nonuse of weather data and services in general. WTP data will be right-censored with a probability spike at zero. Probit analysis therefore can identify some of the factors that account for zero-value responses.

Just et al. (2002, 2003) focused on the importance of three types of heterogeneity in considering demand for economic information in agriculture: heterogeneity with respect to decisionmaker attributes, information sources (providers), and the nature of the information itself. Just et al. (2002, 2003) stressed the difference between raw, unprocessed data and information that intermediaries translate for end users. Jones et al. (1989) pointed out the differences between general and specific information as well as between public and private sources. These forms of heterogeneity are important in studying agricultural demand for weather information as well. In principle, the econometric specification can deal with all these types of heterogeneity.

The cost of acquiring and processing information, as with any cost function, depends on technology. Many providers of climate services have stepped up development of web-based information and decision tools (Pasteris et al. 2004; Jensen et al. 2000; Breuer et al. 2008). But what about farmers and ranchers who do not use computers or have Internet access? According to the most recent USDA Farm Computer Usage and Ownership survey (USDA NASS 2009b), 59% of farms had Internet access (implying that 41% did not), but only 36% of farms used computers for business. Among farms with $100,000 or more in annual sales, 70% had Internet access, and 61% used computers for business. These figures suggest two things. First, a "digital divide" may be opening up between large and small farms. Second, even among larger-scale producers, a significant share is not yet in a position to use web-based meteorological services directly. One question explored in this chapter is whether Internet access has had important effects on farmer and rancher use of weather data.

THE SURVEY AND WEATHER DATA

Data come from Arizona responses to the National Agricultural, Food, and Public Policy Preference Survey (Lubben et al. 2006). USDA's National Agricultural Statistics Service (NASS) administered the survey from late 2005 to early 2006 in 27 states, with support from the Farm Foundation. The survey was a random sample, stratified by farm income. It included questions common to each participating state, optional questions asked in a subset of states, and a set of state-specific questions. Cooperative extension specialists in each state, in conjunction with NASS, developed state-specific questions. Most questions concerned agricultural producer attitudes about current or potential farm legislation. Other questions were about farm program participation, personal characteristics (e.g., age, education), and operation characteristics (e.g., total sales, commodity specialization).

The Arizona survey included two questions related to weather data (Table 11.2). The first asked producers to indicate the importance of different weather data for their production or marketing decisions. Producers could respond on a 5-point Likert scale, where 1 = least important, 2 = less important, 3 = neutral, 4 = important, and 5 = most important. Producers could also choose "don't know/no opinion," although this choice was exceedingly rare. The second question asked producers whether they used weather information to make different production or marketing decisions, such as timing irrigation, moving livestock, or timing harvest.

The Arizona sample of the survey had more than 300 valid responses; here we present results of a subsample of 284 producers with complete responses for weather variables and variables subsequently used in multivariate regressions. Reported producer attributes and responses to weather data questions did not differ significantly between the full sample and this subsample. Comparing both full and subsample responses with Arizona data from the 2002 Census of Agriculture, we found that the policy preference survey modestly oversamples producers with greater agricultural sales.

Figure 11.1 shows the percentage of respondents that indicated a particular type of weather data was either "important" or "most important." For the rest of the chapter, we will use the word "important" to include both categories. Most producers found data on soil moisture (67%), temperature (64%), and frost (56%) important, while relatively few found wind speed (30%) or wind direction (12%) important (Figure 11.1).

Table 11.2 *Weather Survey Questions from the Arizona Sample of the 2005–2006 National Agricultural, Food, and Public Policy Preference Survey*

Please indicate how important the following weather data are for your production and marketing decisions (1 = least important [LI], 2 = less important, 3 = neutral, 4 = important, 5 = most important [MI], X = don't know/no opinion [DK])

	LI				MI	DK
	1	2	3	4	5	X
a. Temperature	1	2	3	4	5	X
b. Precipitation	1	2	3	4	5	X
c. Wind speed	1	2	3	4	5	X
d. Wind direction	1	2	3	4	5	X
e. Soil moisture	1	2	3	4	5	X
f. Soil temperature	1	2	3	4	5	X
g. Frost/freeze conditions	1	2	3	4	5	X
h. Degree days	1	2	3	4	5	X
i. Relative humidity	1	2	3	4	5	X

Do you use weather information (such as temperature, precipitation, humidity, degree days, wind speed, frost/freeze conditions) for any of the following production or marketing decisions? (Check all that apply.)

Crop choice	☐	Crop variety choice	☐
Timing planting(s)	☐	Timing cultivation	☐
Timing pesticides	☐	Timing irrigation	☐
Timing harvest	☐	Crop insurance	☐
Crop/livestock sales	☐	Crop storage	☐
Hedging	☐	Livestock grazing	☐
Moving livestock	☐	Livestock pest control	☐

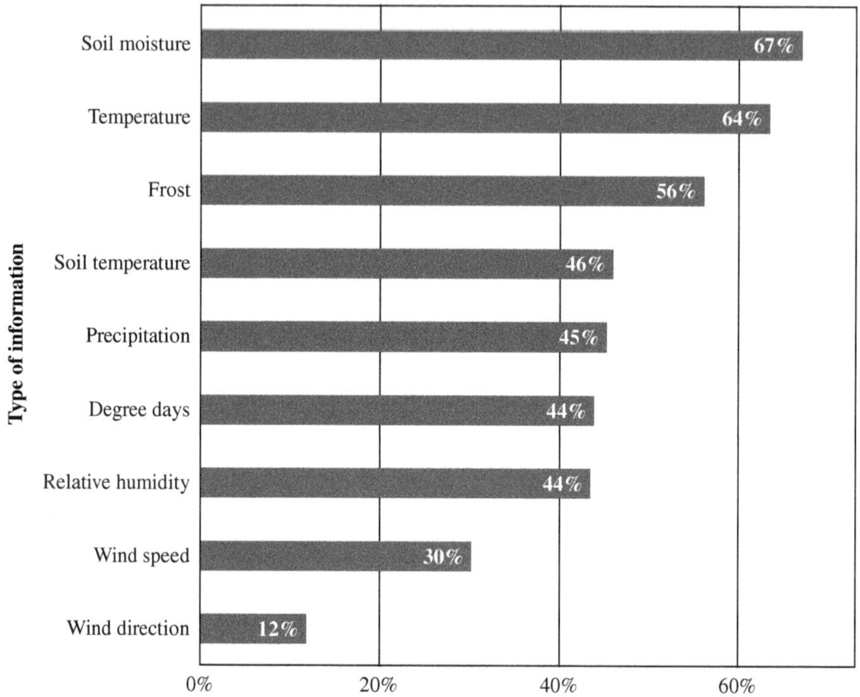

Figure 11.1 *Agricultural Producer Importance Ratings of Weather Data for Production and Marketing Decisions*

Source: Lubben et al. 2006

Figure 11.2 compares assessments of weather data importance separately for three producer groups:

- 105 crop producers, who have crop sales but no reported livestock sales (37% of sample);
- 141 livestock producers, with livestock sales but no reported crop sales (50% of sample); and
- 38 diversified producers, who have both crop and livestock sales (13% of sample).

Diversified producers were more likely to report a given type of data important than were crop or livestock producers. Except for soil moisture, livestock producers were least likely to report a type of weather data important. Here still, diversified producers were most likely to report soil moisture important. Less than half (45%) of all producers found precipitation data important (see Figure 11.1). However, this data was more important for crop producers (60%) and diversified producers (55%) than for livestock producers (32%).

Use of weather data for management decision varied significantly by producer group (Figure 11.3). Largely, this is because of the nature of the data-use questions.

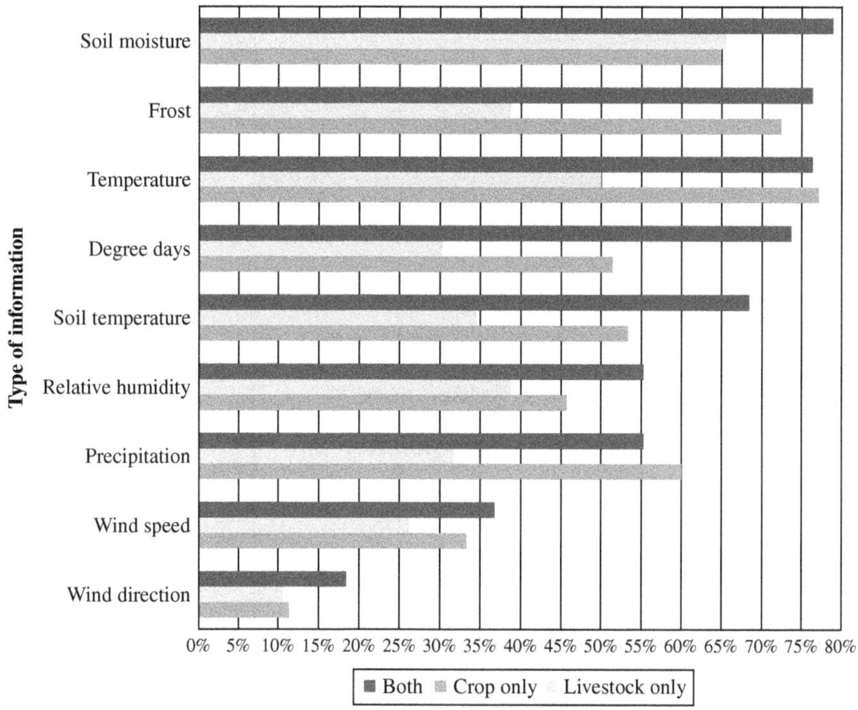

Figure 11.2 *Importance Rating of Weather Data by Producer Type*

Source: Lubben et al. 2006

Nine of the decisions (crop choice, crop variety choice, timing planting, timing cultivation, timing pest control, irrigation timing, timing harvest, crop storage, and crop insurance) relate almost exclusively to crop production, whereas three (livestock grazing, moving livestock, and livestock pest control) are exclusive to livestock production. Crop/livestock sales in principle could apply to any of the three groups, although few crop producers reported using weather data for this decision. Hedging could also apply to either crop or livestock sales, but few producers of any type used weather data to inform this decision. Thus we might expect that diversified producers use weather data for crop-specific decisions in much the same way as crop producers, and that they use data for livestock-specific decisions in much the same way as livestock producers. For the most part, Figure 11.3 bears this out. Notably, however, diversified producers have slightly higher rates of using weather data for timing harvest, timing cultivation, timing planting, crop choice crop variety choice (which also may have a timing element in terms of length of season or planting data), and crop storage.

Figure 11.3 also shows that some crop producers report weather data for some livestock-specific decisions, and some livestock producers use weather data for crop-specific decisions. What might account for this? First, livestock producers could have irrigated pasture, and hence use weather data to time irrigation applications. Second, livestock producers may grow forage crops, not for sale, but

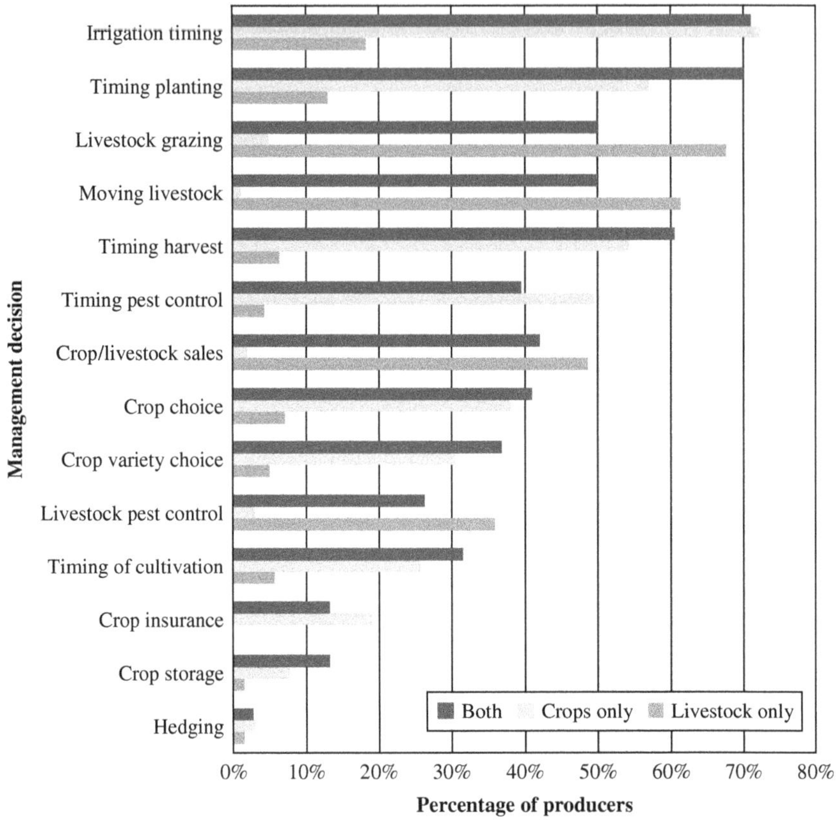

Figure 11.3 *Use of Weather and Climate Data by Producer Type*

Source: Lubben et al. 2006

for use by their animals. Third, some crop producers may have livestock but no livestock sales in the survey year. For example, according to the 2007 Census of Agriculture (USDA NASS 2009a), Arizona had 7,716 operations with cattle and calf inventories, but only 4,078 with cattle and calf sales.

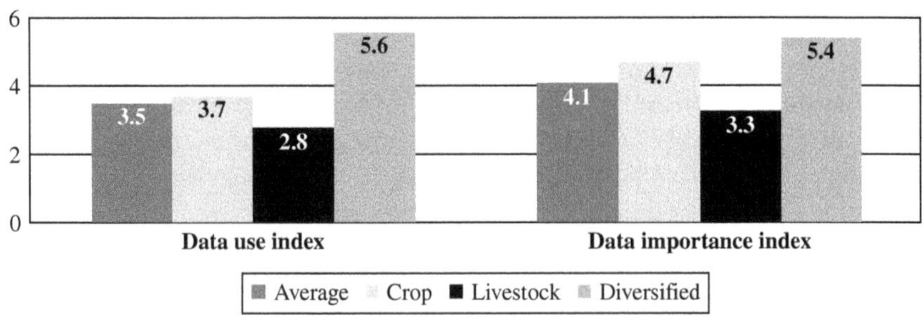

Figure 11.4 *Indices of Weather Data Use and Importance by Producer Type*

Source: Lubben et al. 2006

Two summary indices were created for the weather data. The first is a weather data use index. This is the total number of agricultural decisions (from Table 11.1) for which a producer used weather data. The index could take on an integer value from 0, 1, 2, ..., 14. Similarly, a weather data importance index is the total number of weather data types a producer found important or most important (also from Table 11.1). This could take on integer values 0, 1, 2, ..., 9. The average value of the data use index was 3.5, while the average value for data importance was 4.1 (Figure 11.4). As in Figures 11.2 and 11.3, these indices vary by producer group. Diversified producers use data for more decisions. This is not surprising because, by definition, they have more decisions to make. Diversified producers, however, also use more total types of weather data. In addition, as Figure 11.3 illustrated, diversified producers tended to use weather data for crop-specific decisions more than crop producers did.

ECONOMETRIC MODELS

We conducted multivariate regression analysis on the indices for weather data use and importance. Our interest was to see what factors explain differences in the total number of data types found important and the total number of agricultural decisions for which producers used weather data. We also estimated nine separate probit multivariate regressions for each individual data type. Here the dependent variables equaled 1 if a producer rated a data type "important" or "most important" and 0 otherwise. We ran separate probit regressions for data use variables. Here the dependent variables equaled 1 if a producer used weather data for a particular agricultural management decision and 0 otherwise.

The indices for weather data use and importance are count data. The data use index is a count of the total number of data types a producer deems "important" or "most important." The importance index is the total number of decisions for which producers used weather data. In both cases, the dependent variables (the counts) are integers. Because these variables are discrete integers rather than continuous variables, standard least squares regression is not appropriate. Estimating count data with least squares is problematic for several reasons (Cameron and Trivedi 1998; Gardner et al. 1995). Expected values from the regressions will generally not match observed integer counts and are not even restricted to be non-negative. Regression coefficients, then, do not have any clear interpretation, in terms of how a change in an explanatory variable would alter the predicted integer value. The underlying distribution of the variable can deviate substantially from normality (a crucial assumption in least squares regression). Statistical properties of the variances of the coefficients, and thus standard hypothesis test results, may also be suspect.

Poisson regressions explicitly account for integer-dependent variables. A Poisson regression, however, assumes that the mean and variance of the dependent variable are equal. This assumption can lead to either overestimating the statistical significance of regression parameter estimates when there is overdispersion (variance greater than the mean) or underestimating their statistical significance when there is underdispersion (variance less than the mean) (Cameron and Trivedi 1998). We instead estimate a negative binomial regression that does not restrict the variance of the dependent variable. The negative binomial is more general but includes the Poisson specification as a special case (Cameron and Trivedi 1998; Gardner et al. 1995).

Regression Explanatory Variables

The following explanatory variables were included in the regression. First, there were six (0/1) farm sales class variables. The default (omitted) category was operations below $10,000. The highest sales class included operations with sales of $1 million or more. Next, three dummy variables indicated producer age (45–54, 55–64, 65 and over), and two accounted for education (college degree, advanced degree). Dummy variables also indicated whether a producer was male, Hispanic, had e-mail access, had a satellite television, or owned 50% or more of operated farmland. Three variables also distinguished producers by farm income as a share of household income (26–50%, 51–75%, 76–100%).

As part of the policy preference portion of the survey, producers were asked to rate the importance of maintaining funding for various USDA programs. Another dummy variable equaled 1 if producers rated maintenance of government risk management programs (crop and livestock insurance programs) as "important" or "most important."

To measure farm diversification, dummy variables were included distinguishing crop producers (crop sales only) and diversified producers (crop and livestock sales) from the default, livestock producers (livestock sales only). We also experimented with measures of farm diversification discussed by Pope and Prescott (1980). The one that proved significant in regressions was the total number of crop and livestock categories for which the producer reported sales. The survey asked producers to allocate sales shares between 12 crop and 7 livestock categories. This diversification variable thus accounts for diversification within crop and livestock production categories.

RESULTS AND DISCUSSION

Weather Importance Index

Table 11.3 reports results of the negative binomial regressions indices of weather data use and importance. Controlling for other factors, diversified and crop producers tended to find more types of weather data important. Number of commodities sold did not have a significant effect, however. Producers in the three oldest age categories rated more data types important. Rating risk management programs important and having a satellite TV were positively associated with rating more types of data important, whereas having e-mail access had a negative but insignificant effect. Owning a greater share of one's farmland had a negative effect. None of the effects of sales class, education, or farm income as a share of total income were significant.

Table 11.3 *Negative Binomial Regressions for Weather Data Importance and Weather Data Use Indices*

	Weather Data Importance Index[a]		Weather Data Use Index[b]	
	Coefficient	*|t ratio|*	*Coefficient*	*|t ratio|*
Diversified producer (0/1)	0.298†	2.04	0.560‡	4.52
Crop producer (0/1)	0.339‡	3.40	0.305‡	3.15
Number of commodities sold	0.0856	1.53	0.137‡	2.61
Ag sales ($10,000–$49,999)	0.0789	0.63	−0.119	0.98
Ag sales ($50,000–$99,999)	−0.0354	0.22	−0.118	0.75
Ag sales ($100,000–$249,999)	0.0911	0.59	−0.0837	0.58
Ag sales ($250,000–$499,999)	−0.0151	0.08	−0.462†	2.29
Ag sales ($500,000–$999,999)	0.123	0.57	−0.313	1.56
Ag sales ($1,000,000 and over)	0.151	0.80	−0.399†	2.20
Age (45–54)	0.336†	2.20	0.281†	2.01
Age (55–64)	0.279*	1.79	0.0738	0.51
Age (65 and over)	0.432‡	2.84	0.0873	0.61
Male (0/1)	−0.00679	0.05	−0.037	0.29
Bachelor's degree (0/1)	0.00123	0.01	−0.00847	0.08
Advanced degree (0/1)	0.0225	0.17	0.175	1.37
Farm income/total income (26–50%)	0.146	1.13	0.357‡	2.86
Farm income/total income (51–75%)	−0.00088	0.01	0.351†	2.46
Farm income/total income (76–100%)	0.149	1.06	0.471‡	3.48
Hispanic (0/1)	0.133	0.89	0.245*	1.72
Ag land owned > 50% (0/1)	−0.279‡	2.66	−0.239†	2.39
E-mail access (0/1)	−0.0509	0.57	0.0942	1.07
Satellite TV (0/1)	0.173†	2.10	0.181†	2.27
Risk management important (0/1)	0.145*	1.76	0.135*	1.71
Constant	0.705‡	3.11	1.044‡	3.36
ln (α)	−1.853‡	6.80	−2.455‡	6.25
Observations	284		284	

[a]The dependent variable is an integer count from 0 to 9 of the total number of weather data types a producer rated important or most important for production and marketing decisions

[b]The dependent variable is an integer count from 0 to 14 of the total number production and marketing decisions for which a producer used weather data

*significant at 10% level

†significant at 5% level

‡significant at 1% level

Weather Data Use Index

Both crop and diversified producers used data for more decisions than did livestock producers (Table 11.3). Weather data use also increased with the total number of commodities sold. Considering producer personal attributes, being male had no significant impact on data use, nor did having a bachelor's or advanced degree. Being Hispanic, however, was positively associated with data use, as was being in the 45–54 age group. Producers who felt risk management programs were important used data for more decisions.

Having e-mail access did not have a significant impact on data use, but having a satellite TV had a positive, significant effect. Producers with greater shares of their household income coming from farming used data for more decisions. This is consistent with previous research suggesting that off-farm work discourages use of management information and technology adoption. Somewhat surprisingly, there was a negative association between both farm sales and percent of farmland owned and data use. Coefficients for the highest and third-highest sales class (out of seven classes) were negative and significant.

Importance Rating of Individual Types of Data

Table 11.4 shows the importance ratings for individual types of data, including coefficients and significance levels only for variables that were at least significant at the 10% level in the probit regressions. Livestock producers were less likely to find temperature, precipitation, soil temperature, frost, and degree days important than were crop producers (the default in these regressions). There was no statistically significant difference between crop and diversified producers. Producers in the highest sales class were more likely to find soil temperature and degree days important. Older producers tended to find more types of data important. Greater farmland ownership had a negative impact on importance ratings for a number of data types, whereas satellite ownership and placing importance on risk management programs had positive effects. E-mail access had no significant effect on ratings of any of the individual data types.

Weather Data Use by Producer Type

The regressions in Table 11.5 cover agricultural decisions almost exclusively dealing with crop production. Data for these probit regressions included only diversified and crop producers, excluding producers with only livestock sales. The number of commodities sold increased the probability of using weather data for timing planting, timing harvest, and timing cultivation. When coefficients of sales class variables were significant, they tended to be negative and for the higher sales classes. Although greater age was associated with placing importance on more types of weather data, older producers were more likely to use weather data only for crop choice. The greater importance older producers placed on data did not translate into using data for more decisions. Use of data for several individual decisions increased with the share of household income devoted to farming. Again, this is consistent with findings that greater off-farm work can discourage information use and innovation. As with the importance regressions, when landownership was significant, it had a negative impact, whereas the importance rating of risk management programs and satellite TV ownership had positive impacts. E-mail access had no significant effect on use of weather data for individual decisions.

Regressions in Table 11.6 included only livestock-specific decisions and livestock and diversified producers. Very few coefficients were significant.

Table 11.4 *Results from Nine Probit Regressions of Weather Data Importance Variables*

	Temperature	*Precipitation*	*Wind speed*	*Wind direction*	*Soil moisture*	*Soil temperature*	*Frost*	*Degree days*	*Relative humidity*
Diversified producer (0/1)									
Livestock producer (0/1)	−0.813‡	−0.604‡				−0.445†	−0.980‡	−0.479†	
Number of commodities sold			0.300†						
Ag sales ($10,000–$49,999)									
Ag sales ($50,000–$99,999)									−0.599*
Ag sales ($100,000–$249,999)									
Ag sales ($250,000–$499,999)									
Ag sales ($500,000–$999,999)									
Ag sales ($1,000,000 and over)						0.773*		0.837†	
Age (45–54)			0.571*	7.192‡				0.470*	0.565*
Age (55–64)				6.662‡					0.604†
Age (65 and over)		0.610†		7.433‡	0.702†	0.671†	0.618†		0.603†
Male (0/1)									
Bachelor's degree (0/1)									
Advanced degree (0/1)					0.795‡				
Farm income/total income (26–50%)									
Farm income/total income (51–75%)									
Farm income/total income (76–100%)									0.519*
Hispanic (0/1)							0.563*		
Ag land owned > 50% (0/1)	−0.437*			−0.992‡	−0.415*		−0.443*		−0.471†
E-mail access (0/1)									
Satellite TV (0/1)	0.495‡			0.440*	0.321*			0.405†	
Risk management important (0/1)			0.333*	0.540†				0.525‡	0.299*

Note: Dependent variables = 1 if a producer rated the data "important" or "most important" for production and marketing decisions and = 0 otherwise. Only statistically significant coefficients are reported. All listed explanatory variables were included in each of the nine regression equations

*significant at 10% level

†significant at 5% level

‡significant at 1% level

Table 11.5 *Weather Data Use Probit Regressions: Crop and Diversified Producers Only*

	Crop choice	*Timing planting*	*Timing pesticides*	*Timing harvest*	*Timing cultivation*	*Variety choice*	*Timing irrigation*	*Crop storage*
Crop producer (0/1)					0.675*			
Number of commodities sold		0.320*		0.328†	0.378†			
Ag sales ($10,000–$49,999)					−1.180*	−0.900*		
Ag sales ($50,000–$99,999)					−1.297*			
Ag sales ($100,000–$249,999)					−1.519†			−1.714†
Ag sales ($250,000–$499,999)					−1.489†	−1.741†	−0.918*	
Ag sales ($500,000–$999,999)					−2.540‡			−1.961*
Ag sales ($1,000,000 and over)	−1.044*				−2.330‡			−2.140†
Age (45–54)	0.902†							
Age (55–64)	1.150†							
Age (65 and over)	0.782*							
Male (0/1)								
Bachelor's degree (0/1)								
Advanced degree (0/1)								
Farm income/total income (26–50%)		0.913†			1.849‡	0.800*		1.522†
Farm income/total income (51–75%)	0.879*			0.986*	1.536†	0.994*		1.980†
Farm income/total income (76–100%)	0.904†	0.803*		0.972†	2.258‡			1.821†
Hispanic (0/1)								
Ag land owned > 50% (0/1)			−0.840†	−0.744†				−1.194†
E-mail access (0/1)								
Satellite TV (0/1)					0.548*			
Risk management important (0/1)			0.497*		1.088‡			

Note: Dependent variable = 1 if producer used weather data for the listed production and marketing decision; = 0 otherwise. Producers with no crop sales were excluded from these regressions; 143 observations included. Only statistically significant coefficients are reported. All listed explanatory variables were included in each of the nine regression equations

*significant at 10% level

†significant at 5% level

‡significant at 1% level

Table 11.6 *Weather Data Use Probit Regressions: Livestock and Diversified Producers Only*

	Move livestock	*Livestock grazing*	*Livestock pest control*
Livestock producer (0/1)			
Number of commodities sold			
Ag sales (\$10,000–\$49,999)			
Ag sales (\$50,000–\$99,999)	0.742*		
Ag sales (\$100,000–\$249,999)			
Ag sales (\$250,000–\$499,999)			
Ag sales (\$500,000–\$999,999)			
Ag sales (\$1,000,000 and over)			−1.828†
Age (45–54)			
Age (55–64)			
Age (65 and over)			
Livestock producer (0/1)			
Male (0/1)			
Bachelor's degree (0/1)			
Advanced degree (0/1)			
Farm income/total income (26–50%)			
Farm income/total income (51–75%)			
Farm income/total income (76–100%)	0.731*		
Hispanic (0/1)		1.007†	
Ag land owned > 50% (0/1)	−0.766†		−0.700†
E-mail access (0/1)			
Satellite TV (0/1)		0.379*	
Risk management important (0/1)		0.564†	

Note: Dependent variable = 1 if producer used weather data for the listed production and marketing decision; = 0 otherwise. Producers with no livestock sales were excluded from these regressions; 179 observations included. Only statistically significant coefficients are reported. All listed explanatory variables were included in each of the nine regression equations

*significant at 10% level

†significant at 5% level

‡significant at 1% level

However, those that were had the same impacts on data use as in the crop producer and weather data importance probit regressions.

Discussion of Results

The types of weather data farmers and ranchers most frequently found important were data for soil moisture, frost, and temperature. Producers used weather data most frequently for timing irrigation, followed by livestock management. More specifically, ranchers used weather data most frequently for decisions about moving livestock. Next in frequency, producers used weather data for timing cultivation. There were no other decisions for which more than half of respondents used weather data. The importance of weather data for irrigation timing and frost protection match results of Parker and Zilberman's (1996) study of the California Irrigation Management Information System (CIMIS). While they found that irrigation scheduling was, by far, the primary use of CIMIS information, frost protection was also important.

Consistent with earlier findings, agricultural producer use of weather data varies significantly with commodity specialization and degree of diversification. Consistent with McNew et al.'s (1991) findings from Oklahoma, crop-only and diversified producers relied on more types of data than did livestock producers. In a study of Ohio farms, Jones et al. (1989) reported that dairy-only producers were less likely to rate sources of agricultural information useful than were grain-only or diversified producers. McNew et al. (1991) also found that for decisions involving timing decisions, diversified producers rated weather data more important. In our weather data – use regressions, the total number of commodities sold (a measure of diversification) increased the probability of using weather data for timing planting, harvest, and cultivation. Diversification can reduce risk and possibly decrease demand for risk-reducing weather information. However, diversification increases complexity and the need for optimal timing, which can increase demand for information. Our results, and those of McNew et al. (1991), suggest that this complexity/timing effect may be dominant.

The direct effects of greater farm sales on weather data used appeared to be negative in several cases. Regression coefficients for many of the large farm sales categories were negative. This runs counter to theories and past evidence suggesting positive associations between farm size and information/new technology use (e.g., Artikov et al. 2006; Letson et al. 2001).

Access to e-mail had no significant, positive effect on the importance producers placed on weather data types or on use of weather data for management decisions. In contrast, access to satellite television had positive effects in several cases. Among survey respondents, 64% of crop producers, 68% of diversified producers, and 59% of livestock producers had e-mail access. However, 38% of all survey respondents did *not* have e-mail access. This suggests that platforms to encourage use of weather information and climate services via web-based tools will not directly reach large segments of Arizona's farmers and ranchers.

One way to look at the results is that lack of Internet access (as measured by e-mail access) does not appear to be a barrier to weather data use. Those without access do not appear to rate weather data as less important or to use data less for decisionmaking. One reason might be that producers may be receiving weather data indirectly. For example, irrigation district staff or extension agents may repackage and interpret weather data and information and provide it to producers via print, workshops or other nonelectronic media. Thus, although producers do not use web-based data *directly*, this data may reach them indirectly through intermediaries.

Another possible explanation is, however, that producers do not rely on Internet sources of weather data to make production and marketing decisions. McNew et al. (1991) found that television was the primary source of weather information for most producers, ranging from 63% to 77% of users, depending on the type of information. Carlson (1989) also found television to be a main source of weather information for agricultural decisions and cited earlier studies with similar findings. These surveys predate the rise of the Internet. Yet it is notable that in the Arizona survey, access to satellite television appears to have a more important positive influence on weather data use than Internet access.

Turning to farmer personal attributes, having a college or advanced degree did not significantly increase weather data use, although having an advanced degree increased a producer's rating of the importance of soil moisture data. Producers rating government risk management programs as important also found more weather data types important and used weather data for more decisions. This is consistent with arguments that risk aversion is an important consideration in use of data for agricultural decisions (Bosch and Eidman 1987; Byerlee and Anderson 1982; Feder and Slade 1984; Parker and Zilberman 1996). Women and Hispanic producers were not less likely to find weather data important or to use the data for fewer decisions. In fact, for some decisions, Hispanics were more likely to use weather data.

CONCLUSIONS

This chapter made use of questions concerning weather data included in the Arizona sample of the National Agricultural, Food, and Public Policy Preference Survey to gain some understanding of weather data use in agricultural decisionmaking. We used survey responses from 284 Arizona farmers and ranchers to examine the importance producers placed on different types of weather data for production and marketing decisions; identify which producer characteristics accounted for differences in the importance they placed on weather data; examine producer use of weather data for specific production and marketing decisions; and identify which factors distinguish weather data users from nonusers.

Commodity specialization and household income diversification affect use of information. Livestock producers found fewer data types important. Among crop and diversified producers, those producing more, different commodities used weather data for more decisions. Specifically, they were more likely to use data for timing three activities: planting, cultivation, and harvest. Agricultural diversification can reduce risk and therefore might discourage data use. However, diversification also increases complexity and the value of optimal timing. Our results suggest that these latter effects dominate, increasing demand for weather data. Easterling (1986) found that the timeliness of weather data was the most important factor for discriminating between weather data users and nonusers.

Nonfarm income, in contrast, discouraged weather data use. Use of weather data increased with the share of household income derived from agriculture. This could result from two effects. First, agricultural income risk is a larger share of household income risk for households engaged primarily in agricultural production. Thus returns from using data to reduce agricultural risk are greater. And second, producers spending more time in off-farm employment may have less time to process data. Empirical studies of other agricultural decisions have found that off-farm work can discourage information acquisition or use of time-intensive management techniques (McNamara et al. 1991; Wozniack 1993; Fernandez-Cornejo et al. 1994; Fernandez-Cornejo 1996, 1998, 2007; Fernandez-Cornejo and Jans 1996).

Farms with greater agricultural sales tended to use weather data for fewer decisions, holding all else constant. However, agriculture's share of household income increases with sales volume. As farm sales increase, then, the two effects seem to work in opposite directions. The degree of diversification can also vary with farm sales class. This suggests that further research is needed to sort out the relationships among farm scale, diversification, and off-farm income effects. Another puzzling yet consistent result was the negative relationship between share of farmland owned and weather data importance and use.

Producers rating risk management programs important also found more weather data types important and used weather data for more decisions. This is consistent with arguments that risk aversion is an important consideration in use of data for agricultural decisions (Johnson and Holt 1997; Bosch and Eidman 1987).

Finally, Internet access (as measured by e-mail access) did not increase use of data for agricultural decisions or producer assessments of weather data importance. This could mean that other factors aside from Internet access are barriers to data use. About 38% of respondents did not have e-mail access. These results raise the question of whether farmers and ranchers get weather data from other sources (e.g., irrigation district staff, cooperative extension agents, the USDA, private consultants) instead of accessing data directly. Or does it simply mean that few producers rely on web-based weather information sources? Interestingly, access to satellite TV proved to be a much better predictor of positive weather data attitudes and use.

This preliminary study has some obvious limitations. There is only so much information we can glean from just two survey questions about weather data. While our analysis explores the importance of producer heterogeneity, more work is needed to examine heterogeneity of the sources and quality and technical complexity of the weather data. In addition, more information can be obtained about which types of data are used for which decisions, how and how well producers make use of the data, and why producers do not use weather data to make certain decisions.

We conclude by mentioning two fruitful areas for future research. First, recent research on demand for economic information for agricultural decisionmaking (e.g., Wolf et al. 2001; Just et al. 2002, 2003) could be usefully applied to demand for meteorological information. Particularly important in this work is its appreciation of multiple sources of heterogeneity (producer/production attributes, data sources, data structure/technical complexity) and its emphasis on the role of intermediaries as processors and translators of information. There are two key barriers to use of meteorological information (Artikov and Lynne 2005). First, information providers themselves may lack basic information about the potential users they are trying to serve. For example, what are user needs or constraints? Who is using information and who is not? How do users differ from nonusers? Second, meteorological information providers may not know (or not know well enough) what users think of the information they supply. Is it too detailed or not detailed enough? Is there too little interpretation? Is it timely enough to be useful?

The work of the research team at the University of Nebraska (Artikov et al. 2006; Hu et al. 2006) is also noteworthy. Our Arizona study raised as many questions about agricultural weather data use as it answered. The Nebraska researchers' approach brought concepts from behavioral economics and social psychology to bear on farmer information use, utilizing both focus groups and in-depth surveys to sharpen their questions and hypotheses. Their consideration of social norms and perceived social control, along with farmer attitudes and financial capability, provide a richer perspective of how information use enters into household utility functions.

REFERENCES

Anaman, K.A., and S.C. Lellyett. 1996. Contingent Valuation Study of the Public Weather Service in the Sydney Metropolitan Area. *Economic Papers (Economic Society of Australia)* 15:64–77.

———. 2007. Meteorological Assessment of the Benefits of an Enhanced Weather Information Service for the Cotton Industry in Australia. *Meteorological Applications* 3:127–135.

Artikov, I., S.J. Hoffman, G.D. Lynne, L.M.P. Zillig, Q. Hu, A.J. Tomkins, K.G. Hubbard, M.J. Hayes, and W. Waltman. 2006. Understanding the Influence of Climate Forecasts on Farmer Decisions as Planned Behavior. *Journal of Applied Meteorology and Climatology* 45:1202–1214.

Artikov, I., and G.D. Lynne. 2005. Climate Change and Farm Use of Weather Information. Paper presented at the American Agricultural Economics Association Annual Meeting. July 24–27, 2005, Providence, RI.

Bosch, D., and V.R. Eidman. 1987. Valuing Information When Risk Preferences Are Nonneutral: An Application to Irrigation Scheduling. *American Journal of Agricultural Economics* 69 (3):658–668.

Breuer, N.E., V.E. Cabrera, K.T. Ingram, K. Broad, and P.E. Hildebrand. 2008. AgClimate: A Case Study in Participatory Decision Support System Development. *Climatic Change* 87:385–403.

Byerlee, D., and J.R. Anderson. 1982. Risk, Utility and the Value of Information in Farmer Decision Making. *Review of Marketing and Agricultural Economics* 50:231–247.

Cameron, A.C., and P.K. Trivedi. 1998. *Regression Analysis of Count Data*. Cambridge, UK: Cambridge University Press.

Carlson, J.D. 1989. The Importance of Agricultural Weather Information: A Michigan Survey. *Bulletin of the American Meteorological Society* 70 (4):366–380.

Chapman, R. 1992. *Benefit–Cost Analysis for the Modernization and Associated Restructuring of the National Weather Service*. Washington, DC: U.S. Department of Commerce, National Institute of Standards and Technology.

Dolan, A.H., B. Smit, M.W. Skinner, B. Bradshaw, and C.R. Bryant. 2001. *Adaptation to Climate Change in Agriculture: Evaluation of Options*. Occasional Paper No. 26. Guelph, Ontario: University of Guelph.

Domencich, T., and D. McFadden. 1975. *Urban Travel Demand: Behavioral Analysis*. Amsterdam: North-Holland Publishing.

Easterling, W.E. 1986. Subscribers to the NOAA Monthly and Seasonal Weather Outlook. *Bulletin of the American Meteorological Society* 67:402–410.

Feder, G., and R. Slade. 1984. The Acquisition of Information and the Adoption of New Technology. *American Journal of Agricultural Economics* 66 (3):312–320.

Fernandez-Cornejo, J. 1996. The Microeconomic Impact of IPM Adoption. *Agricultural and Resource Economics Review* 25:149–160.

———. 1998. Environmental and Economic Consequences of Technology Adoption: IPM in Viticulture. *Agricultural Economics* 18:145–155.

———. 2007. *Off-Farm Income, Technology Adoption, and Farm Economic Performance.* Economic Research Report No. ERR-36. Washington, DC: United States Department of Agriculture, Economic Research Service.

Fernandez-Cornejo, J., E.D. Beach, and Wen-Yuan Huang. 1994. The Adoption of IPM Techniques by Vegetable Growers in Florida, Michigan, and Texas. *Journal of Agricultural and Applied Economics* 26:158–172.

Fernandez-Cornejo, J., and S. Jans. 1996. The Economic Impact of IPM Adoption for Orange Producers in California and Florida. *Acta Horticulturae* 429:325–334.

Freebairn, J.W., and J.W. Zillman. 2002. Economic Benefits of Meteorological Services. *Meteorological Applications* 9:33–44.

Gardner, W., E.P. Mulvey, and E.C. Shaw. 1995. Regression Analyses of Counts and Rates: Poisson, Overdispersed Poisson, and Negative Binomial Models. *Psychological Bulletin* 118 (3):392–404.

Hu, Q., L.M.P. Zillig, G.D. Lynne, A.J. Tomkins, W.J. Waltman, M.J. Hayes, K.G. Hubbard, I. Artikov, S.J. Hoffman, and D.A. Wilhite. 2006. Understanding Farmers' Forecast Use from Their Beliefs, Values, Social Norms, and Perceived Obstacles. *Journal of Applied Meteorology and Climatology* 45:1190–1201.

Jensen, A.L., P. Boll, I. Thysen, and B.K. Pathak. 2000. Pl@nteInfo®: A Web-Based System for Personalised Decision Support in Crop Management. *Computers and Electronics in Agriculture* 25 (3):271–293.

Johnson, S.R., and M.T. Holt. 1997. Value of Weather Information. In *Economic Value of Weather and Climate Forecasts*, edited by R.W. Katz and A.H. Murphy. Cambridge, UK: Cambridge University Press, 75–107.

Jones, E., M.T. Batte, and G.D. Schnitkey. 1989. The Impact of Economic and Socioeconomic Factors on the Demand for Information: A Case Study of Ohio Commercial Farmers. *Agribusiness* 5 (6):557–571.

Just, D.R., S.A. Wolf, S. Wu, and D. Zilberman. 2002. Consumption of Economic Information in Agriculture. *American Journal of Agricultural Economics* 84 (1):39–52.

Just, D.R., S. Wolf, and D. Zilberman. 2003. Principles of Risk Management Service Relations in Agriculture. *Agricultural Systems* 75:199–213.

Letson, D., I. Llovet, G. Podesta, F. Royce, V. Brescia, D. Lema, and G. Parellada. 2001. User Perspectives of Climate Forecasts: Crop Producers in Pergamino, Argentina. *Climate Research* 19:57–67.

Lubben, B.D., N.L. Bills, J.B. Johnson, and J.L. Novak. 2006. *The 2007 Farm Bill: U.S. Producer Preferences for Agricultural, Food, and Public Policy.* National Public Policy Education Committee Publication No. 2006-01. Oak Brook, IL: Farm Foundation.

McNamara, K.T., M.E. Wetzstein, and G.K. Douce. 1991. Factors Affecting Peanut Producer Adoption of Integrated Pest Management. *Review of Agricultural Economics* 13:129–139.

McNew, K.P., H.P. Mapp, C.E. Duchon, and E.S. Merritt. 1991. Sources and Uses of Weather Information for Agricultural Decision Makers. *Bulletin of the American Meteorological Society* 72:491–498.

Meinke, H., and R.C. Stone. 2005. Seasonal and Inter-annual Climate Forecasting: The New Tool for Increasing Preparedness to Climate Variability and Change in Agricultural Planning and Operations. *Climatic Change* 70:221–253.

Parker, D., and D. Zilberman. 1996. The Use of Information Services: The Case of CIMIS. *Agribusiness* 12:209–218.

Parry, M.L., and T.R. Carter. 1985. The Effect of Climatic Variations on Agricultural Risk. *Climatic Change* 7:95–110.

Pasteris, P.A., T. Puterbaugh, and R. Motha. 2004. Climate Services: A USDA Perspective. Paper presented at the 14th Conference on Applied Climatology. January 10–15, 2004, Seattle.

Pope, R.D., and R. Prescott. 1980. Diversification in Relation to Farm Size and Other Socioeconomic Characteristics. *American Journal of Agricultural Economics* 62 (3):554–559.

Rollins, K.S., and J. Shaykewich. 2003. Using Willingness-to-Pay to Assess the Economic Value of Weather Forecasts for Multiple Commercial Sectors. *Meteorological Applications* 10:31–38.

Smit, B., I. Burton, R.J.T. Klein, and J. Wandel. 2000. An Anatomy of Adaptation to Climate Change and Variability. *Climatic Change* 45:223–251.

Smit, B., O. Pilifosova, I. Burton, B. Challenger, S. Huq, R.J.T. Klein, and G. Yohe. 2001. Adaptation to Climate Change in the Context of Sustainable Development and Equity. In *Climate Change 2001: Impacts, Adaptation, and Vulnerability*, edited by J.J. McCarthy, O.F. Canziani, N.A. Leary, D.J. Dokken, and K.S. White. Cambridge, UK: Cambridge University Press, 876–912.

Stewart, T.R. 1997. Forecast Value: Descriptive Decision Studies. In *Economic Value of Weather and Climate Forecasts*, edited by R.W. Katz and A.H. Murphy. Cambridge, UK: Cambridge University Press, 147–181.

Stigler, G.J. 1961. The Economics of Information. *Journal of Political Economy* 69 (3):213–225.

Stratus Consulting. 2002. *Economic Value of Current and Improved Weather Forecasts in the U.S. Household Sector.* Boulder, CO: Stratus Consulting.

Teske, S., and P. Robinson. 1994. *The Benefits of the United Kingdom Meteorological Office to the National Economy.* WMO-TD No. 630. Geneva, Switzerland: World Meteorological Organization.

USDA NASS (U.S. Department of Agriculture, National Agricultural Statistics Service). 2009a. 2007 Census of Agriculture Summary and State Data. Volume 1. Geographic Area Series, Part 51 AC-07-A-51. www.agcensus.usda.gov/Publications/2007/Full_Report/index.asp (accessed October 4, 2010).

———. 2009b. *Farm Computer Usage and Ownership Survey.* Washington, DC: USDA NASS.

Wandel, J., and B. Smit. 2000. Agricultural Risk Management in Light of Climate Variability and Change. In *Agricultural and Environmental Sustainability in the New Countryside*, edited by H. Milward, K. Beesley, B. Ilbery and L. Harrington. Winnipeg, Manitoba: Hignell Printing Limited, 30–39.

Wilks, D.S. 1997. Forecast Value: Prescriptive Decision Studies. In *Economic Value of Weather and Climate Forecasts*, edited by R.W. Katz and A.H. Murphy. Cambridge, UK: Cambridge University Press, 109–145.

Wolf, S., D. Just, and D. Zilberman. 2001. Between Data and Decisions: The Organization of Agricultural Economic Information Systems. *Research Policy* 30:121–141.

Wozniack, G.D. 1993. Joint Information Acquisition and New Technology Adoption: Late versus Early Adoption. *Review of Economics and Statistics* 75:438–445.

PART IV

CONCLUSION

CHAPTER 12

Modes of Adaptation and Regional Resilience to Climate Change

Bonnie G. Colby and George B. Frisvold

Regional response to climate variability and change can be usefully viewed from a resilience framework (Nelson et al. 2007). Whereas study of adaptation concentrates on microlevel adjustments, a resilience framework follows a systems approach. Resilience measures a system's ability to undergo changes while still serving its fundamental functions. Here we might consider the southwestern United States or, more specifically, management of water resources in the Colorado or Upper Rio Grande River Basin as the system. A fundamental function of this system is balancing water and energy supplies and demands in the face of population growth and climate change.

In this chapter, we summarize the major modes of climate change adaptation discussed in this volume. We then examine how these adaptation modes contribute to systemwide resilience and identify locations and activities particularly vulnerable to climate change.

MODES OF ADAPTATION

We may divide adaptation strategies into five general categories: government policies and programs, water trading arrangements, changes in production practices, adoption of new technologies, and better use of climate and weather information. These categories are interrelated. For example, government programs can encourage water trading or adoption of new technologies. We could view use of new technologies or information as simply subsets of production practices. However, use of new technologies, especially information technologies, frequently involves pronounced departures from common practices. As such, they often represent novel and long-term changes in behavior and adaptation to new circumstances.

Government Policies and Programs

As Chapters 5 and 6 illustrated, the Colorado River Basin is a heavily managed system. The Bureau of Reclamation along with other federal, state, local, and private management entities manage system water supplies to balance myriad competing interests. These include demands for irrigation, hydropower generation, water-based recreation, ecosystem maintenance, and treaty obligations with Mexico.

Thus far, management of the Colorado has proven quite resilient in the face of recent drought and explosive population growth because of the system's extensive water storage capacity. Climate change will put this resilience to the test, as projections suggest even lower water supplies in the future in the face of continued water demand growth. Chapter 5, however, demonstrated how water rights and production patterns in the region have evolved to limit costs of regional water shortages. Simulations of the U.S. Agricultural Resource Model (USARM) show how the region adapts to a 25% reduction in agricultural water supplies at least cost overall to producers and consumers. It turns out that this least-cost solution involves reduced production of crops grown in central Arizona, which happens to have junior water rights on the Colorado. If a shortage were declared on the Colorado, agricultural water supplies to central Arizona would be the first to face cuts. Our modeling results suggest that this institutional arrangement may also be the economically least-cost arrangement as well.

Turning to federal agricultural pricing policies, Chapters 3 and 4 considered how price supports coupled with current production make rural-urban water transfers more costly and thus can dampen this type of climate adaptation. Chapter 5, however, illustrated that coupled price supports do a poor job of protecting farmers from effects of drought. This occurs because coupled program payments decline with the lower yields and higher prices that drought can bring. Subsidies to agriculture from water supplied by federal surface water projects can also discourage rural-urban transfers (Chapter 3). These subsidies have been criticized on efficiency and equity grounds and have proven politically controversial. Removal of these subsidies, however, is likely to reduce farm income by more than reductions in quantities of water for agriculture (Chapter 5). This occurs because price increases are applied to all water uses, whereas quantity reductions allow producers to limit their most marginal irrigation activities. These results support Moore's (1991) argument that price-based regulation of Bureau of Reclamation water will be more costly to irrigators than quantity-based regulation.

A major federal strategy to encourage water conservation and reallocation to other uses has been government cost-sharing for investments to improve efficiency of on-farm application and water conveyance efficiency. Participation in cost-share programs is greater among larger operations in the Southwest (Chapter 8), suggesting that there may be trade-offs between efficiency and equity. Because large-scale operations account for most of the water use, we could argue that targeting this group would improve program cost-effectiveness (Schaible 2004). However, there are concerns that federal farm programs exhibit a large-farm bias by concentrating financial support to the largest operations (USDA 1998).

Improved on-farm irrigation efficiency can be an effective way for individual irrigators to preserve crop yields and incomes in times of drought. However, results from Chapter 3 should serve as a cautionary tale for policymakers hoping that improved application or conveyance efficiency alone will encourage water conservation or reallocation. There, conditions were identified where improved efficiency may discourage water transfers, increase consumptive use of water, or both. Mitigation requirements (to address third-party effects) along with adaptive management will likely be necessary to prevent technology subsidy programs from having unintended negative consequences for basinwide water management. Chapter 3 illustrated that individual adaptations to water scarcity (improved on-farm efficiency and reduction in conveyance losses) can increase consumptive use of the system as a whole. Adaptations to water scarcity in the United States appear to undermine the resiliency of the Mexico's Colorado River Delta.

Water Transfers and Risk Sharing

The region's agriculture appears quite resilient in the face of water shortages. Model results from Chapter 5 assumed that water can be readily reallocated between different agricultural uses, possibly in different locations. This highlights the importance of water reallocation as a means to reduce adaptation costs. Whereas Chapter 5 examined reallocation of water *within* agriculture, Chapters 2, 3, and 4 considered transfers *out of* agriculture for urban water demand, ecological restoration, or both. Agriculture makes important contributions to the region's resilience. Water demands for urban use or ecosystem maintenance are generally inelastic. Because agriculture maintains a combination of high- and lower-value uses of water, it can serve as a reserve water supply for urban or environmental uses. This book considered different types of agricultural water transfers: short-term leases (Chapter 2); large-scale, multiyear transfers (Chapter 3); and dry-year option contracts (Chapter 4). Dry-year options can provide farmers with guaranteed payments in drought years, while transferring water as needed to urban or environmental uses. Such contracts can increase the scope for agricultural, environmental, and urban demands for water to coexist in a flexible way. They also represent bottom-up, voluntary sharing arrangements as opposed to top-down imposed regulation of water use. A recurring theme in Chapters 2, 3, and 4 was that climate variability has important implications for how and how well water-trading arrangements perform. For water utilities or environmental groups that wish to secure water supplies, an understanding of how climate and climate variability affect agricultural water requirements (and hence water supply) is critical.

Changes in Production Practices

In addition to institutional innovations, such as water trading, agriculture may adjust production practices to adapt to climate change. In areas with sufficient rainfall and soil moisture to permit dryland farming, agricultural producers can switch from irrigated to dryland production. Growers may also switch from more to less water-intensive crops. Yet another option is the practice of deficit irrigation, where growers sacrifice some crop yields but apply less water. These relatively modest changes in cropping patterns and irrigation practices can substantially reduce drought impacts on agriculture (Chapter 5).

Adoption of New Technologies

In some cases, adaption to climate change will require capital investments in new technologies. Ski resorts may rely on snowmaking technology, and farmers may adopt more modern irrigation technologies such as drip or sprinkler irrigation. These technological adaptations, however, are themselves sensitive to climate. Use of snowmaking technology may be limited if climate change involves warmer overnight temperatures and a greater proportion of precipitation falling as rain instead of snow (Chapter 7). Adoption of sprinkler irrigation may be less viable under extremely hot and arid conditions, such as in Arizona and Southern California (Chapter 9).

Information Use and Adaptation

Climate change is more than change in long-run temperatures or precipitation. It also implies changes in climate variability and the probabilities of extremes. Adjusting to climate variability through improved use of climate and weather information is a critical aspect of climate adaptation (Smit and Skinner 2002). The electric utility industry is one climate-sensitive industry that may especially benefit from improved forecasting.

Electric utilities are particularly vulnerable to climate change for two reasons. First, although some fluctuations in electricity demand can be anticipated, the magnitude and precise timing of demand changes can be unpredictable. Second, hydropower generation is sensitive to climate-induced changes in water supply and demand. Climate change is expected to bring not only warmer temperatures, and thus more use of electricity and water at peak periods, but also greater incidence of long-term drought. These water-energy interdependencies amplify the impact of climate change. A key concern is that climate change will increase occurrences of extreme conditions, such as heat waves, that coincide with periods when utilities are operating at maximum capacity. Southwestern electricity use has increased steadily, with summer peak loads in particular increasing. The connection between water and energy can aggravate peak-load reliability issues, because demand for one is closely associated with demand for the other.

Chapter 10 developed and demonstrated a new econometric approach that incorporates weather and climate information in electricity load forecasts. Load forecast predictions improve significantly when climate variables are included in forecast models, especially temperature. Generally, electricity generators make use of forecasting techniques in which climate is assumed to be constant. Chapter 10 tested statistical models that account for weather and climatic information against models that do not incorporate climate information, using load data provided by an electric utility company in the southwestern region. The more comprehensive models clearly and consistently outperform the models that do not employ weather or climatic data. Compared with traditional specifications without climate information, next-day forecast predictions improved by about 1.4% when temperature was included in the model. Augmenting the model with humidity, precipitation, and wind speed reduced the rate of error by another 0.1%.

Although these changes are small in percentage terms, such improvements can save millions of dollars annually. Further, improved load forecasts can reduce pollution and result in more reliable and cost-effective provision of electricity. Better demand forecasts can also aid longer-term planning for electricity supply development. Using information to reduce costs and increase reliability of electricity generation will benefit both utilities and ratepayers and will be an important adaptation to climate change.

Information use may also be an important adaptation strategy in agriculture. While several studies have examined constraints on use of weather and seasonal climate forecasts, many have focused on the properties of the forecasts themselves (e.g., their timeliness or accuracy). Less consideration has been given to heterogeneity among potential users of information for agricultural decisions. Chapters 8 and 11 specifically examined why some producers use information while others do not. Farm scale was an important predictor of use of information for irrigation management. Farms of all types were more likely to rely on lower-cost sources of general information, while larger operations relied more heavily than did smaller ones on private, tailored sources of information. Larger operations were more likely to rely on a greater number and diversity of sources. Smaller farms were less likely to investigate irrigation improvements or use management-intensive methods for irrigation scheduling. These results suggest significant barriers to information acquisition and use of management-intensive irrigation practices for smaller operations.

Chapter 8 also showed that no single institutional source of water management information is relied on by more than half of irrigators. A significant share of irrigators did *not* report relying on public sources of irrigation information. The data also revealed the important role of irrigation districts. In many cases, districts decided the timing of irrigation and were a source of irrigation information, particularly for smaller farms. For public agencies providing water management information, this suggests that intermediaries, such as irrigation district staff, can be an important audience for information and user of decision support tools.

Results from Chapter 11 suggest that income and production diversification of agricultural households have important implications for their use of weather data and seasonal climate forecasts. Effects of diversification are complex. Greater diversification of production *within* agriculture increased climate and weather data use. Specifically, producers who grew a more diverse mix of crops were more likely to use weather data for the timing of three activities: planting, cultivation, and harvest. Diversified agricultural production can reduce risk, so one might think it could substitute for information. However, diversification also increases complexity of management and the value of optimal timing of agricultural operations. Results suggest these latter effects dominate, increasing demand for weather data.

Nonfarm income, in contrast, discouraged weather data use. Use of weather data increased with the share of household income derived from agriculture. This could result from two effects. First, agricultural income risk is a larger share of household income risk for households engaged primarily in agricultural production. Thus returns from using data to reduce agricultural risk are greater. Second, producers spending more time in off-farm employment may have less time to process data.

Therefore, one adaptation to climate variability and change might be to rely less on agriculture as a source of income. For such farms, moving out of agriculture may substitute for more information-intensive agricultural management. Chapter 8 showed that a significant number of growers, especially small-scale producers in Arizona and New Mexico, did not expect to continue farming long, and that this was an important barrier to implementing improvements to conserve water or energy. Smaller-scale operators often rely on off-farm employment for the bulk of their household income. These households may be "too busy" to adopt management-intensive agricultural practices.

CONCLUDING THOUGHTS ON ADAPTATION AND RESILIENCE

Climate-water-energy linkages may be less apparent when considering microlevel climate adaptations. However, these linkages are key building blocks in understanding the overall resilience of the Southwest's economy to climate change. Management of the region's major river systems involves trade-offs between providing water for hydropower and multiple competing uses. More extreme drought events and higher regional temperatures will make balancing competing water demands more difficult. The desire to maintain hydropower generation imposes critical constraints on water management systems. Ironically, if a shortage were declared on the Colorado, surface water deliveries to central Arizona agriculture would be curtailed. A likely response of growers will be to shift to groundwater pumping. Thus restrictions intended in part to protect hydropower supplies could increase electricity demands for pumping and also draw down groundwater reserves, which serve as a drought buffer.

While communities reliant on water-based recreation are vulnerable to declining reservoir levels brought about by climate change, they are also vulnerable to climate change mitigation policies. Future mitigation policies, such as carbon taxes or cap-and-trade systems, will raise gasoline prices. This in turn will discourage recreational visits and the local sales revenues, income, and employment they bring. One way to counteract these negative effects would be to increase the level of federally funded park service employment in the region. The proposed Public Lands Service Corps Act of 2009 (HR 1612) would increase construction and employment in national parks and other public lands through creation of a service corps akin to the Civilian Conservation Corps (CCC), which was instituted as part of the New Deal. A modest increase in jobs under this program could suffice to counter dampened local economic activity from rising gasoline prices.

By increasing energy prices, climate change mitigation policies will raise costs of transporting surface water and pumping groundwater. As highlighted in Chapter 1, energy and water use costs are inextricably linked. Chapter 5, however, suggested that the region's agriculture is resilient to even large (25%) increases in water costs.

To summarize, the chapters in this book suggest that the Southwest as a whole is relatively resilient to climate change, but that several areas and activities are relatively vulnerable. These include central Arizona agriculture (Chapter 5); water- and snow-based recreation (Chapters 6 and 7); outdoor recreation in low desert parks (Chapter 6); hydroelectric power generation (Chapters 5, 6, and 10); urban water providers (Chapters 2, 3, and 4); and the agriculture and ecosystem of Mexico's Colorado River Delta (Chapter 3). We have identified water transfers, technological adjustments, and improved use of information systems as important adaptation mechanisms. However, the chapters acknowledge the limits as well as the promise of these responses.

Climate change presents a cascading array of challenges for water- and energy-using sectors in the U.S. Southwest. This volume has presented a variety of economic techniques and discussed their applications to face these challenges. While economics is an important way to understand these issues, it is by no means the only way. Economics stands alongside the fields of atmospheric science, paleoclimatology, engineering, planning, hydrology, anthropology, rural sociology, and psychology, among others, in contributing to society's collective ability to understand, prepare for, and adapt to climate change.

REFERENCES

Moore, M.R. 1991. The Bureau of Reclamation's New Mandate for Irrigation Water Conservation: Purposes and Policy Alternatives. *Water Resources Research* 27:145–155.

Nelson, D.R., W.N. Adger, and K. Brown. 2007. Adaptation to Environmental Change: Contributions of a Resilience Framework. *Annual Review of Environment and Resources* 32:395–419.

Schaible, G. 2004. Irrigation, Water Conservation, and Farm Size in the Western United States. *Amber Waves* 8 (June):8.

Smit, B., and M.W. Skinner. 2002. Adaptation Options in Agriculture to Climate Change: A Typology. Climatic Change. *Mitigation and Adaptation Strategies for Global Change* 7:85–114.

USDA (U.S. Department of Agriculture). 1998. *A Time to Act: A Report of the USDA National Commission on Small Farms.* Miscellaneous Publication 1545. Washington, DC: USDA.

Index

Note: Page numbers in italics indicate figures and tables.

For Product Safety Concerns and Information please contact our EU
representative GPSR@taylorandfrancis.com
Taylor & Francis Verlag GmbH, Kaufingerstraße 24, 80331 München, Germany

www.ingramcontent.com/pod-product-compliance
Ingram Content Group UK Ltd.
Pitfield, Milton Keynes, MK11 3LW, UK
UKHW010956180926
475701UK00004B/395